THE USER'S GUIDE TO THE
AUSTRALIAN COAST

Greg Laughlin was taught to sail as a young boy and has been an enthusiastic sailor ever since. He holds a PhD in Climatology from the University of New South Wales and is currently Senior Research Scientist at the Bureau of Resource Sciences, Commonwealth Department of Primary Industries and Energy. He is an acknowledged expert on Australia's coastal environment, maritime and weather conditions. In writing *A User's Guide to the Australian Coast*, he has drawn on his own extensive experience, as well as bringing together the expertise of an outstanding group of authorities in the field.

THE USER'S GUIDE TO THE
AUSTRALIAN COAST

A practical guide to our currents, tides, waves, winds and weather for everyone who sails, fishes, hang-glides, surfs – or just sits on the beach

GREG LAUGHLIN

with contributions from

Klara Finkele,

John Finnigan,

Neil Hamilton,

Richard Kenchington,

Jenny Kesteven

and Geof Lennon

Graphics and Artwork

Jeremy Mears

NEW
HOLLAND

A Reed New Holland book
First published in 1997 by
New Holland Publishers Pty Ltd
Sydney • London • Cape Town • Singapore

Produced and published in Australia by
New Holland Publishers Pty Ltd
3/2 Aquatic Drive, Frenchs Forest
NSW 2086 Australia

24 Nutford Place
London W1H 6DQ
United Kingdom

80 McKenzie Street
Cape Town 8001
South Africa

ISBN 0 7301 0515 6

Publisher **Clare Coney**
Designer **Rob Klinkhamer**
Paging **Melbourne Media Services**

Produced by The Australian Book Connection
Bound by M&M Binders

Front cover photographs:
Wave breaking: International Photographic Library,
Inset photographs:
Surfing The Photo Library
Snorkling International Photographic Library
Windsurfing Horizon International
Fishing Horizon International
Sailing Robert Keeley.

To my mother and father, Eileen Laughlin and the late Colonel James Laughlin.

Units of measurement

Throughout most of this book, metric units of measurement — metres, kilograms and seconds (and derivations of these: metres per second, kilometres per hour, and so on) — are used. The exceptions are knots and nautical miles. Some sailors prefer these more familiar units, which were originally derived for navigational purposes (1 nautical mile = 1 minute of earth arc; 1 knot = 1 nautical mile per hour). Knots are also the original basis of the wind feathers (for example, ⇁). Where knots and nautical miles are used, their metric equivalents are also given. For a comprehensive table of units, see Appendix I.

Acknowledgments

This book took more than three years to write. The first draft manuscript was 351 pages without figures! In the time between the first and final draft, there were many attempts to shorten the manuscript and make it more accessible to the average reader. For some of the contributing authors this was an agonising experience; they had gone to great lengths to produce what they saw as *elegant* copy, but in several cases this meant a chapter or section peppered with mathematical equations and technical language. To all the contributors, I am truly grateful for your forbearance and sustained efforts, especially to those who watched with horror as the manuscript continued to be simplified and 'de-technified'.

Dr Richard Kenchington (Great Barrier Reef Marine Park Authority) wrote the Introduction's 'Australia's Coastline: a snapshot' and provided detailed comments on the rest of the Introduction, which, in the first draft, was very much longer than it is now. Dr Jenny Kesteven (Centre for Resource and Environmental Studies, CRES, Australian National University), made a major contribution to Chapters 2 and 6. She provided text and performed the analysis for all of the climate maps with the exception of the wind maps (e.g. pressure, maximum temperature, mid-afternoon humidity) as well as general guidance on both chapters. Dr John Finnigan (CSIRO Division of Environmental Mechanics) wrote Chapter 3's 'Wind and Barriers' and provided the basis for all the figures. Dr Finnigan also made detailed comments on the rest of Chapter 3. Dr Klara Finkele (also CSIRO Division of Environmental Mechanics) wrote the first draft of Chapter 3's 'Sea Breezes' and provided the basis of most of the figures. Emeritus Professor Geof Lennon AO (Founding Professor, National Tidal Facility, Flinders University of South Australia) wrote Chapter 5 and provided detailed comments on all subsequent drafts (there were many). He also provided the basis for many of the figures and detailed, critical comments on the tide maps and related figures.

Jeremy Mear's 'signature' is on all the figures and graphics in this book; he is a freelance graphics consultant in Canberra. We started work on the wind maps in 1993 and he has continued right up to the last figure prior to publishing. His starting point for some of these was a hand-drawn figure on a paper serviette or a bad photocopy of a black and white image. I am most grateful to his quiet and patient approach.

The original manuscript included detailed maps of coastal landform and vegetation, which were excluded in later drafts (we may publish these later as a supplement to this book). Dr Neil Hamilton (CSIRO Division of Wildlife and Ecology) was instrumental in providing access to the CSIRO CAMRIS data for these. Dr Hamilton also wrote the first draft of Chapter 1's 'Beach hazards: a basic understanding could save your life'. While still on the ('missing') landform and vegetation maps, I would like to acknowledge the air-photo analysis work of Dr Bob Galloway (formerly CSIRO Division of Water and Land Resources) which provided the basis of CAMRIS.

Although Dr Michael Hutchinson (Centre for Resource and Environmental Studies, CRES, Australian National University) was not directly involved with this book, you will notice that his name appears on a number of maps in Chapter 2 (e.g. wind maps, raindays) as well as on the digital terrain model in 'Climate Essentials'. Dr Hutchinson has developed a series of powerful techniques for fitting sensible and smooth 'surfaces' to point data. The success of these techniques can be gauged by the large number of disciplines in which they are used, which now includes climatology, ecology, oceanography and many others.

Professor Matthais Tomczak provided detailed comments on Chapter 4 and wrote 'Wave motion and craft'. Dr Neil Lawson (Lawson and Treloar, Consulting Engineers) provided the raw data for the wave height maps and I am indebted to him for allowing me to use this commercially valuable data. Dr Lawson also provided detailed comments on Chapter 4. Dr John Reid (CSIRO Division of Oceanography) also provided critical comments on

Chapter 4 and contributed toward the discussion of the interaction of seas and swell in the open ocean. Mr Paul Tildersley (also from Oceanography) provided additional comments on the chapter.

There were several other people who contributed to Chapter 5: Dr Tad Murty of the National Tidal Facility (Flinders University of South Australia) kindly provided the raw data used for the 'marathon' tide figure (the time series plot). He and Mr Thang Aung provided useful comments on an early draft of the chapter. I am indebted to the NTF for use of the Australian National Tide Tables (ANTT) in the tide range and type figures. Dr George Cresswell (CSIRO Division of Oceanography) provided me with the basis for the map of currents in the Australian region and wrote the first draft of the commentary of the figure.

The severe weather chapter was greatly assisted by the following people: Mr Mark Bedson (author of *SkyChart*) very kindly provided his entire suite of cloud photographs from the publication. I would especially like to thank him for his open and helpful approach to this book. Meteorological and oceanographic consultant, Dr Roger Badham, provided very useful comments on several drafts of the chapter and several original cloud photographs. Dr Graeme Hubbert, also a meteorological and oceanographic consultant, provided considerable information on cyclones and storm surge and comments on the rest of the chapter. From the Commonwealth Bureau of Meteorology, Dr Frank Woodcock, who provided the basis for the world and Australia cyclone maps; Mr Chris Ryan, the basis of the thunderstorm map; Dr Barry Hanstrum, severe weather in Western Australia, especially temperate cyclones; Dr Jonathon Gill, tropical squalls and other tropical severe weather in northern Australia; and Mr Bob Buckley for a final and thorough read of several chapters, especially of Chapter 6.

Finally, to my brother Kit and my mother, many, many thanks for taking the time to go through the early drafts and removing unnecessary and repetitive material.

Contents

Introduction

Boating, beaches and fishing play a big part in the Australian way of life. Approximately three-quarters of Australia's population lives within an hour or so of the coast. Many people who live further inland make at least one recreational or holiday trip to the coast each year.

Among the freedoms most Australians take for granted is the *right* to go fishing, to catch a wave, to take the boat out, to spend a day at the beach, or to retire to the coast. Maritime activities provide a contrast to the pressures of urban and working life.

We have grown up in a world in which the sea has been an infinite source of fish — there have always been 'plenty more fish in the sea'. But in biological terms, most of the open ocean areas are more akin to desert than to mangrove or fertile farming valley. The upper layers are well illuminated but starved of nutrients. The lower layers may be rich in nutrients accumulated from materials falling from above but are without sunlight; the life processes down there are slow. Rich fishing grounds occur only where the right mix of nutrients is brought up from the ocean depths by upwelling currents or down from the land by rivers. Yet Australia is not served by regular upwellings of deep waters, and flow from our rivers is irregular and generally meagre. Local fishery resources may be reasonable from the perspective of a city at the mouth of a river, but nationally, on the basis of experience and much research, we have to conclude that our fisheries are finite. Overfishing is a fact or a highly probable outcome in many inshore and near-city areas.

Most of us have also grown up understanding that our wastes are 'a drop in the ocean', the sea being an inexhaustible receiver of the debris of human activity. The reality is that the sea can accept *some* wastes, but the effect of wastes on the sea varies according to the amount and the type. *Some* added nutrient can lift productivity, but *too much* can lead to overgrowth of algae or red tides. The sea may dissolve and disperse pollutants such as heavy metals and complex synthetic organic chemicals, but these may then be concentrated to toxic levels by marine plants and animals.

The coast reflects a complex interaction of events occurring inland, on the coast or far out to sea. We know about some of the ways in which human actions can change the interactive process, but in most cases we don't know the extent to which the sea's natural self-repair capacity can absorb the impacts of human activity.

With today's technology, much of the skill and the risk involved in navigation and finding fish have been removed by satellite navigation, echo sounders and sidescan sonar. Add to this the fact that more and more people are producing more and more wastes that find their way into the sea, and it's easy to see why the balance has changed dramatically. The sea and its creatures are no longer invulnerable or infinite; they can be gravely affected by human activity.

In Australia, given our small population and long coastline, we might expect to be free of the problems of the crowded and polluted coasts of Europe, Asia and America. This does apply, except in the areas that we use most.

AUSTRALIA'S COASTLINE: A SNAPSHOT

The length of Australia's coast is approximately 36,000 kilometres. Typical landforms on the coast are sand, rock, mud and water (inshore waters). These are what would be seen by an observer standing on the foreshore looking toward the land.

For Australia as a whole, sandy foreshores are by far the most common, comprising a little over half the total Australian coastline. Most are backed by sandy dunes. Rocky and muddy shores are also common, each comprising about one fifth of the Australian coastline. Gravel beaches are rare in Australia.

The states of Australia have quite different

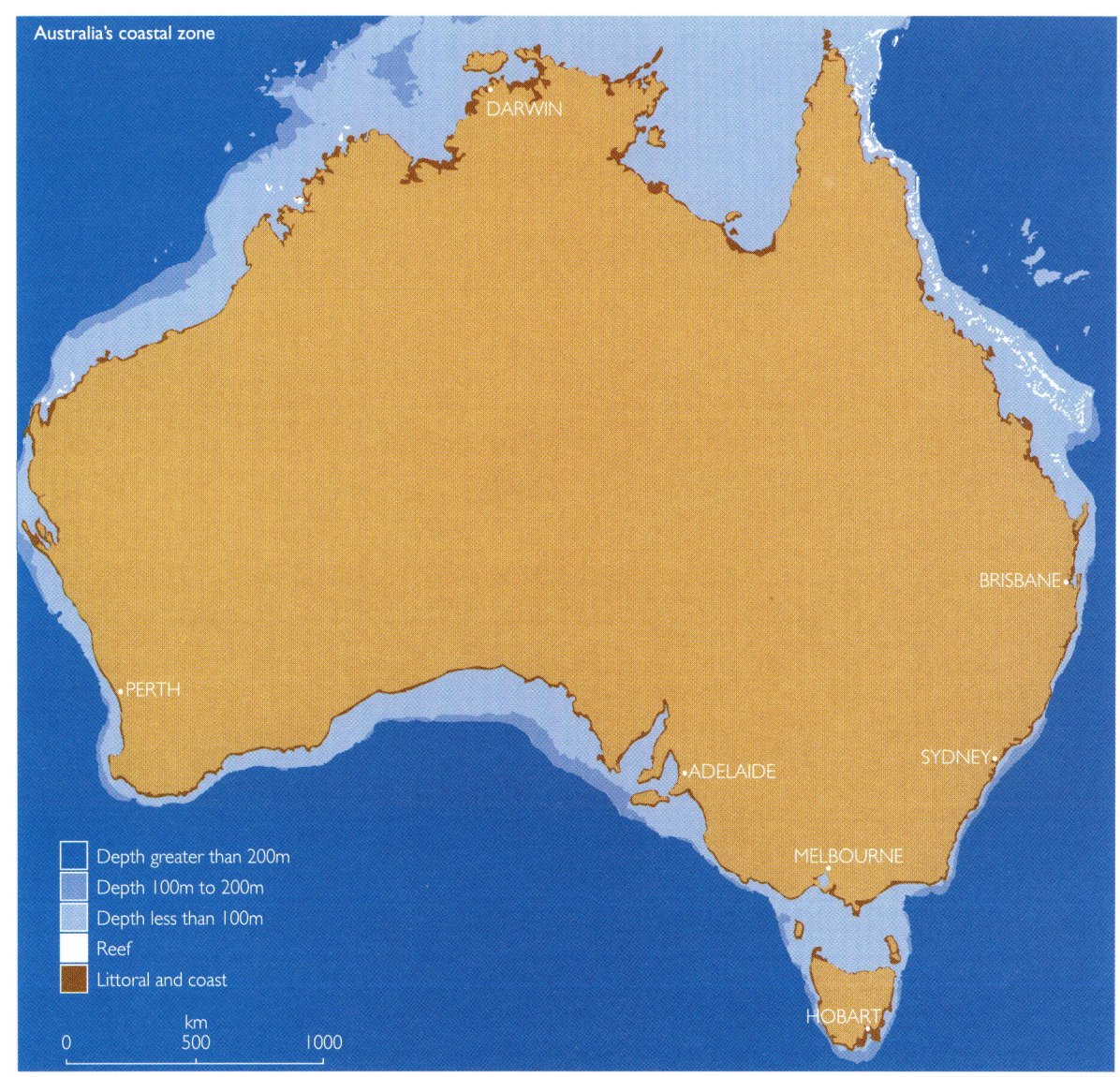

Australia's coastal zone

Depth greater than 200m
Depth 100m to 200m
Depth less than 100m
Reef
Littoral and coast

km
0   500   1000

Figure Int.1
The physical environments of the Australian coast and location of the continental shelf and reef systems. (Data supplied by CSIRO Division of Wildlife & Ecology, Canberra.)

proportions of the foreshore types; for example, nearly half the coastline in the Northern Territory is comprised of mud, which is in stark contrast to Tasmania, with only one per cent mud coasts.

The littoral zone

The littoral zone is the area affected by waves, or alternately exposed and submerged by tides, or both. The width of this zone reflects both the size of the tide and the slope of the land in the zone; in areas which have both a large tidal range and slight gradients in the land, the sea will penetrate further inland than in areas that have small tide ranges, steep land near the coast (or cliffs) or both.

Figure Int.1 clearly shows the extensive littoral zones in north and north-west Western Australia and in much of the Northern Territory and parts of Queensland, particularly in the Gulf of Carpentaria. There are also more localised areas including, for example, Shallow Inlet near Wilsons Promontory in Victoria and near the Derwent estuary in Tasmania. Such extensive littoral zones are generally low-lying wetland areas and are very important biologically and ecologically.

The continental shelf

The continental shelf marks the outer boundary between the relatively shallow coastal waters and deep ocean waters. The formation of the shelf relates to a distant time some 30 to 90 million years ago. More recently — perhaps 15,000 to 20,000 years ago — sea levels were at least 100 metres lower than they are today and parts of the outer edge of the continental shelf — as defined by sea level today — would have been the littoral zone. In Figure Int.1, the outer edge of the shelf is suggested by the 200 metre depth contour.

In Australia, the continental shelf is, on average, about 50 nautical miles wide and has a gentle gradient of about 10 feet per nautical mile (3 metres per kilometre). The width of the shelf varies greatly, from just a few nautical miles near South West Cape in Tasmania to over 100 nautical miles in the Great Australian Bight.

The continental slope

Beyond the continental shelf is the continental slope, which is 12 to 60 nautical miles wide and has steep slopes of about 300 feet per nautical mile (57 metres per kilometre).The continental slope continues to the deep ocean floor, which has an average depth of 12,000 feet (3.6 kilometres) or more in the case of the Pacific Ocean.

Surf beaches are usually found where the outer edge of the continental shelf is relatively close to the present-day coast and where there are no reef systems to dissipate the energy of approaching swell waves. Consistently large surf often occurs in areas where the shelf edge is unusually close to the shore and local underwater topography 'funnels' the approaching swell, especially if that swell is large and has travelled over long open water distances. Shelf morphology (shape, depth and slope) can be a more important factor than the width of the shelf.

Reefs

The major reef systems around Australia are also shown in Figure Int.1. Although most of these are within the continental shelf, there are exceptions (such as the Rowley Shoals off Western Australia). Apart from possessing great beauty and ecological richness, the reefs also profoundly influence the currents, tides and wave conditions of the littoral zone.

> Surfers and boaters might note that offshore reefs, especially when numerous and dense like the Great Barrier Reef, partially obstruct approaching swell waves. It is partly for this reason that much of the Queensland coast does not regularly experience large surf.

Australia's great natural wonder, the Great Barrier Reef, is about 1200 nautical miles long and comprises more than 2900 individual reefs, approximately 900 islands and numerous sand cays. Like parts of the outer edge of the continental shelf, the present reefs were formed during and since the last major sea level rise. However, sea level has risen and fallen many times over geological history, causing some of the present reefs to be built on top of much older 'relic' reefs.

THE MARINE AND COASTAL ENVIRONMENT

In 1995, an official report on the state of our marine environment concluded: 'Australia's marine environment is, on average, good, but the state of the marine environment near where urban Australia lives is often not good.' The report listed five main areas of concern:
- declining marine and coastal water/sediment quality
- loss of marine and coastal habitat
- unsustainable use of marine and coastal resources
- lack of marine science policy and lack of long-term research and monitoring of the marine environment
- lack of strategic, integrated planning in the marine and coastal environments.

We are degrading our coasts and the sea. This is apparent on Australia's eastern and southern coasts from Port Douglas to Port Lincoln, in the west from Geographe Bay to Jurien Bay, and at more isolated locations around human settlements. Records show poorly planned development, erosion, loss of mangroves and seagrass, over-fishing, threatened fish stocks, pollution, littering and the degradation of beaches and waters. Radio stations broadcast warnings of water pollution for swimmers and fish pollution for anglers. We bemoan poor catches, loss of fishing spots and the loss of unspoiled coast to inexorable development.

This state of affairs will continue — and get worse — if we don't collectively accept respons-ibility for the protection and maintenance of the rights mentioned earlier. That is not a remote matter to be whinged about until 'they' 'do something' to fix it. It is an immediate issue that *we* must *deal with*.

The first step we must take is to accept that human activity is now the major factor affecting the condition and productivity of coastal and shallow sea environments. Occasionally deterio-ration occurs as the result of a single big decision. More usually it results from many small actions and many small decisions taken without thought for the consequences. The attitudes of individuals will make a difference.

Oily bilge water, emissions from a poorly maintained marine engine, furtively dumped sump oil, turpentine and hydraulic fluid, together with the runoff from suburban streets and stormwater drains account for most of the hydro-carbons that find their way into coastal waters. In urban areas much of the pollution comes not from industry and agriculture but from the fertilisers and pesticides used on lawns and gardens and washed off into stormwater drains.

The other major source of pollution is sewage. In the days when we still believed the sea had infinite capacity, we thought it was the cheap-est dumping ground for our wastes. Now we know that there is a price to pay for such a conve-nience: degradation of the coast and coastal waters. However, politicians see no votes in sewage. The problem will be fixed only when it becomes politically smart to fix it and when passive concern and vague demands about the quality of the coast and coastal waters sharpen into action.

In the case of fishing we may consciously accept that there are no longer plenty more fish in the sea, yet it takes time to change the mindset of generations. It is easy to blame commercial opera-tors who land tonnes of fish; but for most of the populated coast the recreational fin fish catch is as great as, or greater than, the commercial. Not only that — when the fish stock is too low for the catch to repay the costs of taking it, commercial fishing goes bankrupt or moves on. Recreational fishing, by contrast, continues to use the same places and to apply high effort because catching fish is only part of the attraction of a day's fishing.

WHAT CAN WE DO?

There are several simple but powerful things we can do to turn this situation around, either acting within an organised structure, or on our own.

Get involved

Australia's Commonwealth, State, Territory and local governments are working together on a national coastal action program. Key endeavours within this are Coastcare, Clean Up Australia, Beachcare — voluntary programs that form the basis for coastal communities and visitors to become involved in addressing the problems.

Take action

Although the problems are large, individuals can do much on their own.

The first step is to remove unnecessary impact, such as littering of beaches and water. Any containers of food, drink, bait, sunscreen and insect repellent that we take to the beach or out in the boat can be brought back and disposed of properly. Broken glass is a hazard on beaches, yet it is easy to put an empty stubby back in the carton and recycle both. Plastic bags can kill the turtles that eat them, having mistaken them for jellyfish, and they can also cause serious damage to motors when they are sucked in to cooling systems; yet it is easy enough to rinse a bag (if necessary) and put it in a waste container.

The second step is to reduce pollution. Although it may require some effort, it is efficient as well as environmentally responsible to main-tain engines so that they don't discharge oil. Put sump oil, oily bilge, paint waste, turps, cooking oil and grease into containers for proper disposal onshore; it is grossly irresponsible not to do so.

Leave some fish in the sea

Fish will still keep biting after we have enough for a meal or two. We must therefore reset our expectations of what is reasonable and what is a good day's fishing, and leave some fish for tomorrow. Attitudes to recreational fishing are changing fast but there is still a way to go. A decade ago any suggestion of limiting catches was greeted as an attack on a fundamental democratic right. Now individual recreational fishers and their organisations recognise the need to do so.

Find out more

Few Australians would admit to being wilful destroyers of their natural heritage. Rather, many people's crime is apathy born out of ignorance — or just plain ignorance itself. The suggestions made above are advice of a general sort. To apply them effectively, we need to do one last thing: we need to bone up on the subject of Australia's coastal environment. Then, and only then — armed with knowledge and understanding — can we become responsible users of a priceless, irreplaceable resource.

This book has been written to help in such a quest.

CHAPTER ONE

The Beach and You

Australians see the beach as one of their most important cultural icons, spending many summer holidays and weekends along the coast. However, we generally know little of how beaches function, leaving ourselves vulnerable to their hidden dangers while surfing or swimming. Equally we lack knowledge of our own physiology, and how it works in a beach environment.

Lifesavers rescue about 10,000 people from the surf each year in Australia. Almost all are young and fit but lack understanding of simple natural processes that affect the coast. Groups featuring prominently in rescues are:
- country children (who lack familiarity with common beach conditions)
- overseas tourists (nearly half of those rescued cannot swim).

It isn't just inexperience that contributes to accidents; certain patterns of beach shape and waves are also important. Regular beach users, such as surfers and fishers, know these patterns, and use them to their own advantage. Other beach users can reduce the risks by learning how beaches work. Beach safety experts, such as the local surf lifesaving club, are the best people to ask about local conditions.

Beaches are dynamic, natural, high-energy systems, with changing waves, tides, winds, currents and shape. The three-dimensional shape of a beach will alter to compensate for wave, wind, and tide, and this in turn will modify the water circulation near the beach (the 'nearshore' zone), so swimming conditions can change rapidly without any obvious signal. However, the conditions at any beach are broadly predictable, if you have some knowledge and take time to make observations.

BEACH HAZARDS – A BASIC UNDERSTANDING COULD SAVE YOUR LIFE

To describe in detail how a beach works would take a whole book (and Paul Komar has already done this in his classic *Beach Processes and Sedimentation*). The following pages introduce some basic concepts and hazards of which any beach goer should be aware.

Circulation patterns near the shore

Beaches are areas of the coast where waves travelling from deeper water reach the land. They act as cushions for wave energy, gradually absorbing and sometimes reflecting it. Beaches change shape in response to the incoming waves, and in turn modify the waves themselves as they break and wash up onto the beach. The water that is moved towards the beach by the waves forms currents that sweep along the beach (longshore currents), and then head back out to sea (rip currents). This circulation pattern is common to all surf beaches.

Simply put, the water has to go somewhere; and the larger the seas, the stronger the currents may become. In detail, however, the circulation is quite variable, depending on the shape of the beach, the size of the waves yesterday as well as today, the particle size of the sand on the beach, and the state of the tide. But if you remember this fundamental pattern of water circulation you will find it easier to understand the details of beach processes.

Beach states

Most Australian surfing beaches lie in areas with a tidal range of less than 1.6 metres. Research by the Coastal Studies Unit at the University of Sydney has shown that these beaches have a fairly predictable 'state' at any given time. In essence, just remember that beaches and waves are tied together in a feedback mechanism, so that an increase in wave height causes changes in beach profile, which in turn modifies the incoming waves.

Beach state also changes with changing wave conditions. Understanding the states of a beach requires regular observation — and it provides users with the key to beach safety.

Surfing beaches are classified as three main types, according to their predominant state.

Reflective beaches

The lowest energy beach state is called 'reflective'. It is characterised by a very steep beach face with prominent cusps, coarse sand, small waves (0 to 1 metre high) that break right on the beach face, no real trough or bar, but deep (1 to 2 metres) water very close to the shore. Reflective beaches are commonly found in very sheltered sites, sometimes even inside harbours. When large waves appear at a reflective beach (for example, after a storm), the beach will usually modify its shape to form the next state, called 'intermediate'. Swimming is generally safe at reflective beaches — apart from the sudden drop into waist-deep (or even deeper!) water.

Intermediate beaches

The 'intermediate' state occurs in a number of varieties, which are sometimes found at different locations on the same beach at the same time. The varieties are called (from lowest waves to highest) low tide terrace, transverse bar and rip, rhythmic bar and beach, and longshore bar and trough. The names indicate the characteristic shapes seen on intermediate beaches over long periods of time. These shapes occur on beaches with wave heights between 0.5 and 2 metres. Intermediate beaches commonly change their shape even within one tidal cycle, so they can quickly become hazardous.

Intermediate beaches have a shallow sand bar (often with a wave-like shape) and trough in the surf zone, rip currents returning water offshore at various intervals, and more complex underwater terrain than is found on reflective beaches.

Intermediate-state beaches are the most common type of surf beach in Australia. They also commonly feature in rescue statistics because of their irregular topography, changeable nature and circulation patterns. Unless you are an expert, take the advice of surf clubs about where to swim on these beaches.

Dissipative beaches

The state that produces the highest waves (consistently over 2.5 metres) is called the 'dissipative' state. Dissipative beaches have multiple breaker zones stretching up to one kilometre offshore, a gentle beach slope, usually several straight offshore bars, fine sand, and rips that appear and disappear. Dissipative beaches are common on the south coast of Australia (Goolwa beach near Adelaide is the classic example), but occur on the east coast only after big storms or cyclones. They can be extremely dangerous, and most swimmers don't — or shouldn't — venture far from the shore at such a beach.

Characteristics of rips

Rip currents put fear into the hearts of most occasional beach users. However, most rips are not as terrifying as they at first seem. They are simply the path that water takes from the beach back out through the breaking waves, completing the nearshore circulation cycle. Rips normally begin to gather strength in the trough near the beach, accelerate over the bar in the surf zone, and disappear just behind the breaker zone. Complicated beach morphologies (intermediate beach states) can confuse these circulation patterns, but generally rips may be seen from the beach, and therefore avoided.

When arriving for a swim the first thing you should do is make a general survey of the whole beach and surf area. (Serious surfers do this religiously.) Rips can be difficult to recognise when you are in the water — or even from the beach! Try to stand as high as possible above the beach, perhaps on a headland or dune. Rips usually appear as disturbed areas of water or disruptions to the regular patterns of incoming waves. They can look dark and deep, with no breaking waves (because some rips flow through deeper channels), or might be covered with white foamy wavelets (as the current passes over sand bars). Watch for plumes of sandy water spreading out on the seaward side of the surf zone. If you remember the basic pattern of water circulation in the nearshore zone and have a suitable vantage point, it is quite easy to see rips on most beaches.

If you happen to be caught in a rip (despite having thoroughly surveyed the beach before swimming!), again remember the water circulation pattern. If you simply float in the water, you will be carried out behind the breaking waves, where the current dissipates, and will be able to catch a wave back to the shore. If you don't like that idea, swim alongshore (parallel to the beach), and you will soon be out of the current, and will easily reach the beach again. Most rips are quite narrow (often less than twenty metres), so you don't have to swim far. People tend to feel themselves losing control when 'caught' in a rip, and panic, when in fact they are quite safe if they keep calm.

The worst thing to do is to try to swim for shore (our immediate instinct) — against the rip. If the rip current speed is close to, or greater than, your swim speed, you'll exhaust yourself trying to get to shore.

Compounding factors

Almost all beach hazards are associated with the morphology and beach states described above. The safety of any beach is influenced by its unique situation, but the following factors make some beaches more hazardous.

- **Headlands and rocky outcrops** are commonly associated with rips and unusual current patterns. Surfers often use these headland-associated 'shoots' as a fast lane out through the waves, but they can be dangerous to inexperienced or young swimmers.
- **Tides** must always be considered by swimmers. Deeper water and sometimes stronger rips occur at high tides. Low tides, on the other hand, may make rips more easily visible, but may also strengthen them as they become restricted to channels or pass over shallow bars.
- **Wave and wind characteristics.** Above all, wave height is a good indicator of the safety of conditions. Waves over two metres high should always be considered hazardous, and on some beaches even smaller waves might indicate that it is dangerous to swim. For example, waves approaching the beach at an angle cause stronger longshore currents, strange wave patterns and rips that may not otherwise be expected. Rising seas may have similar effects. Winds can enhance the currents, and produce confusing wave appearances. There is no consistent pattern relating wind to beach hazard, although strong winds can create locally difficult swimming conditions.
- **The topography of the sand** in the surf zone can also be hazardous. Sand bars can move in response to changing wave patterns or tidal levels, and mega-ripples (large ripples up to one metre high) migrate along the beach or onshore in troughs. On very rare occasions, entire bar–trough systems can move rapidly to adjust to wave heights and water levels.

> Above all, beach safety is the responsibility of each individual swimmer or surfer. The main factor that contributes to your safety is knowledge of the natural systems around you. Visit your local surf club, or the club where you are holidaying, to find out about the types of conditions you may expect. Spend time watching the waves before you swim. You may even notice some of the more subtle effects, like systematically rising and falling water levels ('surf beat'), or different waves arriving at regular intervals. The more you know about your beach, the more respect you will have for it.

THE HAZARDS OF BEING HUMAN

We all know of the dangers of bumping into certain living creatures — particularly sharks, crocodiles and box jellyfish — in shallow waters. But we are probably less aware of the serious threat that we humans can pose to ourselves in coastal situations. Human physiology makes us vulnerable to two very real hazards: heat stress and cold stress.

Heat stress (hyperthermia) and dehydration in air

If you are active in warm weather, you should study Figure 1.1. It shows how air temperature combines with relative humidity to influence human comfort and the hazard of hyperthermia.

Human skin is not dry. When conditions are hot we rely heavily on evaporation (sweating) to

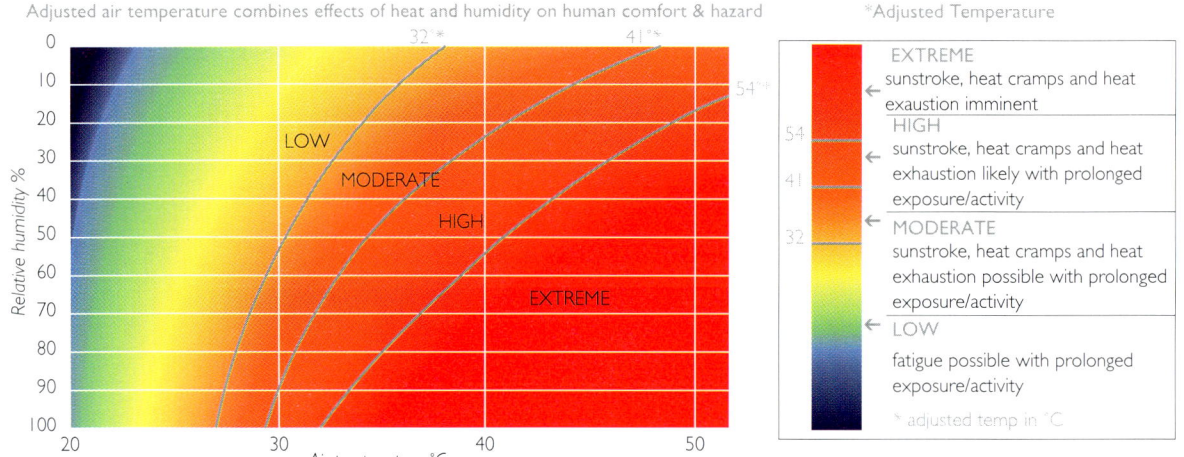

Figure 1.1

Potential heat stress hazard posed by combinations of air temperature and relative humidity. (Based on data published by US National Weather Service, NOAA.)

remain cool. Hence, anything that reduces or enhances this evaporation will affect how we feel and possibly our health. The figure shows that, for any given air temperature (across the bottom of the figure), the effect of increasing humidity is an increase in the *adjusted* air temperature. Adjusted air temperature is an estimate of the equivalent dry air temperature that the skin (and you) would sense.

Let's take an example. Suppose it's a warm day of 30 degrees Celsius with humidity of 20 per cent (very dry); by entering air temperature and humidity on the figure, you will find yourself with an adjusted air temperature of about 30°C, the 'actual' air temperature. As shown, this is still in the low range of hazard. But because this is very dry air you will be losing water at a great rate, so it is wise to keep the fluids up to your body. However, if the humidity is above 50 per cent (still at the same air temperature, 30°C) sweating will be less effective and your body is less able to keep cool; as a result sunstroke, heat cramps and exhaustion are now possible with prolonged exposure. If the humidity is really high — and this is not uncommon in some parts of Australia — you could find yourself in the high danger category with an adjusted air temperature of 40°C! Put another way, although the air temperature is 'only' 30°C (as measured by a shaded thermometer), your body will be experiencing the level of stress it would if exposed to 40°C.

If the air temperature is much warmer, say 45°C (an extreme day), you will experience a moderate level of stress and discomfort even if the humidity is zero (the water loss from your body will also be extreme). At 45°C any increase in humidity above 35 per cent or thereabouts will put you in the extreme category and real trouble will be just around the corner.

Although Figure 1.1 is based on data that has been published in many places, it should only ever be used as a guide. The purpose for quoting it here is to draw to your attention how easily you can become heat stressed under conditions that you may not consider to be extreme. So, when sailing on a warm day, be wary if the air is humid; even if it's dry and you feel comfortable, be aware that your body will be losing water at a considerable rate and you may become dehydrated if you do not replace this loss.

Chapter 2 includes maps of average maximum temperatures and mid-afternoon humidity for January, April, July and October (that is, the 'mid-season' months) around Australia. From these you can get some idea of the heat stress and dehydration potential of the area you are intending to visit.

Cold stress (hypothermia) and wind chill

In warm conditions we rely on sweating to remain cool, but in the cold, exposing moist skin to moving air enhances cooling and can chill us rapidly.

Figure 1.2 shows how air temperature combines with wind speed to become a potential

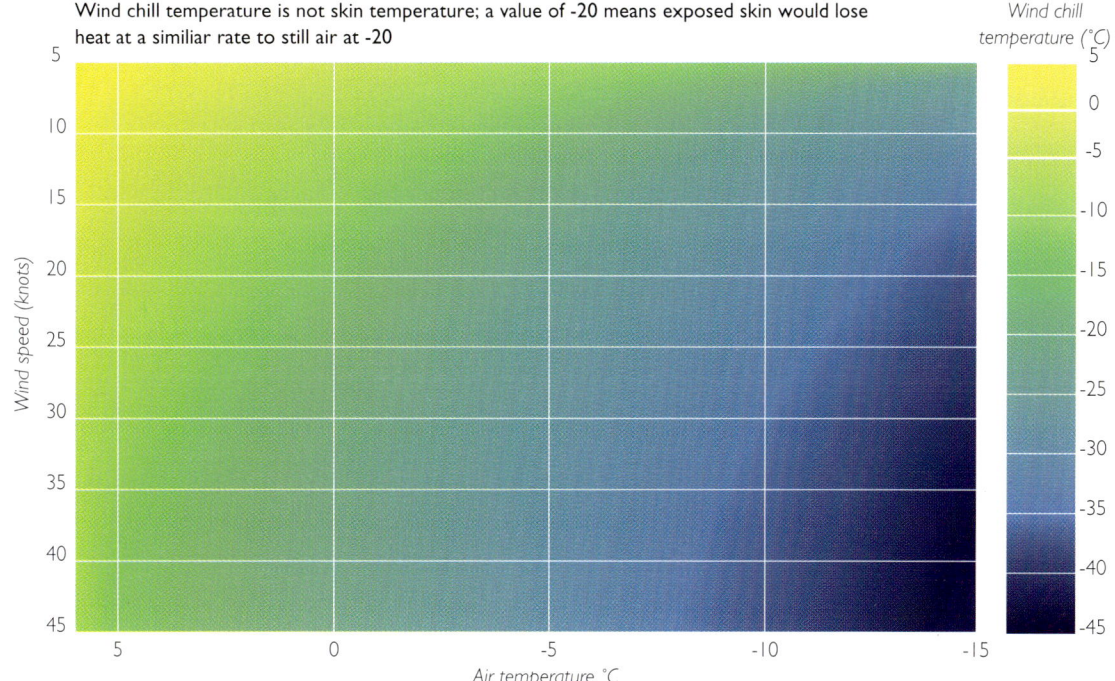

Wind chill temperature is not skin temperature; a value of -20 means exposed skin would lose heat at a similiar rate to still air at -20

Figure 1.2
Potential cold stress hazard (wind chill) posed by combinations of air temperature and wind speed. (Based on data published by Moran & Morgan, 1986.)

hazard, known as hypothermia. This figure shows that, for any given air temperature (across the bottom of the figure), the effect of increasing wind speed is a decrease in skin surface temperature. This is expressed as the 'wind chill' temperature, which is an estimate of an equivalent still air temperature of the skin.

To take an example, with an air temperature of 0°C and a wind speed of 5 to 10 knots, there would be a wind chill temperature of about –10°C, well below the 'actual' air temperature. With the same air temperature and a wind speed of 30 knots, the wind chill temperature is equivalent to –20°C.

As with Figure 1.1, Figure 1.2 should only ever be used as a guide and is intended to draw your attention to how easily you can experience wind chill under conditions you may not consider to be extreme. When sailing on a cool day, for example, be aware that if the wind is blowing your body could be losing heat very rapidly.

Cold stress (hypothermia) by immersion in cold water

An unprotected person immersed in cold water can lose body heat at an alarming rate. Although there is considerable variation in people's response to hypothermia, there are generally agreed immersion/temperature thresholds. Figure 1.3, based on a graph provided by the Australian Oceanographic Data Centre (AODC) in Sydney, shows approximate survival times for various water temperatures. The boundary lines on the original AODC figure have been deliberately blurred and overlapped; for example, in the original figure, the boundary line between safe and intermediate hazard at 5°C water is approximately half an hour — in the figure here, there is virtually no safe immersion time at this temperature. This is done so as to dispel any belief that there is *any* period of time for which you can plunge into cold water without a worry. Depending on your fitness, fat level, tiredness and the timing of your last meal, you may find yourself in trouble very quickly. At 0°C survival times are very short generally. Even in water at a relatively balmy 15°C, immersion for more than an hour can be dangerous.

These are not absolute times; certain people will survive longer and others may not survive as long. Chapter 5 covers the average water temperatures around Australian coasts in more detail. The figure here suggests that an unprotected

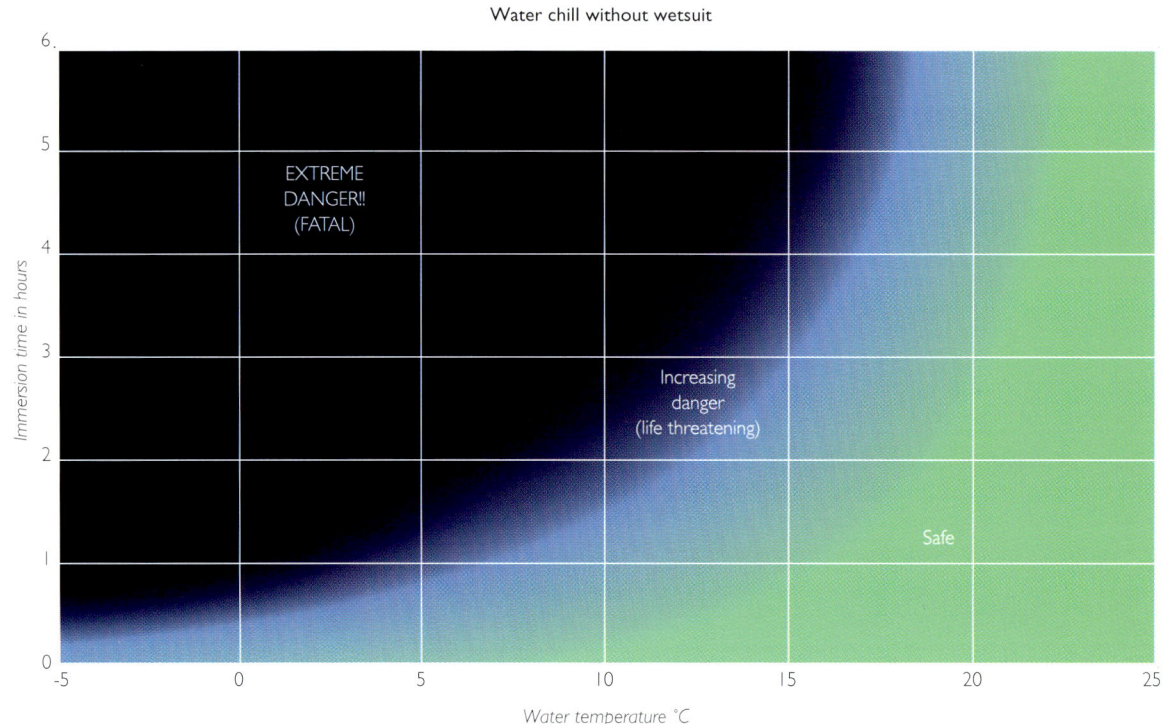

Water chill without wetsuit

EXTREME
DANGER!!
(FATAL)

Increasing
danger
(life threatening)

Safe

Immersion time in hours

Water temperature °C

Figure 1.3
Potential cold stress hazard for immersion in cold water without body protection. (Based on nomogram supplied by the Australian Oceanographic Data Centre, Sydney.)

person (that is, without a wetsuit) could face real danger of hypothermia if immersed for much longer than about an hour at 15°C and half an hour at 5°C.

Climate Essentials

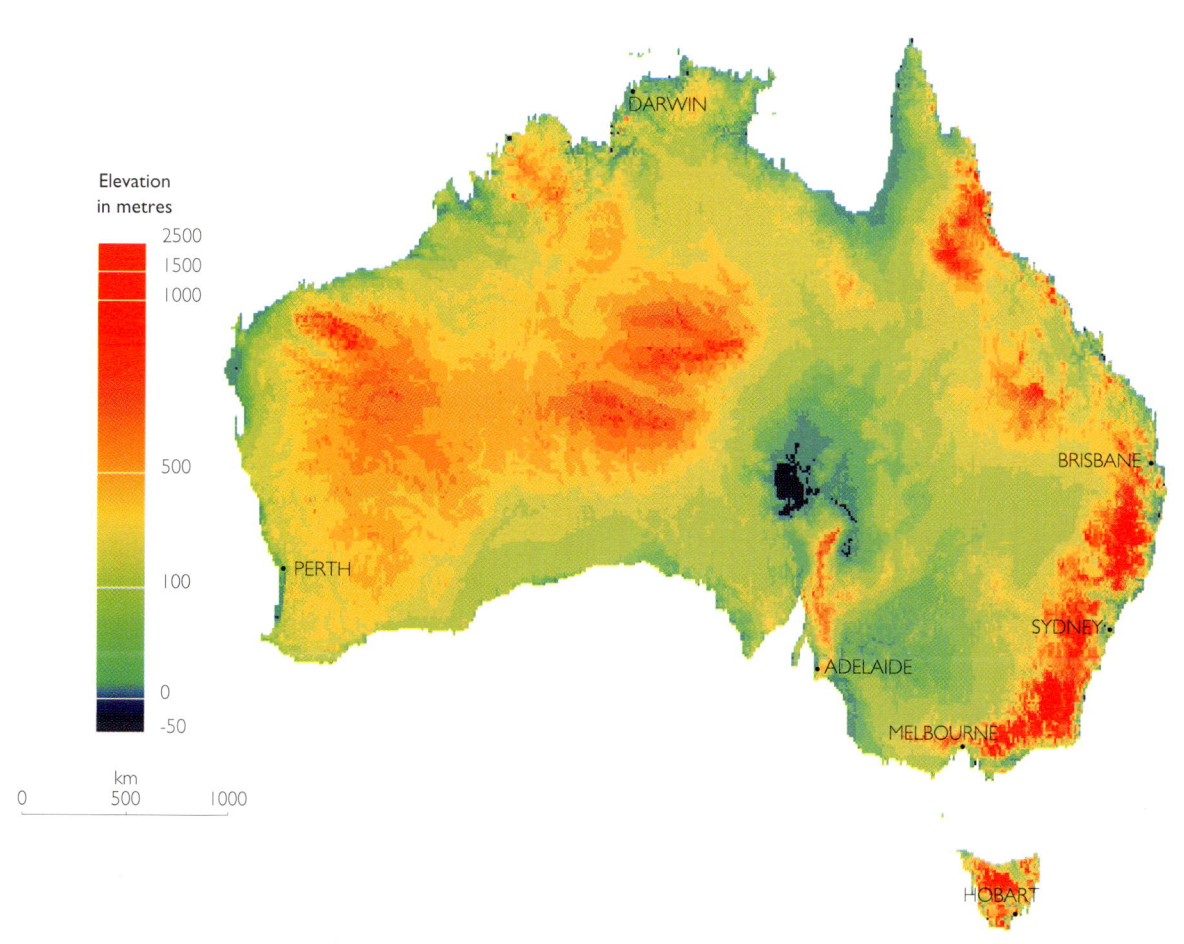

Figure 2.1
A digital terrain model of Australia. (Courtesy M.F. Hutchinson, CRES, Australian National University.)

Coastal weather patterns are greatly influenced by the nearby land. In Figure 2.1, which is a map of Australia showing elevation, the areas below sea level are designated deep blue, sea level green and the highest areas shades of red. The low-lying coastal areas in northern Australia are immediately obvious, as are hilly and mountainous coastal areas in Queensland, New South Wales, Victoria and Tasmania.

This model helps to give a terrestrial context to Australia's climatic conditions, and to explain the patterns in the large-scale weather maps featured later in this chapter.

KEY FACTORS

The key ingredients in climate are air pressure, air masses and fronts, and global air circulation.

Air pressure

Air pressure (or barometric pressure) is the weight of the column of air above a particular location. Air pressure normally decreases with increasing altitude. A key reason for measuring atmospheric pressure is to help forecast the weather. Maps are drawn to show the air pressure at a horizontal level, the most common being mean sea level, or MSL. Air pressure at sea level normally varies between about 970 hecto-Pascals (hPa) and 1040 hPa, but values as low as 870 hPa and as high as 1084 hPa have been recorded. Pressure maps for several levels in the atmosphere are useful for experts because, among other things, they allow them to draw conclusions about the horizontal and vertical motion of the atmosphere.

> A change of pressure at any particular place indicates an alteration in the amount of air above. For example, a falling barometer reading implies a net outflow of air from the region – and this is frequently associated with deteriorating weather. On the other hand, a rising barometer reading implies a net inflow of air into the region – and this is frequently associated with improving weather.

Air masses and fronts

Air that originates in the Antarctic will obviously be quite different from air from the tropical waters to the north. Apart from differing in temperature, the air masses will differ markedly in the amount of moisture they carry, and thus in their ability to form clouds.

Australia is affected by most of the main types of air masses: cold subpolar maritime air originating over the Southern Ocean, hot dry subtropical continental air originating over the Australian continent, and warm moist Pacific and Indian subtropical maritime air originating over the Pacific and Indian oceans.

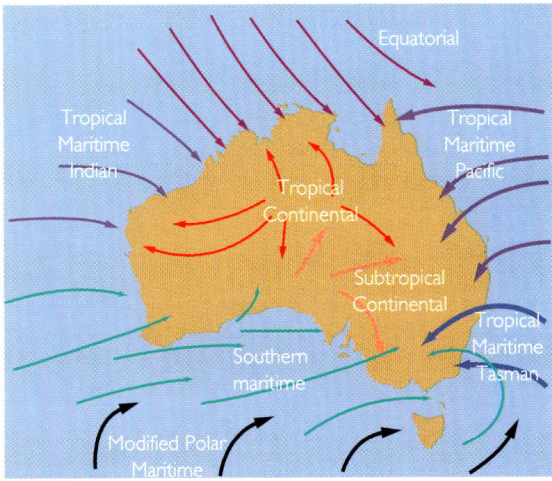

Figure 2.2
The influence of the major air mass sources affecting the Australian climate. (Based on Tapper & Hurry, 1993, *Australia's Weather Patterns.*)

As Figure 2.2 shows, the limits of penetration of the air masses into the continent vary according to the time of year.

In the Southern Hemisphere, winds from high pressure systems rotate anticlockwise and low pressure systems rotate clockwise (the reasons for this are explained in the next chapter). We thus know what type of surface the wind has been passing over, and hence its likely characteristics. For example, if a large high pressure system is bringing onshore winds to eastern Australia, then the air reaching the coast will have travelled over hundreds of kilometres of the Pacific Ocean bringing subtropical maritime air to the coast. On the other hand, if air is being driven over hundreds of kilometres of land, as is often the case in the northern part of the continent when there is a large high over Central Australia, the air is dry tropical continental.

When looking at weather maps, don't forget the enormous size of pressure systems; it is not uncommon for the whole of the Australian continent to be covered by a single high. Eventually, the air will reflect the characteristics of the surface over which it has been blowing.

The transition zone between different air masses is called a front. Fronts are most active in the high latitudes and mid-latitudes and mostly affect the southern parts of Australia. At the

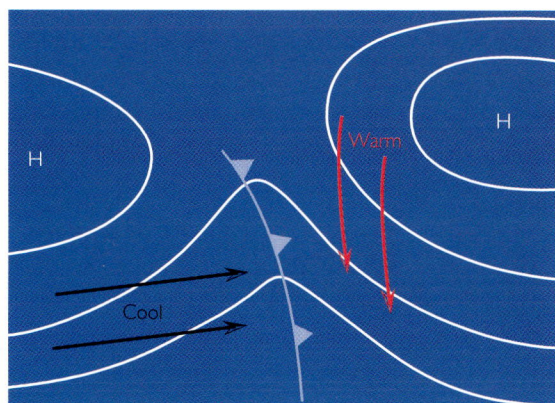

Figure 2.3
A typical pattern of a trough between two anticyclones, conducive to the development of cold fronts.

Earth's surface, the passage of a front is generally marked by a fairly abrupt change in wind and temperature. The three main types of fronts are cold, warm and occluded, although the latter two are seen infrequently in the Australian region.

The formation of fronts around Australia is largely due to the interaction of subtropical and polar air. This produces a continual cycle of fronts developing in the prevailing westerly airflow over the southern parts of the continent, although they may penetrate into the subtropical north.

Certain circulation patterns encourage the formation of fronts. In the Australian region the most common is between two subtropical anti-cyclones (see Figure 2.3).

A front may be many kilometres wide and extend to the upper levels of the atmosphere; it moves at an average speed of 35 kilometres per hour, but can reach more than 50 km/h. Cold fronts tend to move faster than warm fronts. Fronts are not vertical, but slope with height, and much of the characteristic weather and cloud patterns associated with fronts are a consequence of air moving up or down these frontal zones. To understand the patterns of weather from a diagram it is necessary to have some idea of the global processes involved in these patterns.

Global circulation

Although weather changes from day to day, there are basic underlying patterns of pressure and wind.

The main source of energy for the atmospheric circulation of the Earth is its orientation to the Sun. Due to the orientation of the Earth, the tropics receive more energy (heat) than the polar regions.

At ground level in the tropics, warm air from the southern tropics meets warm air from the north. A zone of surface convergence is formed, where the warm air rises and results in low surface pressures. This is called the Intertropical Convergence Zone (ITCZ). Air from above the ITCZ moves towards the polar regions in the upper atmosphere at an altitude of approximately fifteen kilometres. As it moves it radiates heat, cools, becomes denser and subsides, descending around the mid-latitudes (25–30°) to produce a belt of high pressure, known as the subtropical highs (often just referred to as 'highs'). To the south of this, the air ascends again in a belt of low pressure characterised by strong westerly winds — the 'roaring forties' (~40° latitude) — before finally descending as cold dry air toward Antarctica (Figure 2.4).

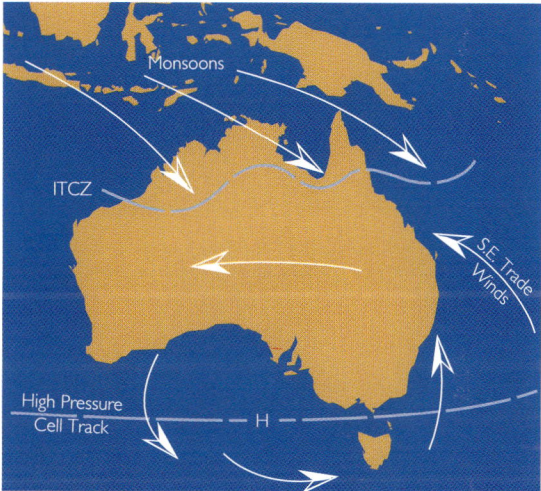

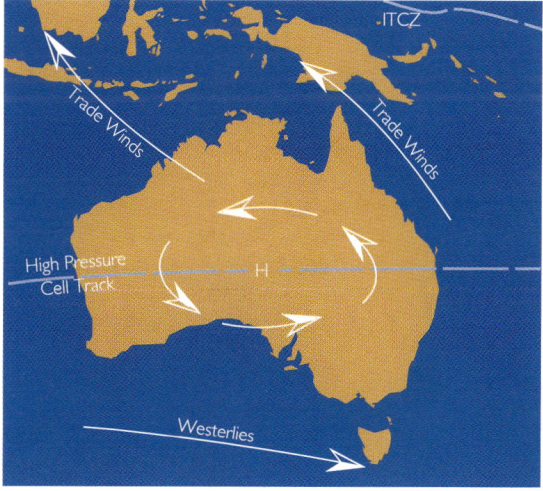

Figure 2.4
A schematic representation of the Intertropical Convergence Zone (ITCZ). Summer (top), Winter (bottom).

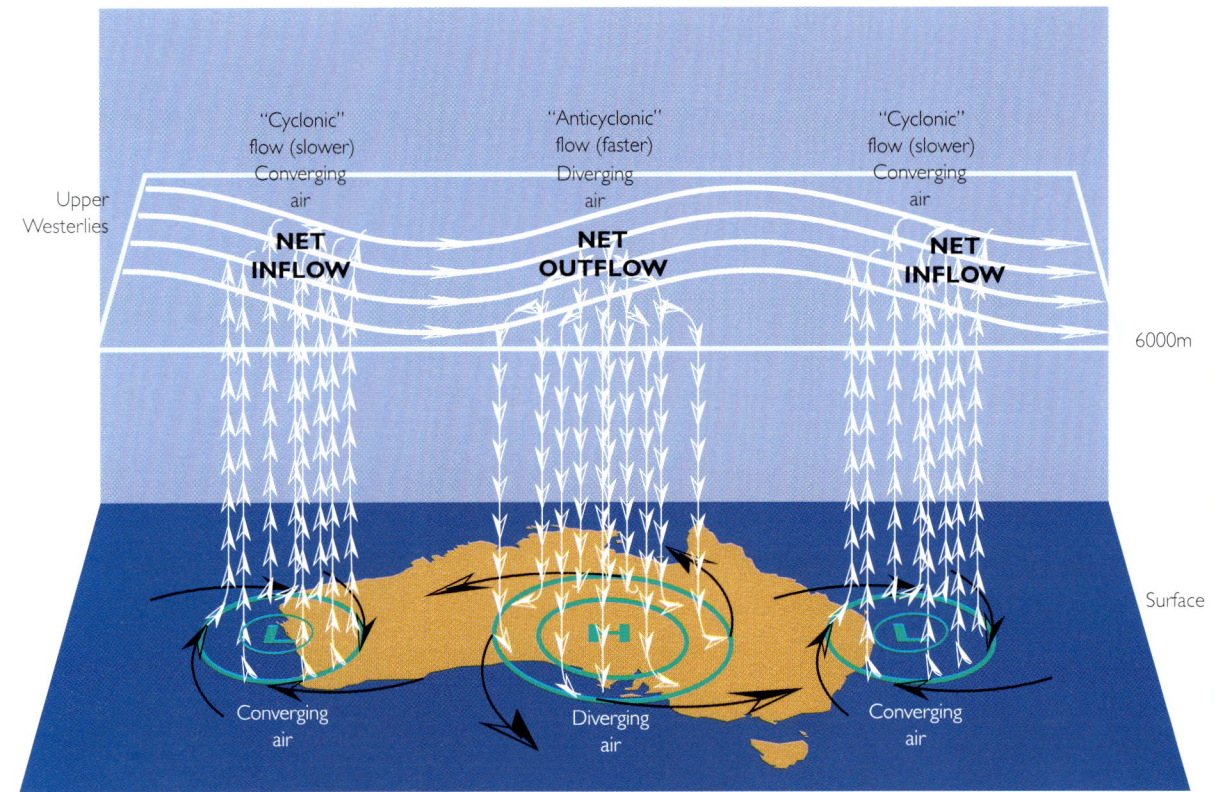

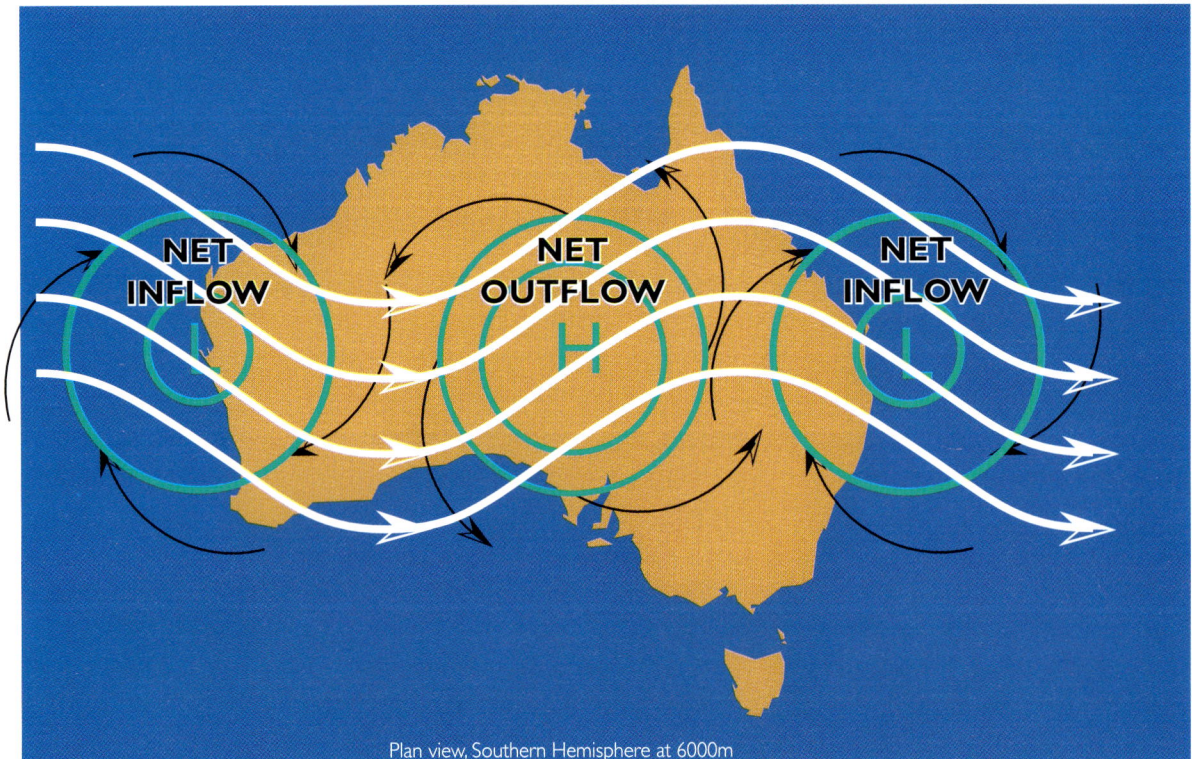

Figure 2.5
A schematic representation of the three-dimensional structure of the atmosphere, highlighting the relationship between the upper westerlies and the surface pressure features.

The vertical and horizontal movements of air form cells of circulation. These cells shift north or south as the latitude of maximum incoming sunlight varies throughout the year.

Air is in constant motion, not only horizontally, but also vertically — it is rising and falling. There are thus intimate links between its flow at high altitudes and ground-level disturbances. Winds at the 500 hPa level (approximately six kilometres above the surface) and above are the counterparts of the surface winds, balancing equator-ward surface flows by pole-ward winds, and vice-versa. In the Southern Hemisphere, the average flow of air is generally westerly, particularly in southern latitudes, and there is a strong westerly flow at 30–45°S.

The westerlies are stronger and more extensive in the Southern Hemisphere than in the northern, due to there being fewer land obstructions to slow down the wind. They are also stronger in winter than in summer as the temperature gradient between the equator and the Poles increases. On any given day, however, the air's circulation pattern may include many regional-scale disturbances that are travelling easterly.

Higher in the atmosphere (around 200 hPa, or about 12 kilometres) the subtropical jet stream meanders over the continent throughout the year, although there are marked seasonal variations. The jet stream is a small bank of extreme wind in the upper part of the general westerly flow. Its existence is a direct result of the mechanism responsible for transfer of heat from the equator to the pole. Both the seasonal and latitudinal features and the year-to-year variations of the jet stream are reflected in the behaviour of the pressure systems of the Australian region. The magnitude and latitude of an anticyclone are closely related to the speed and latitude of the subtropical jet stream, while the pressure trough between successive anticyclones is associated with an equator-ward meander of the subtropical jet stream and its subsequent acceleration.

GENERAL CIRCULATION PATTERNS IN THE AUSTRALIAN REGION

Australia spans latitudes from the tropics (10°S) to the lower middle latitudes (45°S) of the Southern Hemisphere. Its weather is dominated by the descending and diverging air that is characteristic of the belt of subtropical high pressure, generally comprising vast areas of light winds and gently subsiding air. The weather is usually fine, apart from the east coast where winds from highs are onshore.

Figures 2.6 to 2.13 show the average mean sea level (MSL) pressure and inferred large scale (geostrophic) wind direction for January, April, July and October. The wind direction figures show patterns which would exist well above the ground (i.e. away from friction effects). As for most of the maps in this book, the pressure maps show higher values as warmer colours (e.g. orange to red) and lower values as cooler colours (e.g. green to blue).

A simplified summer pattern

In summer, the noon sun is overhead in the Southern Hemisphere and the ITCZ is around 15°S, allowing warm moist air to move over northern parts of the continent, with an alternation of monsoonal north-westerlies and tropical easterlies. Air from above the ITCZ sinks in the vicinity of 34–40°S, and central and southern Australia are influenced by the dry variable winds of the high pressure belt. As the air sinks it warms, and the clear skies associated with subsiding air allow sunlight to heat the land, thus further raising the temperature of the air. With the subtropical highs to the south of Australia, easterly winds predominate over most of the country, although mid-latitude westerlies may be experienced in the south. Sea breezes are common during summer, when the land is strongly heated by the sun and weak horizontal pressure gradients exist (that is, the isobars are wide apart) — these gradients generally occur with large high pressure systems.

In the tropics, the Trade Winds blow from the south-east. When travelling in the Trade Wind belt, especially if you are inland, you will quite commonly experience wind from the same direction (SE) for days on end. During this time the skies are mainly clear and humidity is low. Toward the coast, however, local effects such as sea breezes can overshadow the Trades, especially in the afternoons.

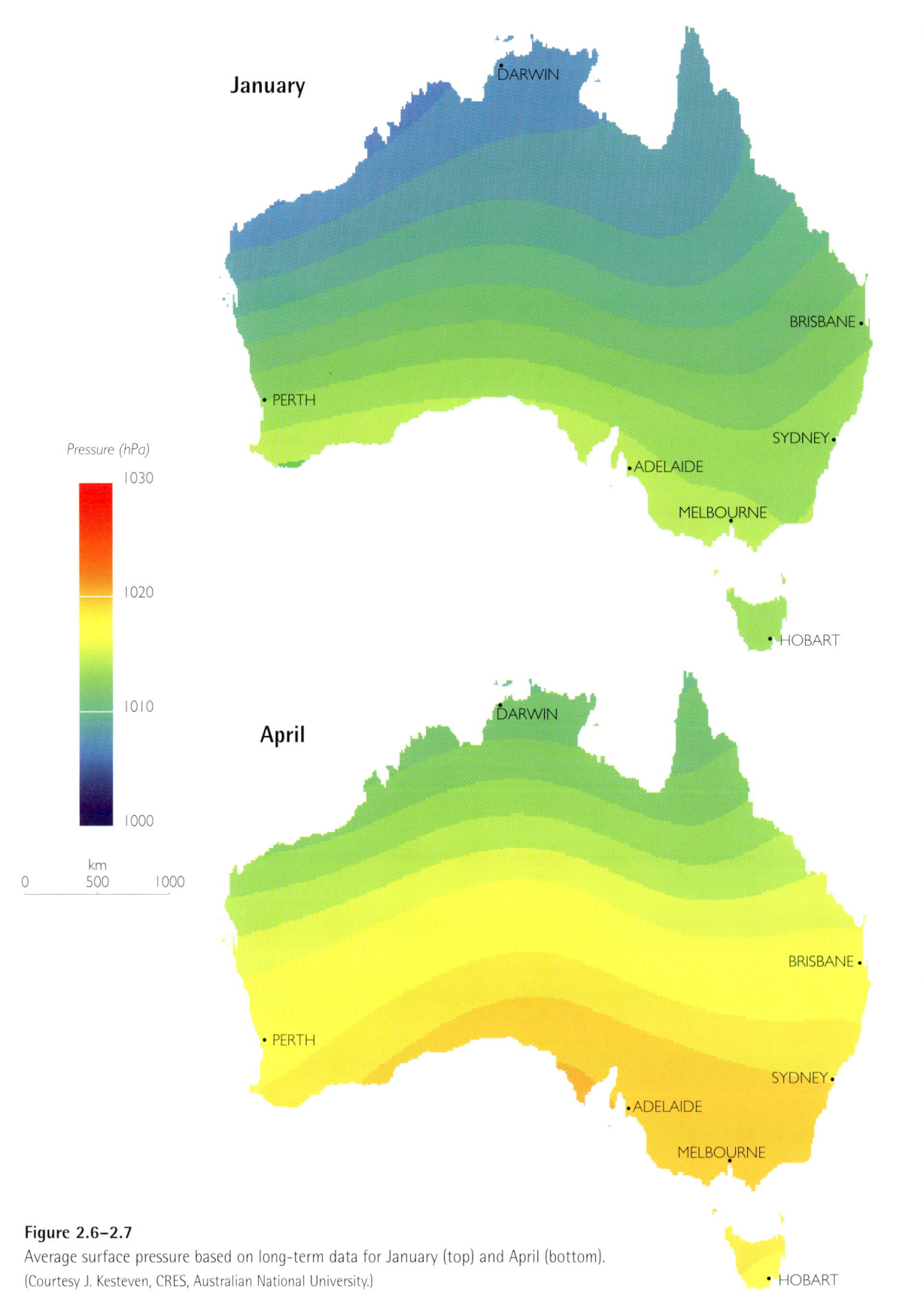

Figure 2.6–2.7
Average surface pressure based on long-term data for January (top) and April (bottom).
(Courtesy J. Kesteven, CRES, Australian National University.)

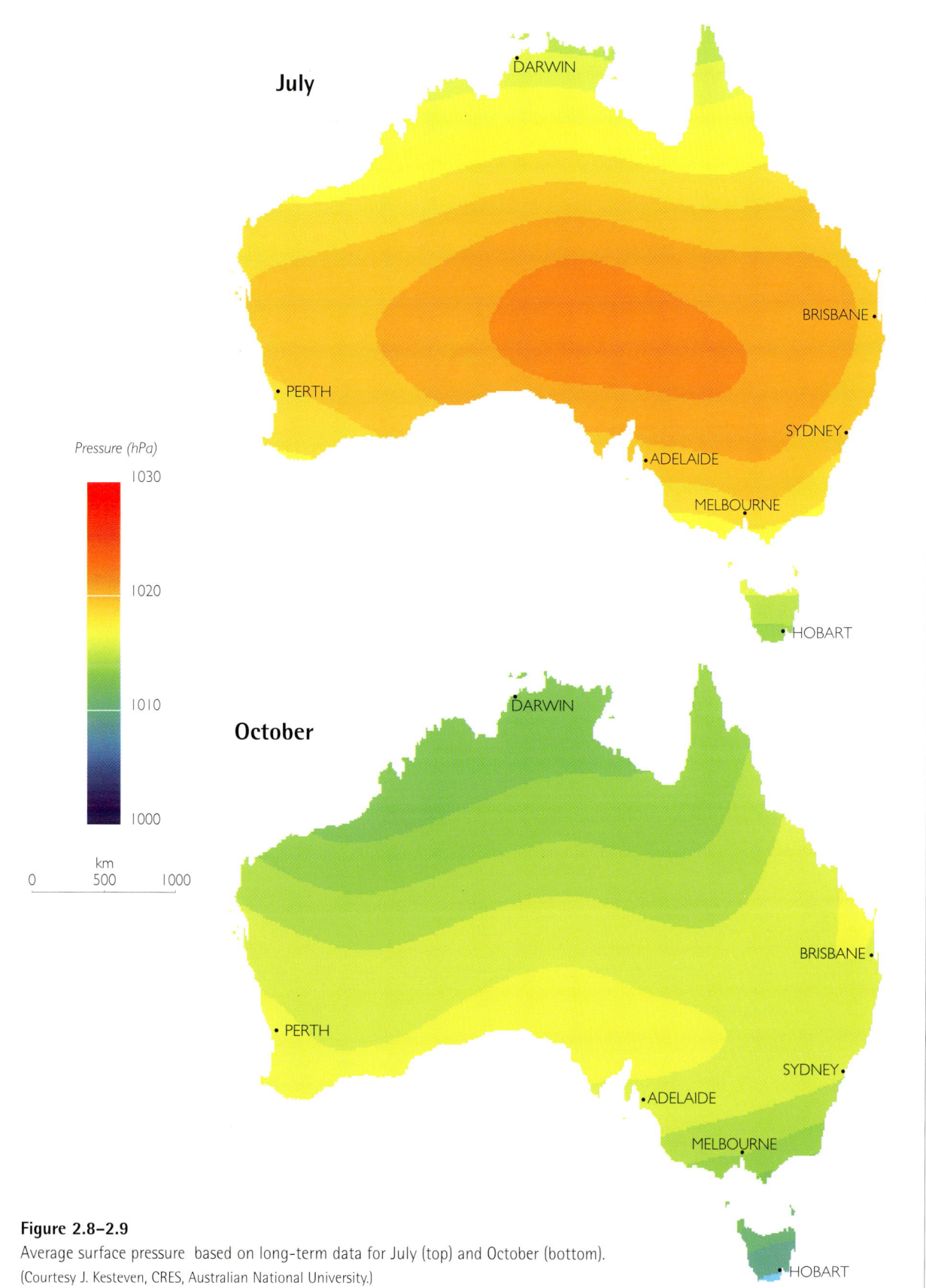

Figure 2.8–2.9

Average surface pressure based on long-term data for July (top) and October (bottom).

(Courtesy J. Kesteven, CRES, Australian National University.)

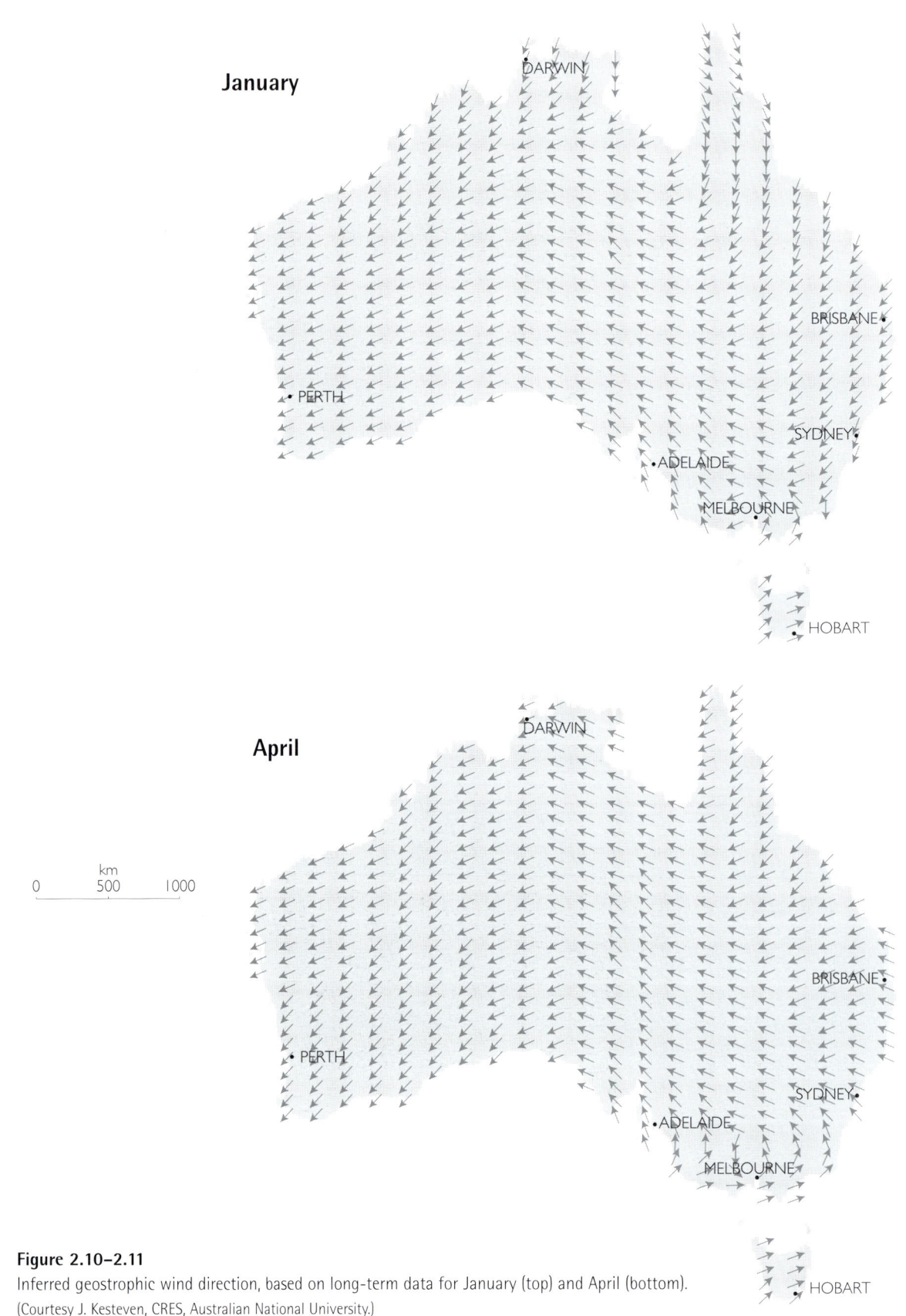

Figure 2.10–2.11
Inferred geostrophic wind direction, based on long-term data for January (top) and April (bottom).
(Courtesy J. Kesteven, CRES, Australian National University.)

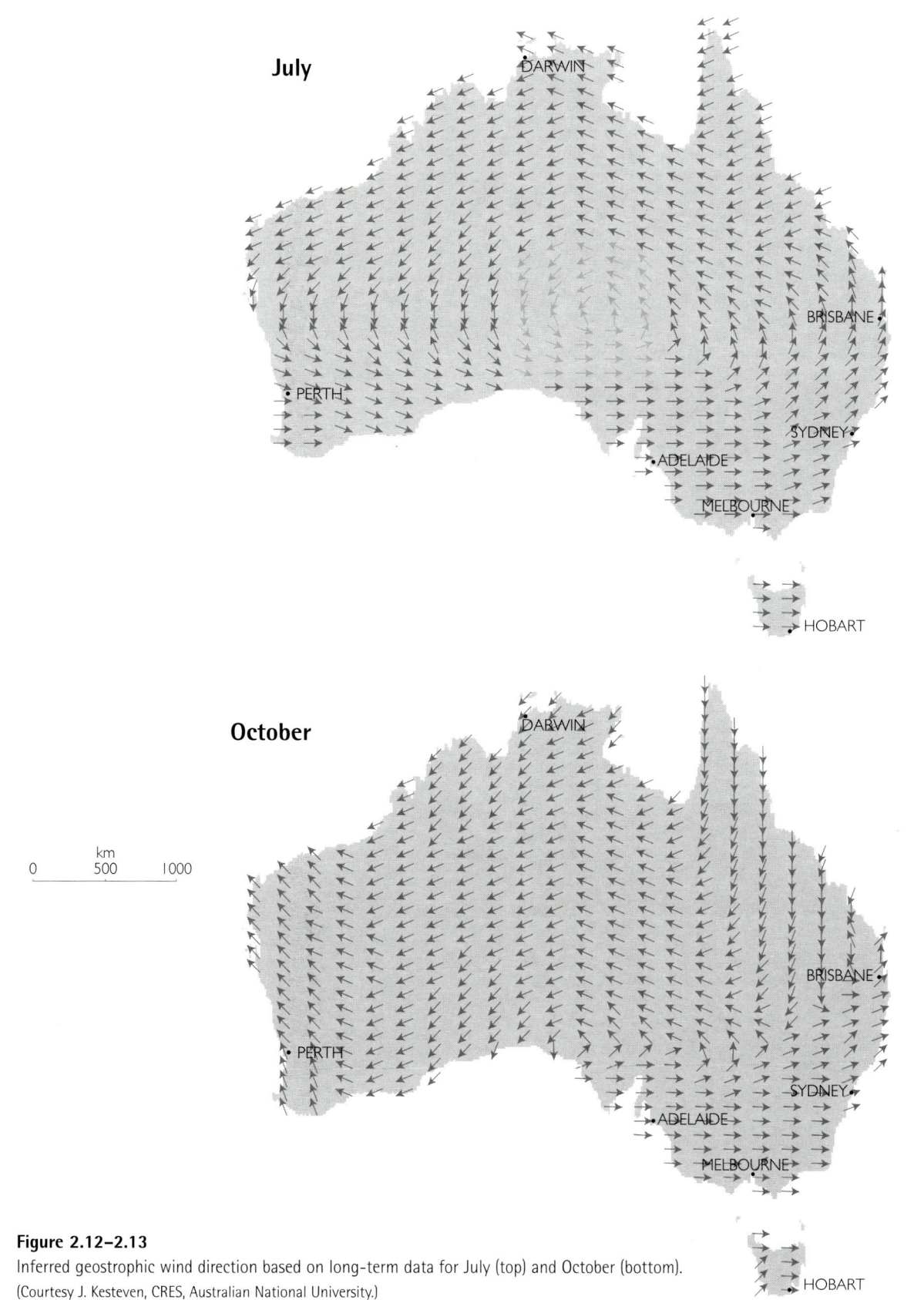

Figure 2.12–2.13

Inferred geostrophic wind direction based on long-term data for July (top) and October (bottom).

(Courtesy J. Kesteven, CRES, Australian National University.)

The northern Australian wet season, sometimes referred to as the 'Monsoon', usually extends from December to March. The west season starts when the ITCZ moves over northern Australia, and warm moist air moves in from the north-west, replacing the drier air of the south-easterly winds that prevail in the dry season.

> Geostrophic wind, or U_g, is a theoretical wind observed well above the Earth's surface, where friction effects are negligible and the pressure gradient (as measured by isobars on weather maps) is balanced by the Coriolis force. This is explained in detail in Appendix V. At 1000 metres or more above the Earth's surface, observed winds are frequently in close agreement with U_g; the term *theoretical* indicates that it is based on a theoretical equation.

The January maps show a large area of relatively low pressure over the northern part of the continent — the arrows certainly suggest the presence of the ITCZ. To the south, pressure increases, more or less, with increasing latitude. In April (Figure 2.7), MSL pressures are considerably higher than in January, especially in the southern and south-eastern parts of the mainland. The mid-winter pattern (July, Figure 2.8) is quite different: MSL pressures no longer simply increase with latitude, but instead the subtropical high pressure belt is over Central Australia and pressure decreases to the north and south. Bearing in mind that we are talking about long-term average pressures (and hence the maps do not show individual high or low pressure cells), the mid-winter period resembles a single large high pressure system centred on the continent. October resembles April, where the high pressure centre is displaced to the south and with lower pressures than in the mid-winter period.

A simplified winter pattern

In winter, the overhead noon sun has moved to the Northern Hemisphere and there is a build-up of solar energy in the vicinity of 15°N, resulting in the ITCZ being around 15–20°N. Over northern Australia fine weather prevails under the influence of the tropical easterlies. The winter cooling of the land causes an increase in the density of the overlying air, and a consequent alteration in the barometric pattern. The high pressure belt moves north with the seasonal change, generally being between 29°S and 32°S. Usually the track is at its furthest north over most of Australia in June–July, but, due to the greater lag in water temperatures off the east coast, the northward deflection can be retarded until August–September. The subtropical high pressure belt is narrower than in summer and is broken into cells; areas under the influence of the centre of the system experience settled weather, and westerly winds cover the southern parts of the landmass.

During winter, the temperate or extra-tropical cyclones (also called 'mid-latitude depressions' or 'temperate cyclones'), which usually track well south of the continent, move closer (normally they are closest in July). These deep and complex lows maintain a westerly airflow. Any large depression well to the south of Australia sends a strong southerly flow toward the land, which may be as cold as 2–3°C when it reaches the southern shores. If these depressions move north with their centres passing near Tasmania, the rapid passage of cold fronts gives surges of very cold southerly air, and, if upper level moisture is high, general precipitation results (rain or snow usually). As the trough moves eastward and the next anticyclone becomes dominant, winds abate, skies clear, nocturnal radiation increases and widespread frosts result. Between latitudes 40°S and 50°S is a zone of strong westerly winds, which develop in these latitudes as the oceans extend around the globe practically unimpeded by land.

Southern Australia is usually on the northern margin of the westerlies and their influence is felt most in winter and spring, when storms in the belt of mid-latitude depressions are further north than usual. Western Tasmania is particularly affected by these westerlies, which result in windy, wet winters.

And so it is that Australia's weather in winter is characterised by a more or less regular progression of large high pressure cells (also called anticyclones) from the west to the east, bringing (typically) three to five days of fair weather. In between the highs you get cold fronts (polar air), warm fronts (rarely), and occasional low pressure systems (usually to the far north and south of the continent). These inter-high periods are typically unsettled with some cloud and rain.

MAIN FEATURES OF AUSTRALIAN WEATHER

The pattern of surface pressure on any given day is composed of a series of areas (or cells) of high and low pressure, known as pressure systems. The pressure cells interact with each other. The extent of the interaction depends on factors such as the distance, and the strength of the pressure gradient, between pressure centres. Pressure systems tend not to be stationary. Generally they move from west to east in southern Australia. Over northern Australia, the systems tend to move from east to west, except during the wet season, when they tend to move from the north-west to the south-east. The direction and speed of a pressure system are influenced by the rotation of the Earth, variations in topography and inter-actions with other pressure cells.

High pressure systems

A high pressure system (sometimes called a 'high' or an anticyclone) is a downward moving mass of air that rotates anticlockwise (in the Southern Hemisphere) and diverges at the Earth's surface (see Figure 2.5) under the influence of the pressure gradient and the Earth's rotation (due to the Coriolis force, which is explained in Chapter 5 & Appendix V).

High pressure systems generally form in one of two ways: by the convergence and descent of cold dense air in the upper atmosphere, or by the movement of cold dense air into the lower layers of a relatively warm atmosphere (resulting in highs or 'cold highs'). The latter form in the first 2000 to 3000 metres of the atmosphere and commonly happen in winter (in Australia), when modified polar air moves in behind a northward moving front.

High pressure systems are characterised by subsiding air and weak pressure gradients and can extend over several thousands of kilometres. Because the air is subsiding, the air mass tends to be stable (that is, it will resist vertical movement) and high pressure systems are generally associated with fine weather (apart from the east coast as mentioned earlier). The troughs that form between the high pressure centres play an important role in the formation of fronts, as the two adjacent anti-cyclones draw contrasting tropical and polar air masses into these troughs (see Figure 2.3).

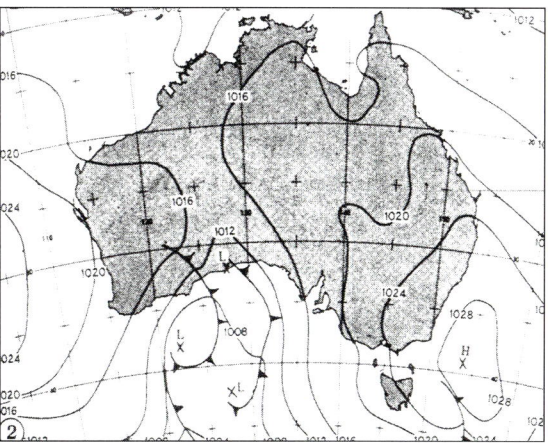

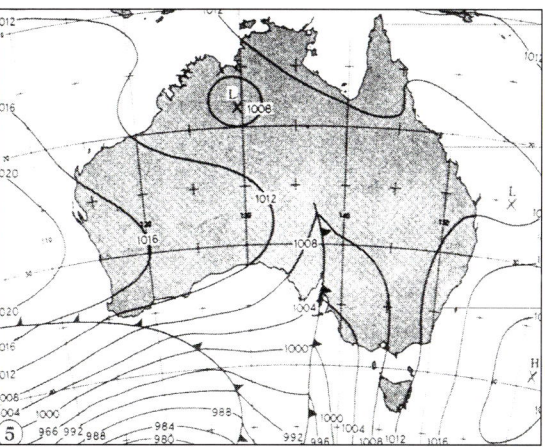

Figure 2.14
Examples of weather maps showing a blocking high.
(Courtesy Commonwealth Bureau of Meteorology, Melbourne.)

Normally, high pressure systems progress from west to east across Australia at a slower rate than the travelling depressions (lows) further south. Sometimes, though, a slow moving anti-cyclone may occur with strong pressure gradients around its perimeter, which steer approaching pressure systems around its perimeter. These are known as 'blocking highs'. In the vicinity of Australia such anticyclones frequently occur in the eastern Tasman Sea in the region of New Zealand.

A region under the direct influence of a blocking anticyclone experiences relatively low rainfall. The blocking anticyclonic zones are generally pole-ward, although it is quite common for a high pressure ridge to extend into southern parts of Australia, causing persistent dry weather there (Figure 2.14).

Low pressure systems

A low pressure system (sometimes called a 'depression', 'cyclone' or 'low') is upward moving air that rotates clockwise (in the Southern Hemisphere) and converges at the Earth's surface (see Figure 2.5). The rising air causes adiabatic cooling (cooling by expansion of the air mass), which often results in condensation, cloud formation and rain. In contrast to a high, the low pressure system may consist of unstable air — that is, air which moves vertically — so that lows are often associated with unsettled weather.

Low pressure systems may form in one of two ways: with the coldest air at their centre or core, or with a warm core. The former occur mainly in the high and mid-latitudes and result in temperate or extra-tropical cyclones, or mid-latitude depressions, while the latter form mainly in the low latitudes and result in tropical cyclones, heat lows, cut-off lows and monsoon depressions. The different temperature structures of the two kinds of low pressure systems produce different wind patterns, with the warm-cored cyclones having the potential to be more destructive by far.

Temperate or extra-tropical cyclones, or mid-latitude depressions

Large cyclones move regularly across the world's temperate latitudes, mainly between 40–65° north and south. These low pressure systems, with diameters often larger than 1000 kilometres, are also sometimes known as 'mid-latitude lows'. They usually contain both cold and warm fronts, although the latter are seldom shown on Australian weather maps. Although normally well south of Australia, these depressions exert a considerable influence on the weather of southern Australia, particularly in winter, when they may be associated with very strong winds, as described in Chapter 6. The passing of cold fronts is associated with overcast conditions, frequently accompanied by rain, drizzle or snow.

Tropical cyclones

Small, intense low pressure cells sometimes occur along the northern Australian coast and are known as 'tropical cyclones' if their winds reach gale-force strength. They are the most destructive of the tropical weather systems, being associated with stormy weather, extremely strong winds, heavy rainfall and high seas, which can result in severe damage to property and loss of life. All

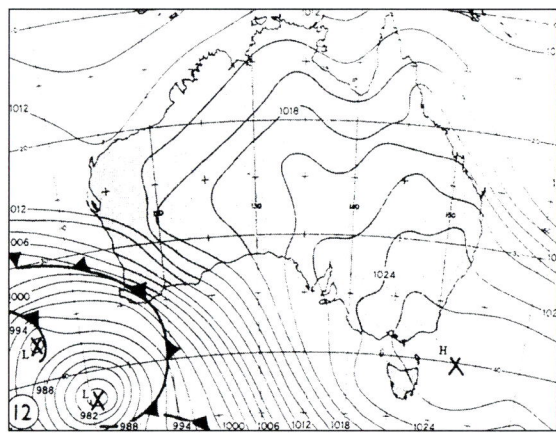

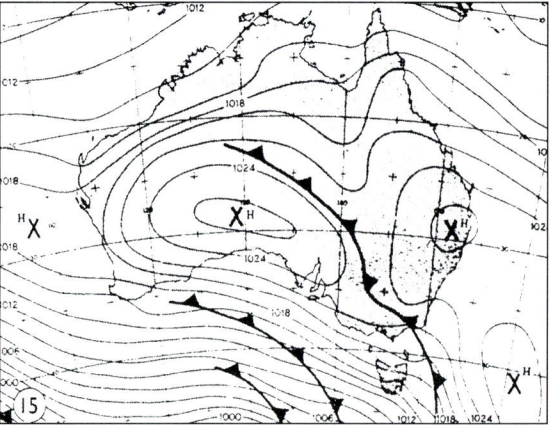

Figure 2.16

Examples of weather maps showing a temperate cyclone.

(Courtesy Commonwealth Bureau of Meteorology, Melbourne.)

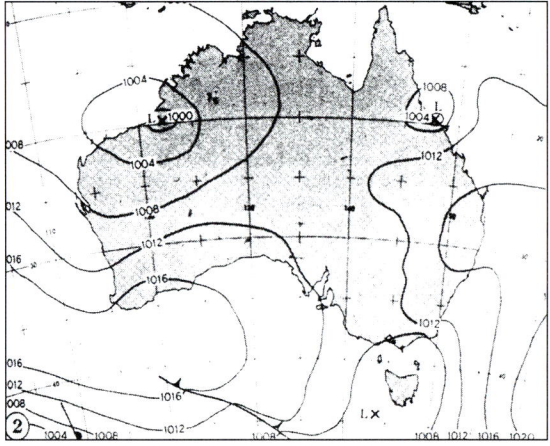

Figure 2.15

An example of a weather map showing a series of lows.

(Courtesy Commonwealth Bureau of Meteorology, Melbourne.)

tropical cyclones pass through a life cycle of immaturity, maturity, and finally decay (when they can degenerate into tropical rain depressions, usually after crossing the coast). These depressions are very important in northern Australia, bringing much needed rain to pastoral lands. Tropical cyclones are discussed in more detail in Chapter 6.

Heat lows

Lows also form in Australia from strong heating of the surface during summer, resulting in rising air and falling surface pressure — commonly called 'heat lows'. Heat lows with a light cyclonic circulation often occur over northern Western Australia ('Pilbara heat low') and north-western Queensland ('Cloncurry heat low'). They are important as they help draw onshore moisture-laden oceanic air from the north and west. Although quite extensive and persistent in summer, heat lows are generally quite shallow (that is, they have a weak pressure gradient and do not have very low pressure at their centre).

Cut-off lows

Cut-off lows are formed when a part of a low pressure system — initially in the upper air — becomes isolated from the main low pressure system by a high pressure system that may be ridging rapidly eastwards. This frequently occurs in the lower temperate latitudes of the Australian region (for example, 30–40°S), particularly along the east coast. The cut-off low often deepens and may cause extensive rains and even floods. Southern

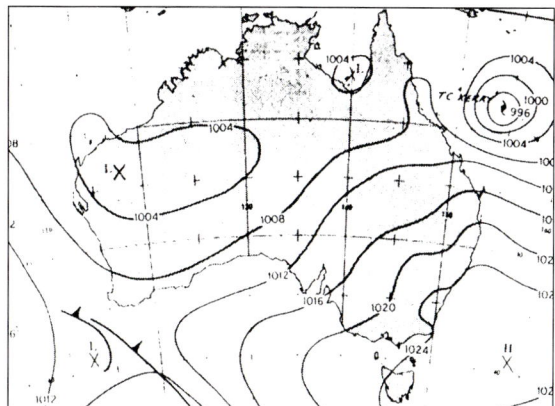

Figure 2.17
An example of a weather map showing a tropical cyclone.
(Courtesy Commonwealth Bureau of Meteorology, Melbourne.)

Australia receives a substantial portion of its winter and spring rainfall from cut-off low pressure systems, especially in the wheat growing areas of western Victoria and southern New South Wales. Rain-producing cut-off lows occur on average slightly more often than once per month from August to November and produce approximately 30 per cent of the total rainfall during that period.

Monsoon depressions

The northern Australian heat lows are necessary forerunners of the formation of the monsoon circulation, as they draw moist tropical air onto the continent and provide the base state that is then triggered into monsoon depressions by

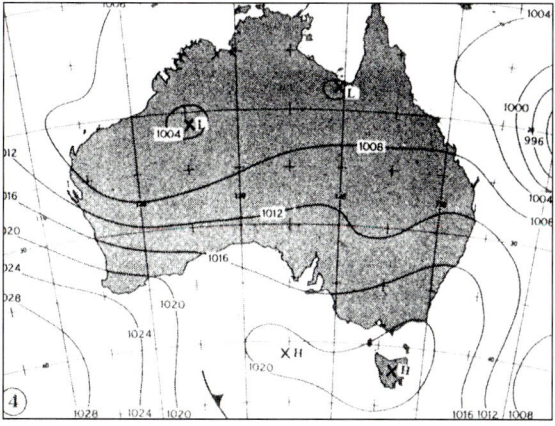

Figure 2.18
An example of a weather map showing a heat low. (Courtesy Commonwealth Bureau of Meteorology, Melbourne.)

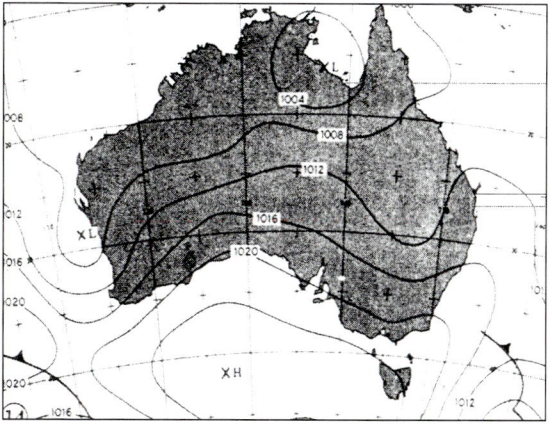

Figure 2.19
An example of a weather map showing a monsoon depression.
(Courtesy Commonwealth Bureau of Meteorology, Melbourne.)

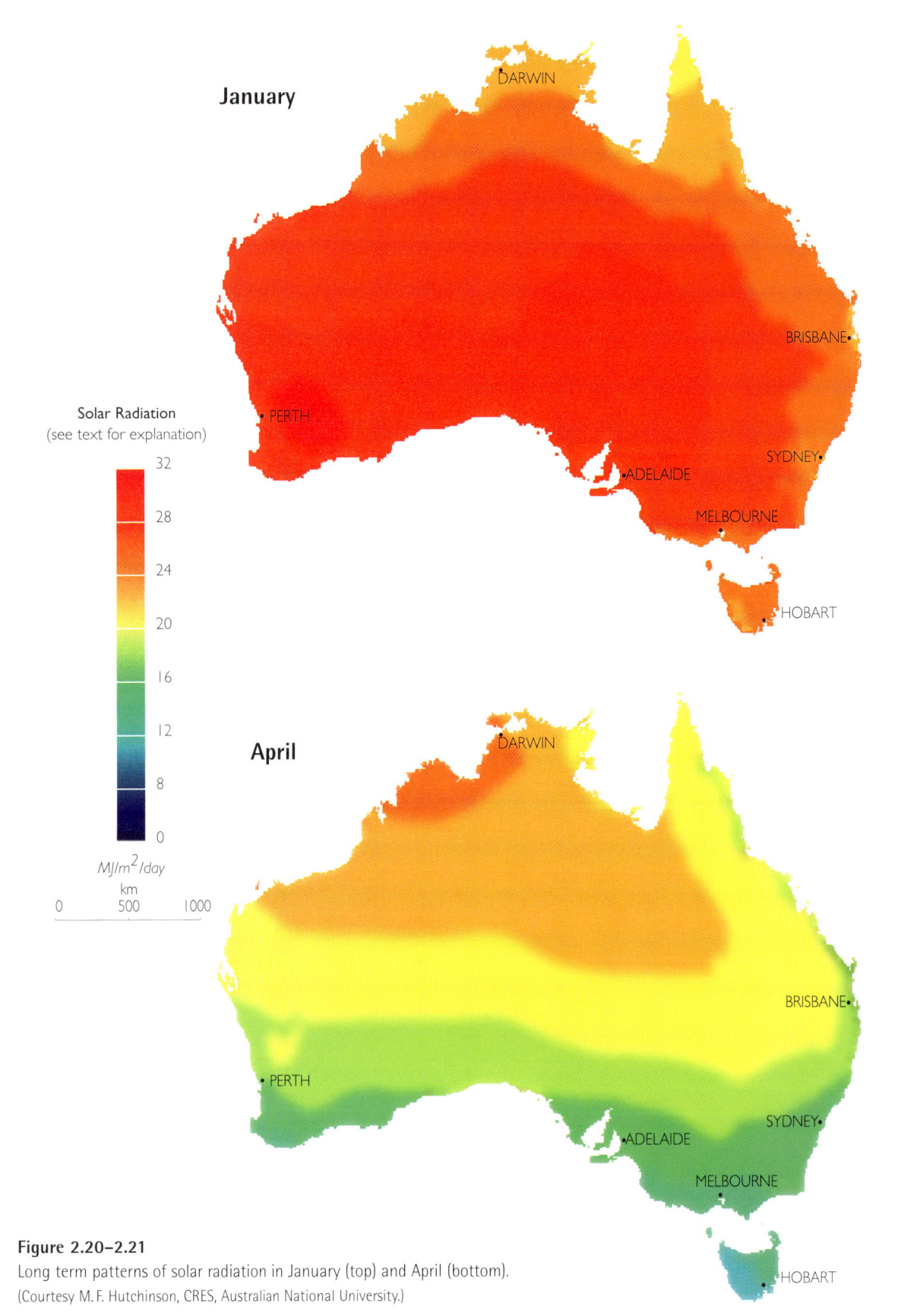

Figure 2.20–2.21
Long term patterns of solar radiation in January (top) and April (bottom).
(Courtesy M. F. Hutchinson, CRES, Australian National University.)

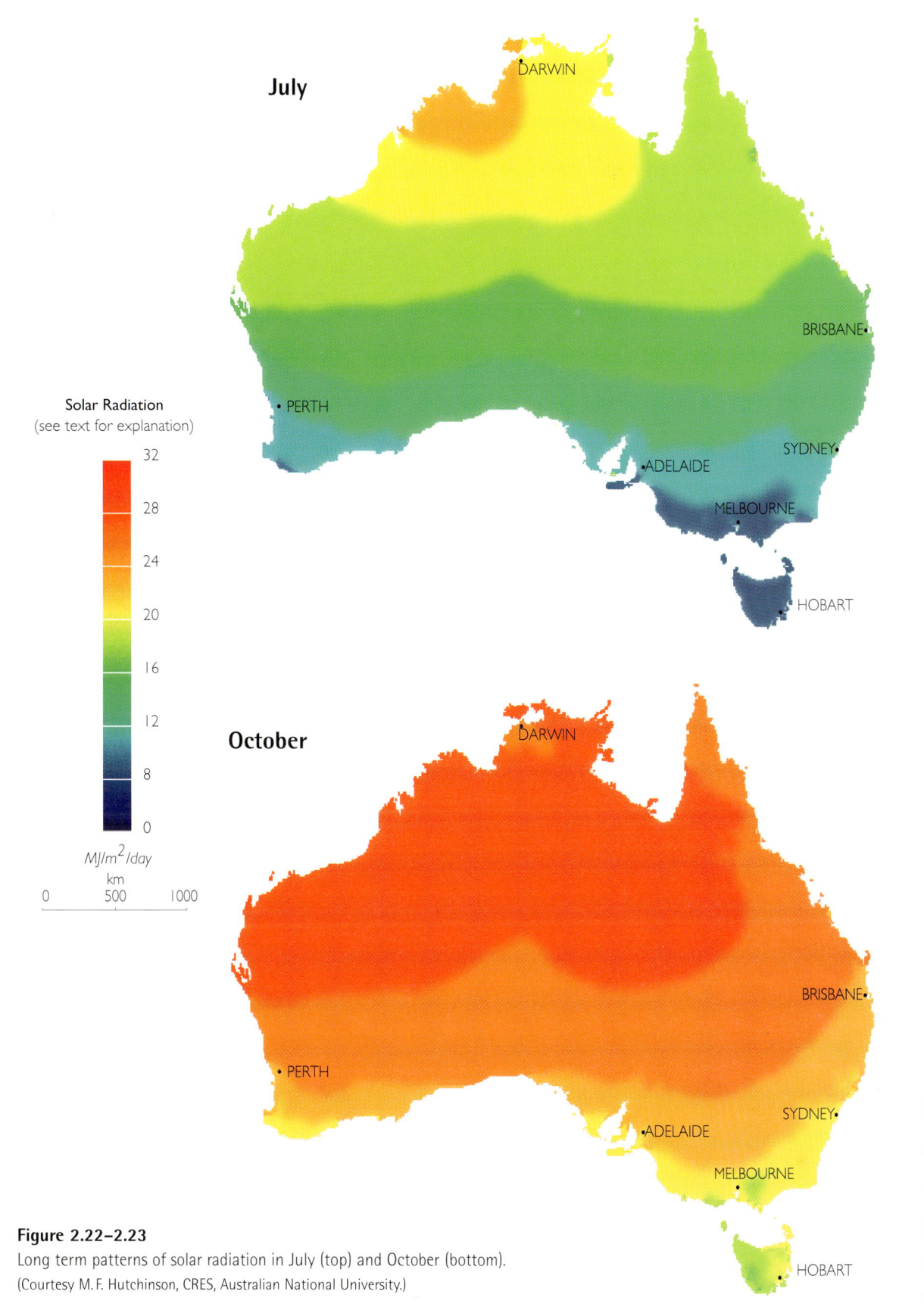

Figure 2.22–2.23

Long term patterns of solar radiation in July (top) and October (bottom).

(Courtesy M. F. Hutchinson, CRES, Australian National University.)

higher latitude air movements.

Monsoon depressions occur in the Australian tropics about five times a year and are responsible for a significant portion of the seasonal rainfall. They are normally located over land, but may intensify into tropical cyclones when they move over warm ocean water. Although less destructive than tropical cyclones, monsoon depressions can often be quite strong, with winds approaching 40 knots at the Earth's surface.

MAPPING THE CLIMATE

Solar radiation; highest and lowest expected temperatures; rainfall; humidity; wind speed and direction — these are all significant concerns for people who spend time in coastal situations. The seasonal patterns are shown as maps for January, April, July and October, the 'mid-season' months. Other months can be 'guestimated' — for example, February would resemble January, March might resemble April more than January, and so on. Maps are based on all available data.

Solar radiation (and sunburn potential)

One of the most persistent hazards of water based activities is sunburn. If there is no breeze to cool down the exposed skin, you will normally feel solar radiation as warmth and 'see' it as brightness. If there is a breeze and the air is cool, the strength of the radiation can be masked. Without realising it, you can get severely burned.

As we have seen, the surface of our skin is wet and relies on evaporation for cooling; anything that increases evaporation (low humidity, air movement) will cool the skin. In dry and cool (and moving) air, the body feels cool even though it may be intercepting very strong radiation. People on the water are particularly vulnerable to sunburn because the direct solar radiation (which could already be at high levels) is 'enhanced' by reflection from wet surfaces such as the deck or the water itself.

There are very few sites in Australia that measure the 'burning' ultraviolet (UV) component of solar radiation — certainly not enough to produce reliable maps. Figures 2.20 to 2.23 map 'solar radiation' rather than sunburn potential because they are based on computer models of *total* solar radiation rather than just the UV component of it. Many published maps are for an atmosphere-less Earth, but these maps have been corrected for altitude, dust, cloud cover, rainfall and latitude, to reflect conditions at the ground accurately.

Surprisingly, the effect of latitude on solar radiation is not straightforward, in the sense that the closer you get to the tropics the stronger the sun will become. The tropics also experience more cloud cover and rain than many temperate areas, and this reduces average radiation levels. Also, as one moves away from the tropics, a higher proportion of solar radiation is reflected and scattered by the atmosphere, rather than arriving at the surface directly. This effect is very noticeable in Tasmania, for example, where solar radiation can still be very high even though the centre of Tasmania, at about 42°S, is a long way from the tropics. Anyone who has bushwalked in Tasmania's Cradle Mountain region in January knows you can get badly burned very quickly on a clear or slightly hazy day.

> Sometimes high, wispy or hazy cloud acts like a magnifying glass, concentrating radiation. The effect of dense, dark cloud is the opposite. Advice: use sun protection, especially on clear days and where there is high, wispy cloud.

Figures 2.20 to 2.23 map the solar radiation for January, April, July and October. In these maps, the higher the potential for sunburn, the 'warmer' the colour; hence red shows the most extreme solar radiation levels. It is safe to say that the potential to be sunburnt is very high at the medium to high solar radiation levels.

> Remember that the maps are based on total solar radiation, or loosely, sunlight. The burning UV component is a small fraction of this total and one that varies. Under certain atmospheric conditions it is possible for human skin to burn very quickly, even in regions with relatively low levels of total solar radiation. Always take precautions

against sunburn.

Also, these maps merely identify areas where radiation is high (long, clear days, high solar angle) and where it is low (shorter days, cloudy or overcast, low sun angle). There is NO simple correlation between total solar radiation and burning UV levels. Nevertheless, you should find the maps useful for locating areas that have lots of sun and areas that do not.

Maximum air temperatures

Maximum air temperatures should be of interest to readers of this book because the time when these highest temperatures are reached — usually 2 to 3 pm — often corresponds to the peak period of coastal usage and the time of maximum health hazard or discomfort. You might be surprised by just how high some of these averages are, especially given that, on a daily basis, much higher (and lower) temperatures will occur.

The hottest time of the year becomes progressively later from north (November) to south (February), corresponding in a general way with the time of most effective heating by the sun. There is a great diversity of temperatures across Australia, stemming from the effects of latitude, altitude and the degree of 'continentality' (the tendency for large landmasses to experience higher maxima and lower minima than smaller landmasses). Temperatures range from humid tropical heat in the north through the dry heat of the centre to alpine winters in the highlands of the south-east.

Remember Figure 1.1, which illustrated the potential heat stress hazard posed by combinations of air temperature and relative humidity? If you examine the maximum temperature maps alongside the mid-afternoon relative humidity maps later in this chapter, you will be able to identify areas and times of year where and when conditions could be hazardous to your health. There are many examples of high temperatures being combined with high humidity levels (in the same month) and you should be aware of the dangers these represent.

Australia occupies a larger extent of the drier latitudes (those under the influence of the subtropical high pressure belt) of the Southern Hemisphere than either South America or Africa. As such, Australia may be considered the hottest continental mass, despite the fact that higher temperatures occur in both Asia and Africa. This large longitudinal span in the dry latitudes causes modifications in the air masses that travel over the surface of the land, so that the degree of continentality is greater than might be expected.

Briefly, very high top temperatures can be expected over large areas of Australia in January, especially to the north-west at or around the latitude of Port Hedland.

In some areas, very high maximum temperatures occur quite close to the coast, so you should not assume that the weather will be relatively 'mild' by the sea. Even Tasmania — which many people expect to have mild temperatures during summer — experiences average top temperatures in the mid-to-high 20s (the extreme maximum for Hobart, for example, is 40.6°C).

Figures 2.24 to 2.27 map daily maximum air temperatures, averaged for the 'mid-season' months and based on a long period of record. (As an aside, it is interesting to compare these maps with the solar radiation maps, Figures 2.20 to 2.23.)

In April, maximum temperatures are usually much lower than January, but there are still many parts of Australia that experience an *average* of 35°C! The mid-winter pattern sees much cooler temperatures in the southern parts of the continent, especially in elevated south-eastern areas. Nevertheless, there is no doubt that the eastern states have generally milder temperatures. The October map, again, may surprise some people, since the 'extreme' maximum temperatures are once more evident in the northern and north-western parts of the continent.

Minimum air temperatures

Minimum air temperatures are usually reached around an hour before dawn. This does not usually correspond to a time of high coastal usage, yet minimum temperatures will influence how well (or whether) you sleep and this is important if you are camped for the night or moored close-in in your boat. Local conditions can strongly influence minimum temperatures.

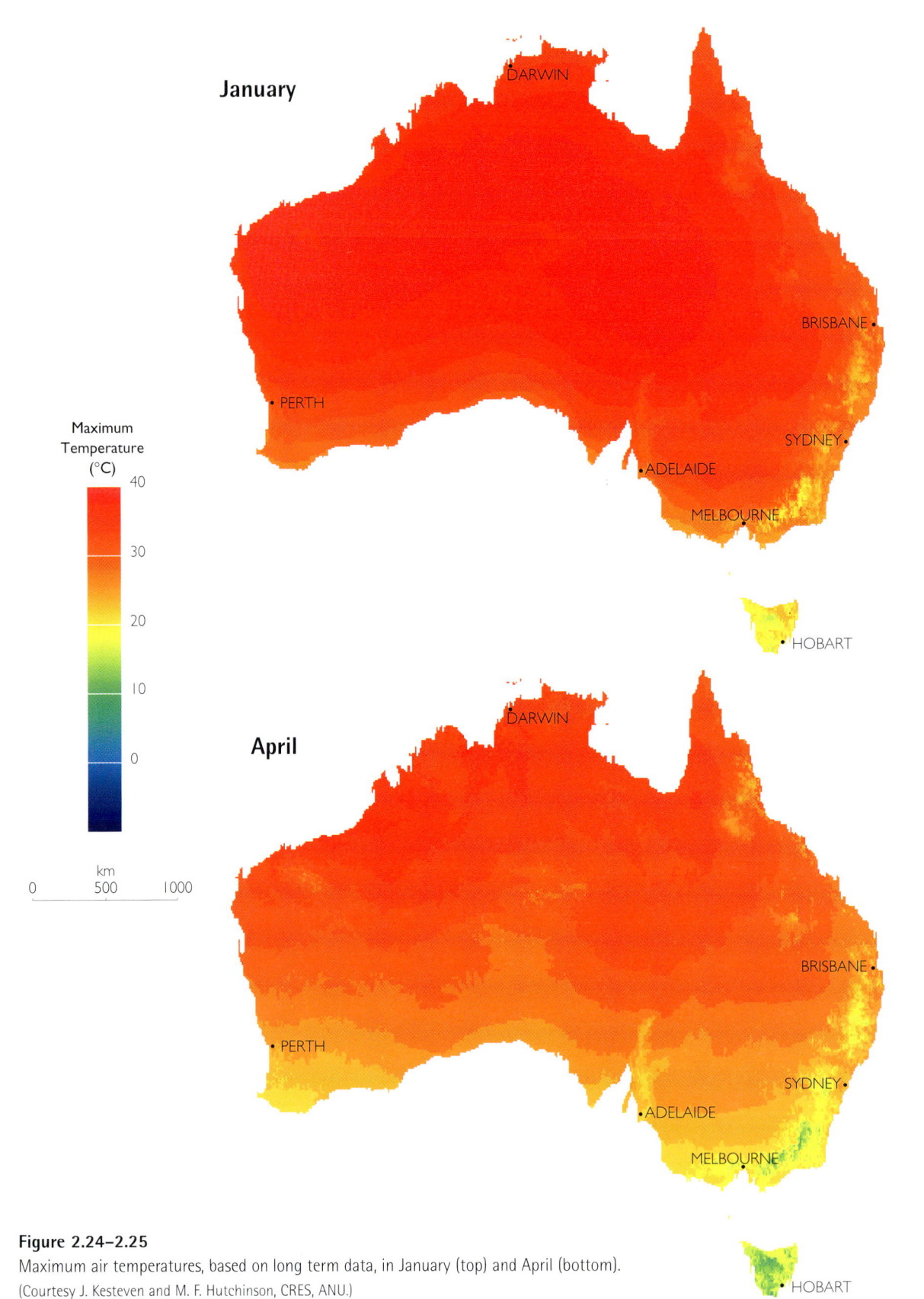

Figure 2.24–2.25

Maximum air temperatures, based on long term data, in January (top) and April (bottom).

(Courtesy J. Kesteven and M. F. Hutchinson, CRES, ANU.)

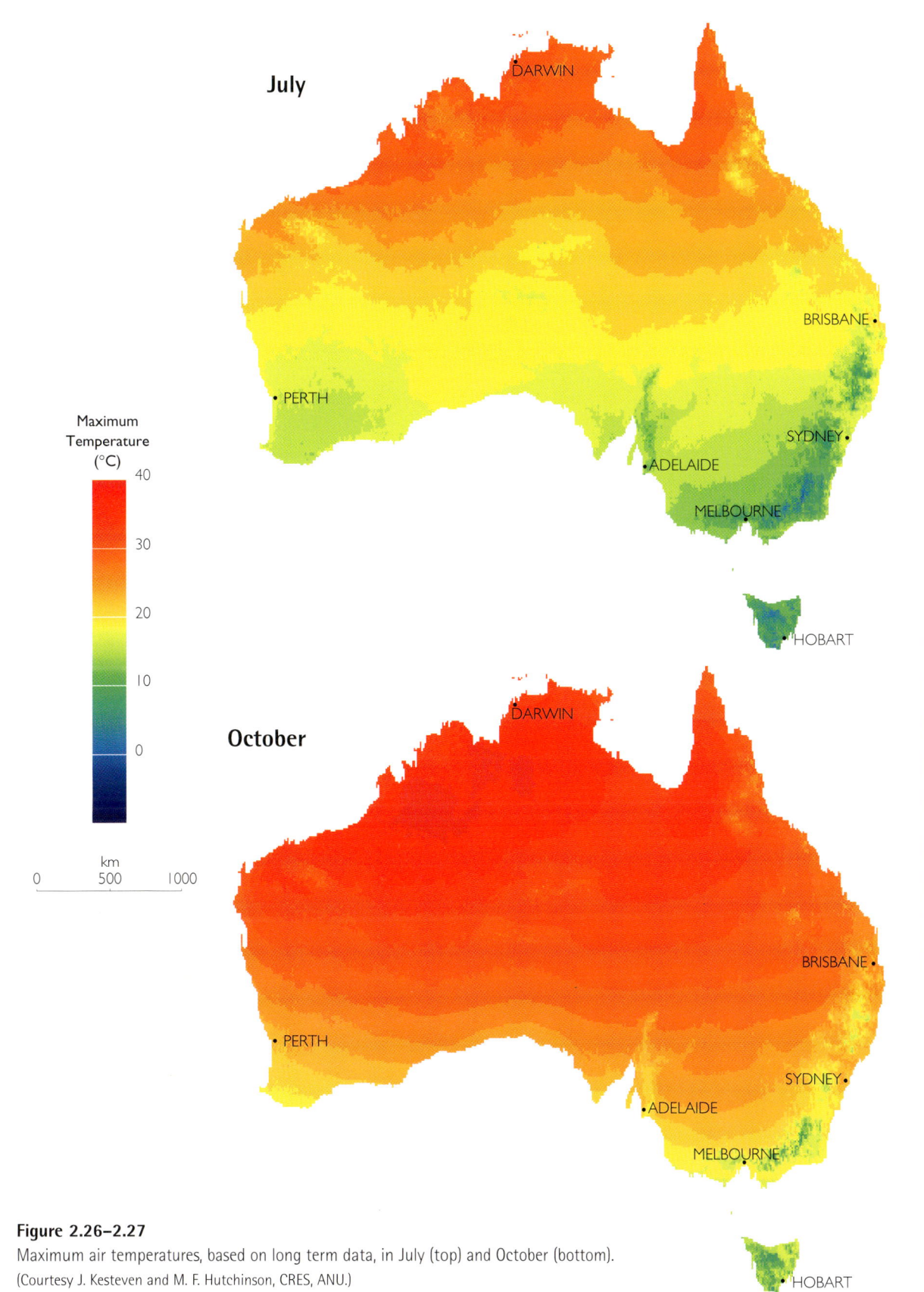

Figure 2.26–2.27

Maximum air temperatures, based on long term data, in July (top) and October (bottom).

(Courtesy J. Kesteven and M. F. Hutchinson, CRES, ANU.)

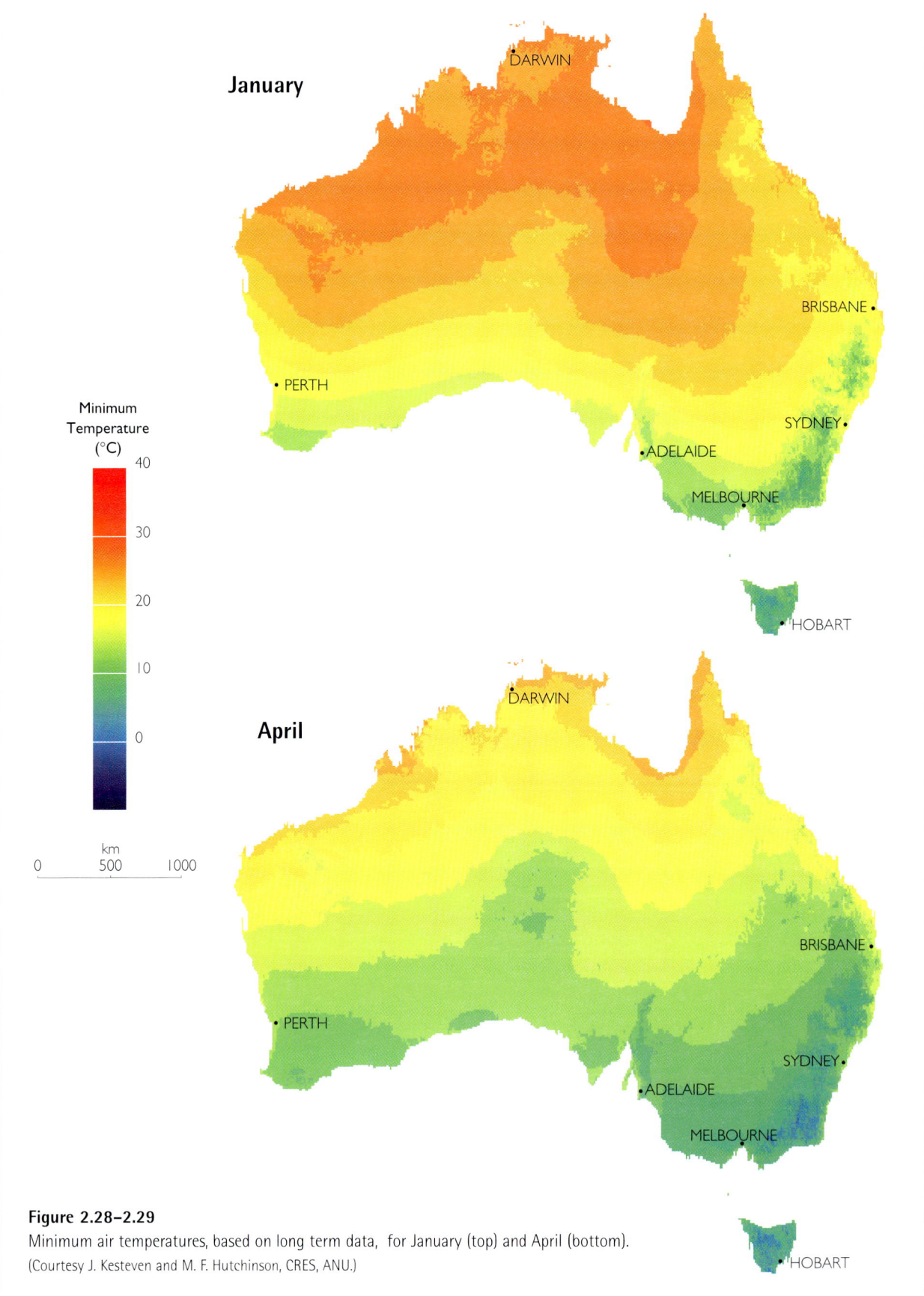

Figure 2.28–2.29

Minimum air temperatures, based on long term data, for January (top) and April (bottom).

(Courtesy J. Kesteven and M. F. Hutchinson, CRES, ANU.)

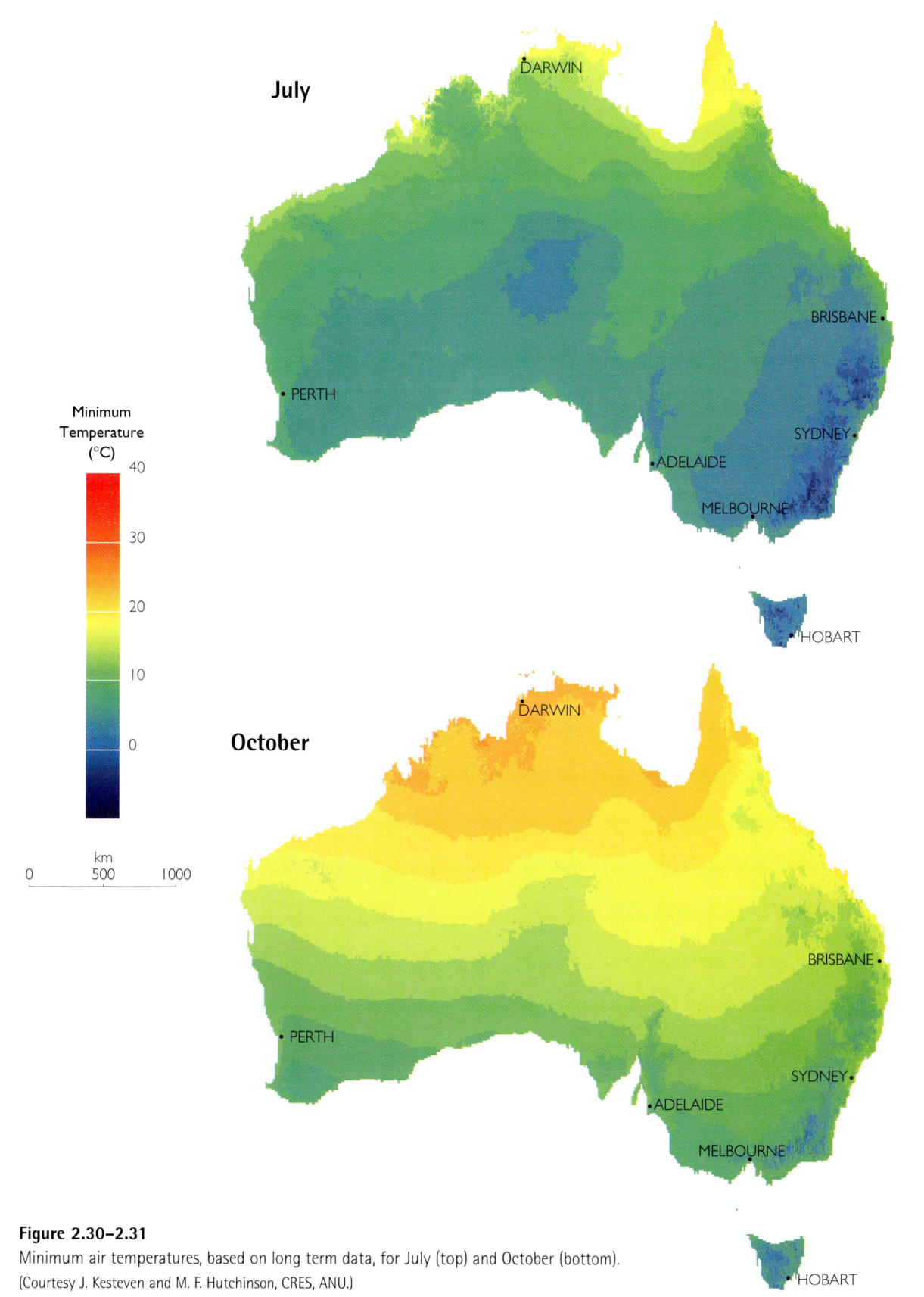

July

October

Minimum
Temperature
(°C)

40

30

20

10

0

km
0   500   1000

Figure 2.30–2.31
Minimum air temperatures, based on long term data, for July (top) and October (bottom).
(Courtesy J. Kesteven and M. F. Hutchinson, CRES, ANU.)

A brief commentary on the minimum temperature maps: recalling that the maps show average minimum temperatures, some readers might be surprised by how cool some of the January temperatures are in elevated areas like Canberra and the Blue Mountains (although this is hardly 'coastal'). For example, there are quite a few areas which experience an average of 10°C and of course much lower temperatures will occur from time-to-time. Further south, and definitely coastal, is Hobart with an average 11.5°C and a lowest on record of 4.5°C—quite a contrast to Darwin, for example, with a January average of 25°C and a lowest on record of 20°C. April minima, as expected, are generally lower, and July minima are generally lower still, with quite extensive areas experiencing 3–4°C average minima (including large areas in Central Australia). The October pattern is very interesting in that the northern part of the continent seems to have warmed up considerably, whereas many areas to the south are still experiencing quite low average minimum temperatures. This is particularly evident in southern New South Wales and much of Tasmania.

The difference between the maximum and minimum temperature on any particular day is called the 'daily temperature range'. The average temperature range for any month can be determined by comparing the corresponding maximum and minimum temperature maps (making sure they are for the same month). For example, let's take Perth in mid-summer (January): the average maximum temperature is about 30°C and the average minimum is about 18°C; the average daily temperature range (for January) is therefore about 12°C.

Figures 2.28 to 2.31 map average daily minimum air temperatures for the 'mid-season' months.

Rainfall (raindays)

Australia is a dry continent, with more than half its land considered to be desert. Less than 30 per cent of the continent has sufficient rainfall to support general agriculture and the relatively narrow coastal strip receives most of the rainfall. The dryness of the interior is mainly due to the subsiding air of the subtropical high pressure belt.

When the subtropical high pressure belt moves over southern Australia in summer, the summer monsoon moves in over northern Australia. These moist northerly winds bring showers, thunderstorms and the threat of tropical cyclones, which generally dump large amounts of rain in a short period. As the subtropical high moves north in winter, dry south-easterly winds prevail over northern Australia, and southern Australia comes under the influence of the winter westerlies. Cold fronts associated with depressions to the south of the continent move eastwards across southern Australia and are often accompanied by strong winds, heavy rain showers and snow in the alpine areas. Areas exposed to the westerlies — such as the south-west corner of Western Australia, south-western Victoria, and particularly western Tasmania — have heavy winter rainfalls, but most of the south-east part of the continent is more sheltered and rainfall is more evenly distributed throughout the year.

Precipitation patterns are always complex in mountainous areas and rainfall varies enormously from one region to another, depending on the characteristics of the place (for example, height, orientation of the mountains) and the landscape's interaction with the prevailing winds and storms that occur in the region. Australia is relatively flat and its mountains are small compared with many other parts of the world. Nevertheless, the Great Dividing Range along the eastern seaboard, and the Alps of the south-east, are important factors in determining the climate of the continent and the distribution of our water resources.

The higher rainfall along the east coast of Australia is primarily due to the inflow of moist air from the ocean meeting the mountain chain. The heaviest rainfalls occur where moist winds are forced to flow up mountain slopes, and very arid areas can occur in the lee or shadow of mountains.

The ranges also enhance the very heavy rain from tropical cyclones, particularly when these storms move very slowly over the ocean close to the coast. Thus the heavy rainfalls around the coast region of far north Queensland are the result of the moist south-east trade winds blowing across the Pacific and releasing large amounts of water as they strike the coast and move up the slopes of the mountains. Very heavy falls also occur in winter on the mountains of western Tasmania that are exposed to the westerlies.

The great variability of Australian rainfall, both in terms of amounts and the time of year in which most of the rain falls, is reflected in Figures 2.32 to 2.35, which map the number of days with significant (measurable) rainfall.

Although there is great seasonal variation, if

you compare the rainday figures between, say, Sydney and Melbourne, you might be surprised. Take January (the peak of the cricket and sailing season) — Sydney 13, Melbourne 8; April — Sydney 13, Melbourne 11; July — Sydney 11, Melbourne 15 and October — Sydney 12, Melbourne 14. Take the whole year — Sydney 149, Melbourne 143! Regional comparisons aside, the rainday maps should be useful for helping you decide when and where to go for that trip to the coast.

> If you are planning to head toward one of the 'red' areas (where there are very few raindays) you would be advised to carry plenty of water if you are away from town water supplies.

Mid–afternoon relative humidity

As mentioned in Chapter 1 (see especially Figure 1.1), the combination of high air temperature and high relative humidity can pose a real hazard to your health, or even be life threatening.

Figures 2.36 to 2.39 map humidity more or less at the time of maximum air temperature — frequently the time when coastal activity is at its peak. By comparing the maximum temperature and humidity maps, you will get an idea of the areas and times of year likely to be linked to hazardous combinations.

The humidity maps show very large variations across the continent. Most coastal areas are considerably moister than adjacent inland areas. Some areas — including most of the Great Australian Bight, the Perth–Geraldton region and much of the north-west — do not show a widespread ameliorating effect at the coast. The air in these places is very dry (on average) and if you are there during warm weather and are active as well, you will be perspiring at a great rate and will need to drink lots of fluid.

On the moister end of the scale, most areas on the east coast and in the tropical north (especially during the wet season) experience high humidity. If you look at the map for April, you might be a bit surprised to see that, despite common impressions to the contrary, there is little difference between the average humidity of Brisbane, Sydney, Melbourne and Hobart. This shows how

strongly the human body reacts to the combination of high temperature and humidity — low temperatures and high humidity seem to go unnoticed, in the sense that the combination causes no discomfort.

Mid–afternoon coastal wind speed and direction

Figures 2.40 to 2.43 map the probability of experiencing 'light winds and above' — in other words, 8 knots and above — at key Australian coastal cities in January, April, July, and October.

If the probability shown on one of these maps is, say, 40 per cent (which loosely equates to 3 days per week), it follows that the probability of *less* than 8 knots is approximately 60 per cent (4 days). A value of 10 per cent would mean that, on 10 per cent of the days in that particular month (that is, about 3 days), the wind would exceed the specified speed.

This can be looked at another way. If the probability of days with wind exceeding a particular value is 10 per cent, then the probability of days with less than that wind speed must be about 90 per cent. In such places the wind obviously does not blow very often. A value of 90 per cent would mean that a particular wind speed would be exceeded on about 24 days per month or perhaps between 5 and 6 days per week (90 per cent of 7 days). The probability of days with less than that wind speed is only about 10 per cent. In such places the wind obviously blows most of the time.

> As a rule of thumb, a value of 10 per cent in the wind maps would usually correspond to less than one day a week (that is, 10 per cent of 7 days), and a value of 90 per cent would usually correspond to 5 or 6 days a week (that is, 90 per cent of 7 days).

It is important to bear in mind that the maps provide a picture of average conditions at 3 pm, based on about 30 years of observations. Useful as this is, it obviously cannot tell us what will happen tomorrow! However, if you are in one of the 90 per cent areas shown on the map and the wind isn't blowing, then the chances of it blowing the next day are good indeed.

The maps cannot be used to estimate wind

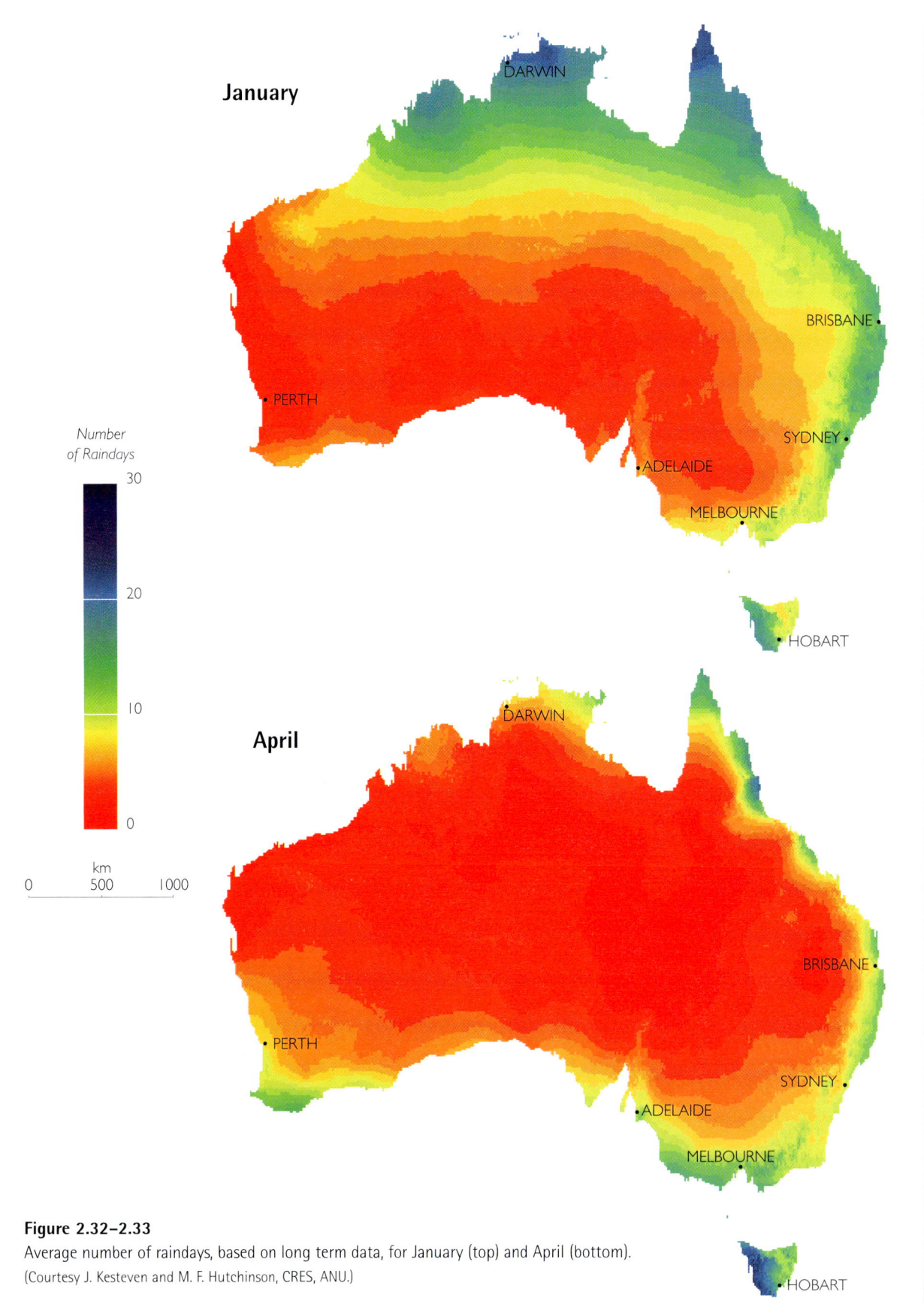

Figure 2.32–2.33

Average number of raindays, based on long term data, for January (top) and April (bottom).

(Courtesy J. Kesteven and M. F. Hutchinson, CRES, ANU.)

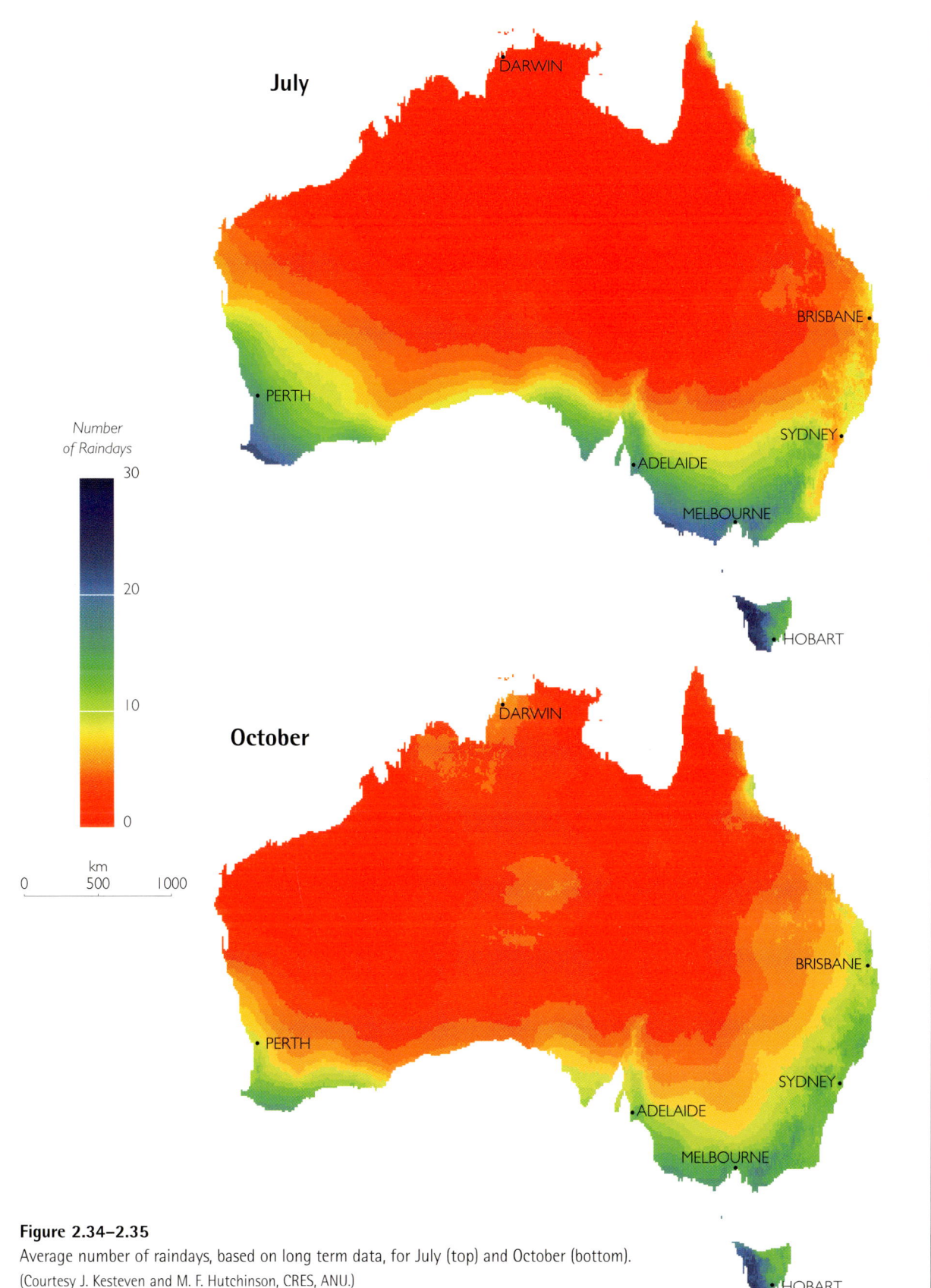

Figure 2.34–2.35
Average number of raindays, based on long term data, for July (top) and October (bottom).
(Courtesy J. Kesteven and M. F. Hutchinson, CRES, ANU.)

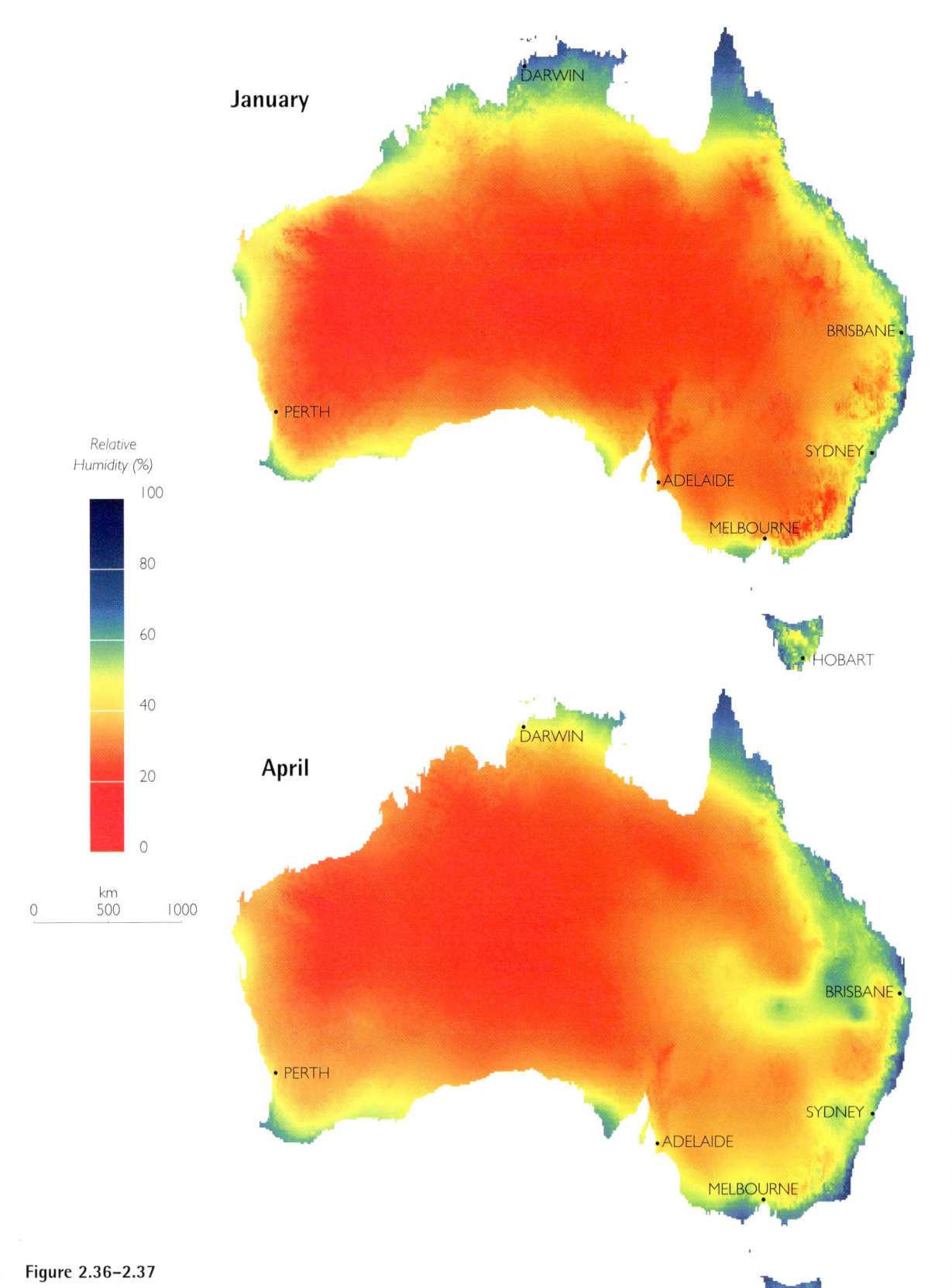

Figure 2.36–2.37

Estimated mid-afternoon humidity, based on long term data, for January (top) and April (bottom).

(Courtesy J. Kesteven, CRES, ANU.)

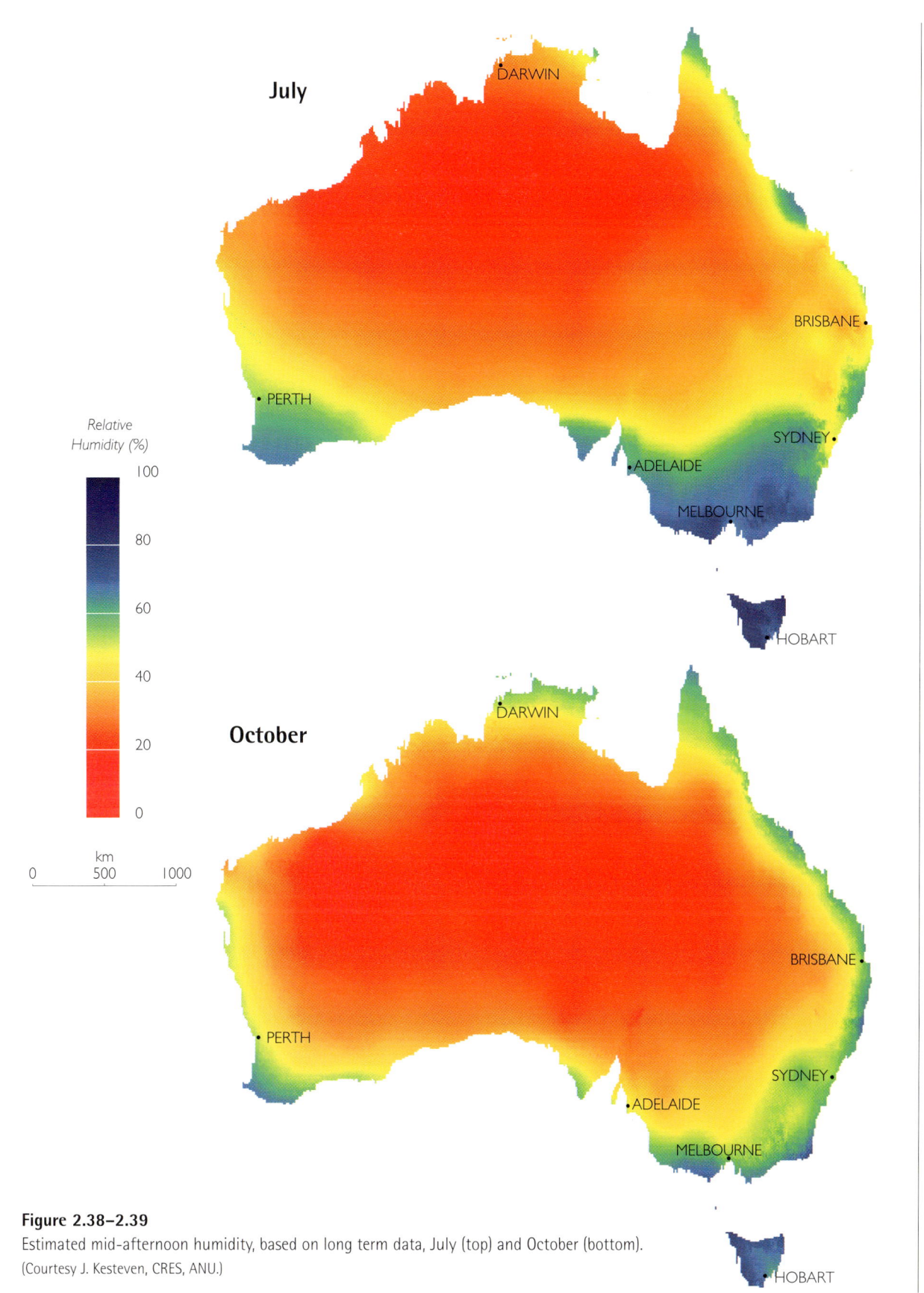

July

October

Relative
Humidity (%)

Figure 2.38–2.39
Estimated mid-afternoon humidity, based on long term data, July (top) and October (bottom).
(Courtesy J. Kesteven, CRES, ANU.)

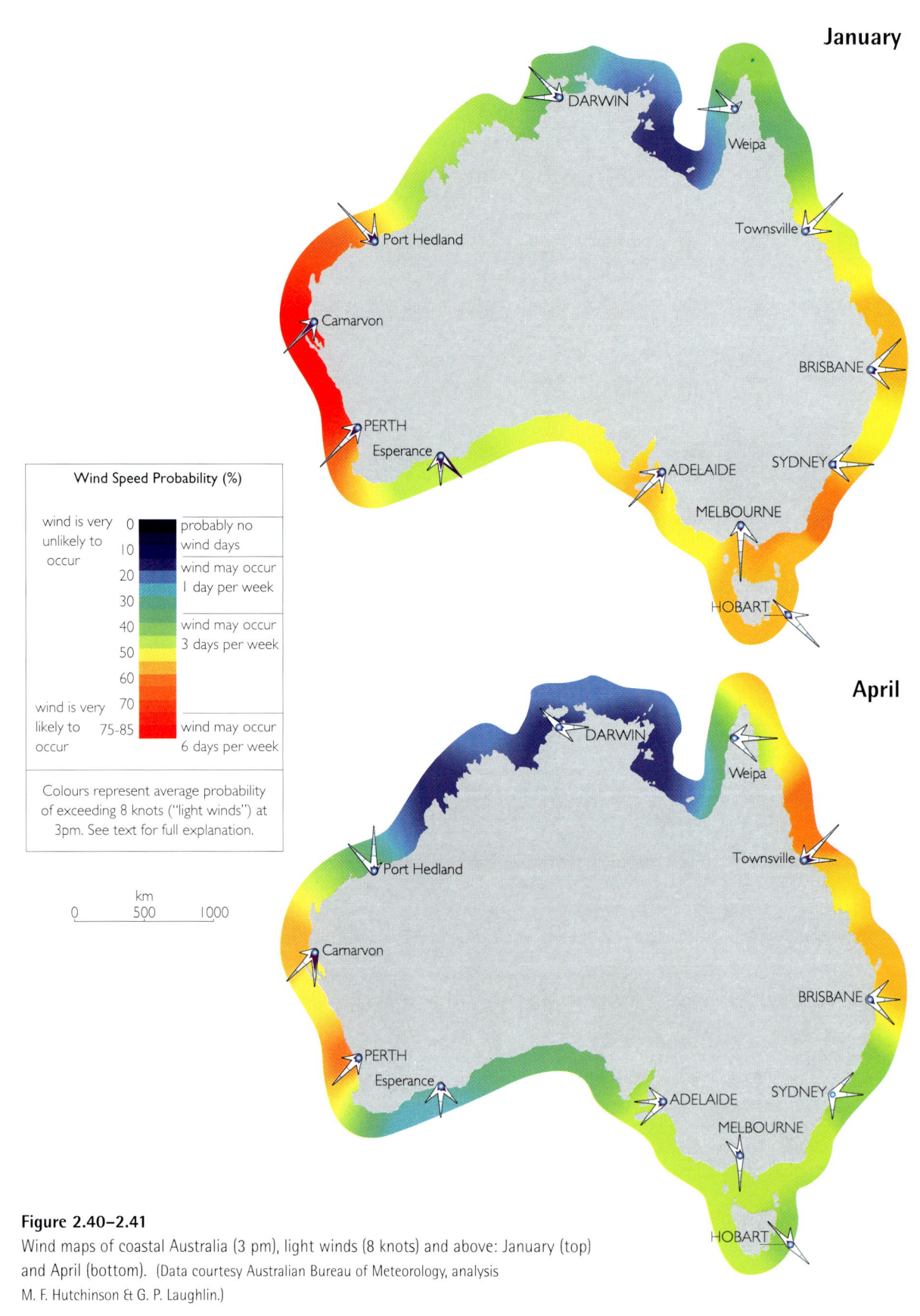

January

April

Wind Speed Probability (%)

wind is very unlikely to occur	0	probably no wind days
	10	wind may occur 1 day per week
	20	
	30	
	40	wind may occur 3 days per week
	50	
	60	
wind is very likely to occur	70	
	75–85	wind may occur 6 days per week

Colours represent average probability of exceeding 8 knots ("light winds") at 3pm. See text for full explanation.

km
0   500   1000

Figure 2.40–2.41
Wind maps of coastal Australia (3 pm), light winds (8 knots) and above: January (top) and April (bottom). (Data courtesy Australian Bureau of Meteorology, analysis M. F. Hutchinson & G. P. Laughlin.)

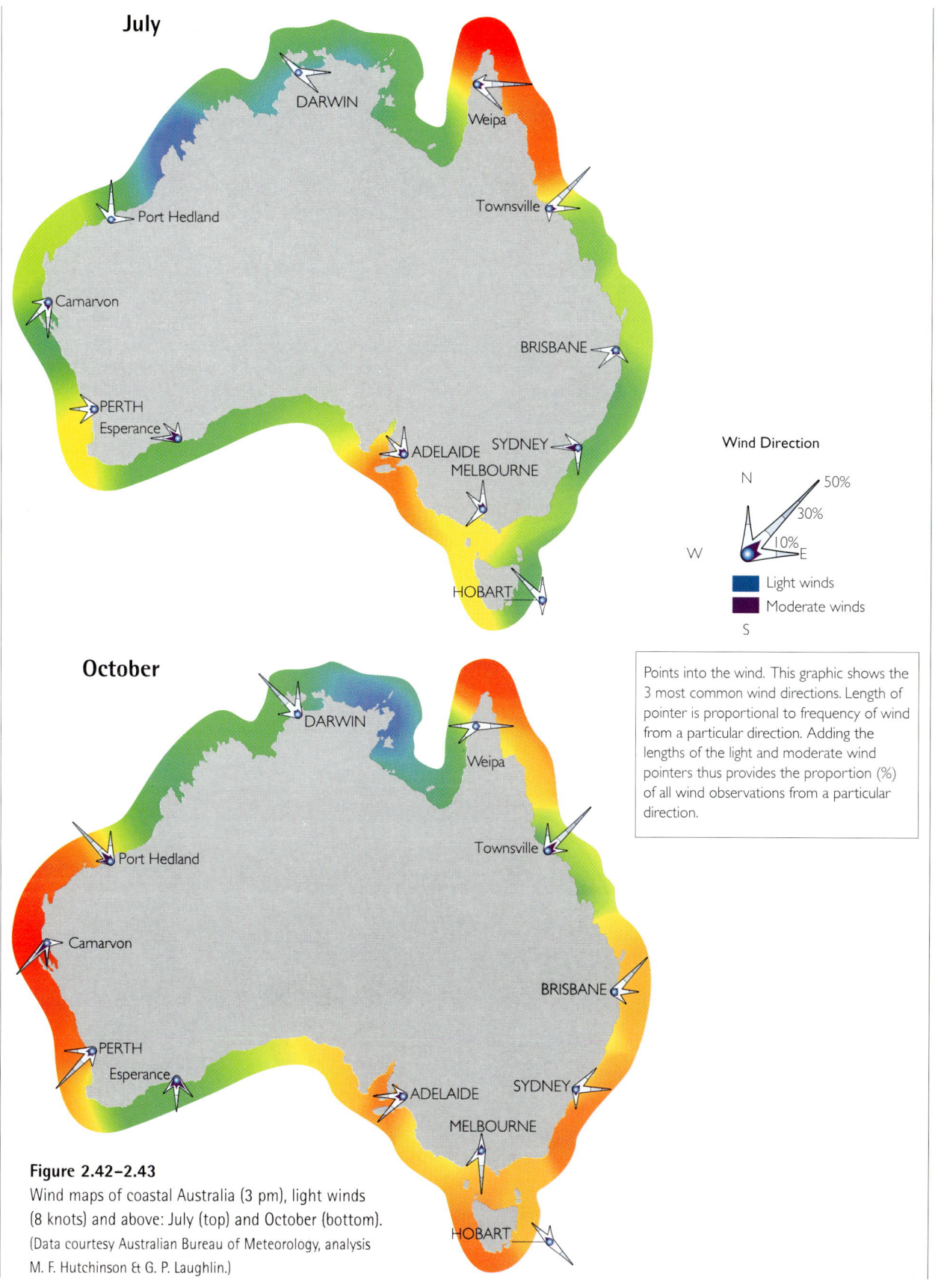

July

DARWIN
Weipa
Townsville
Port Hedland
Carnarvon
BRISBANE
PERTH
Esperance
ADELAIDE
SYDNEY
MELBOURNE
HOBART

Wind Direction

N
50%
30%
W
10%
E
Light winds
Moderate winds
S

Points into the wind. This graphic shows the 3 most common wind directions. Length of pointer is proportional to frequency of wind from a particular direction. Adding the lengths of the light and moderate wind pointers thus provides the proportion (%) of all wind observations from a particular direction.

October

DARWIN
Weipa
Port Hedland
Townsville
Carnarvon
BRISBANE
PERTH
Esperance
ADELAIDE
SYDNEY
MELBOURNE
HOBART

Figure 2.42–2.43

Wind maps of coastal Australia (3 pm), light winds (8 knots) and above: July (top) and October (bottom). (Data courtesy Australian Bureau of Meteorology, analysis M. F. Hutchinson & G. P. Laughlin.)

conditions out to sea or very far inland. They are an estimate of conditions close to shore, and are based on all available sites within 50 kilometres of the coast (that is, about 200 sites). The data from each site have been corrected objectively for its distance from the coast.

The maps also show a number of summary wind 'roses' for the three most frequent wind directions for that location (less frequent wind directions are not shown). A wind rose is a graphical representation of wind direction; note that *the pointer of the rose faces into the wind*. The length of each pointer is proportional to the frequency of wind from a particular direction. The inner pointers — in shades of blue — show wind direction for moderate winds and above (16 knots and above) and the (usually) longer, outer pointers — in shades of green — show wind direction for all winds up to 16 knots. If, for any given direction, the moderate wind pointer was 15 per cent and the light wind pointer was 60 per cent, this would mean that 75 per cent of all winds come from that direction. (The enhanced wind roses are based on a series published in *The Atlas of Australian Resources* (NATMAP 1986).)

These wind maps show a great deal of variation, which probably reflects the distribution of sea breezes, the Trade Winds and the Monsoon. For example, the January map ranges from about 15 per cent probability in the Gulf of Carpentaria to over 80 per cent in Carnarvon, Western Australia.

The combination of wind speed — shown as probability — and wind direction — shown as enhanced wind roses — allows readers to gain a detailed insight into wind conditions anywhere on the Australian coast and at any time of the year. Recall that the length of wind roses is proportional to the frequency of wind from a particular direction and that we have separate pointers for light and moderate winds in each wind rose. If all directions — not just the three most common — were shown, they would add up to 100 per cent.

January

January shows the influence of sea breezes over extensive areas of the coast as well as the effect of the Monsoon in northern Australia. Generally speaking, this results in light and infrequent winds in the north and increasingly reliable winds to the south. It can also be seen that reliable winds (> 45 per cent) can be found virtually anywhere south of about 18°S latitude. (Here and below, the term *reliable* means highly probable in percentage terms.)

The most reliable winds (> 60 per cent) are found over a very broad area in Western Australia from Port Hedland to Perth. It can be seen, however, that the wind is reliable well north and south of this area. At 3 pm the winds at Carnarvon are most commonly from the south-west and south. Although most of the east and south coasts of Australia experience reliable winds in January, it can be seen that the most reliable winds (> 60 per cent) are around the area Jervis Bay to Gabo Island (Victoria). Here the predominant direction is north-east and east.

The areas with the least reliable winds are those influenced by the Monsoon (or wet season) in the north. For example, see the area to the east of Darwin and around to Weipa where probabilities are around 10 per cent — that is, the winds are unreliable.

April

Wind probabilities in April are generally lower than those in January. Also, the general north–south pattern has contracted, leaving more isolated 'pockets' of reliable wind. Very large areas of unreliable (infrequent) wind are evident in northern Australia.

The most reliable winds (> 55 per cent) are found in Western Australia (again), with centres north of Perth (south-west is the most common direction), and in Queensland, particularly around Townsville and areas to the north.

The least reliable (around 15 per cent) winds are found over a broad area of the Northern Territory and the north of Western Australia (< 15 per cent), but also note the relatively low probabilities (approximately 25 per cent) in Western Australia around Esperance and Albany.

July

In mid-winter the south-east Trade Winds are most dominant over north-eastern Australia, and many inland sites show a very strong dominance of winds from the south-east. For example, Cloncurry (Queensland) has nearly 65 per cent of winds from the south-east in July. However, very few coastal sites have a predominant south-east wind, although most have a very predominant onshore wind direction; this departure from what might be expected (since the Trades *do* blow from the south-east) is probably due to coastal topography and sea breeze influences. Notable exceptions are Groote Eylandt (Gulf of Carpentaria), with 80 per cent of winds from the south-east, and Thursday Island (Cape York), with 75 per cent.

The mid-winter pattern shows the highest probabilities to be in northern Queensland and a quite definite north–south gradient in these probabilities. To a lesser extent, but still reliable, are southern Victoria, most of the west coast of Tasmania (no predominant direction), the area around Adelaide in South Australia (from the north, north-west and west) and the Perth region in Western Australia.

The least reliable winds are found in a less extensive area centred on King Sound, Western Australia.

October

The October pattern resembles the January pattern far more than it does July. The most reliable winds (> 75 per cent) are found in a relatively narrow band centred around Carnarvon in Western Australia; here the predominant direction is from the south-west. Very reliable winds are also found at the top of Cape York (> 65 per cent), and over much of the east coast south of about Bundaberg (> 50 per cent), with direction ranging from east to north-east. Notice that the area of fairly reliable wind in Queensland is now very much smaller than for January. Reliable wind is also found over much of the coast of New South Wales (particularly its south, which has a probability greater than 60 per cent), Victoria, Tasmania (especially the south-west) and South Australia.

The least reliable winds (< 35 per cent) are found in the Gulf of Carpentaria, particularly in the Gove region and along parts of the south and north coasts of Western Australia.

CHAPTER THREE

Wind Essentials

Wind is the movement of air. Air moves because the sun heats the Earth's surface more at the equator than at the Poles; it is also influenced by the Earth's rotation. The result is a pattern of rotating, continent sized masses of air called the 'general circulation'. On a smaller scale, wind patterns are modified by local influences such as the differential heating of the land and sea, which gives rise to sea and land breezes, and the blocking and channelling of winds by mountain ranges. On an even smaller scale, winds in the lowest few hundred metres of the atmosphere respond directly to local features of the ground, such as small hills and various types of vegetation.

In Australia, wind is monitored at about 540 sites. About 68 of these are fully equipped 'synoptic' sites — sites that measure a wide range of weather variables — 200 are automatic weather stations (AWSs) with cup anemometers, and 430 are 'cooperative', where wind is estimated by observing the effect it has on objects (including the sea). At most cooperative sites, estimations are made only at 9 am and 3 pm local standard time.

THE BEAUFORT SCALE

The estimation procedure used at cooperative sites is referred to as the Beaufort Scale. It is actually a misnomer, since, in his original scale (circa 1806), Sir Francis Beaufort, a rear admiral and a trainer in the Royal Navy, did not mention wind speeds except in the first three classes. His intention was to communicate information on ship speed, sail carrying ability and survival to those involved in the blockade of Europe.

For each Beaufort number, an equivalent wind speed at a height of ten metres above the ground is given.

It takes some experience to get consistent results with the Beaufort Scale since it is based on recognising the effect wind has on the land or sea surface. In the table below, for example, one can find that when large branches are in motion and whistling is heard in powerlines, the wind speed is *probably* in the range covered by Beaufort class 6. Long-term experience has shown that such wind is commonly 22–27 knots.

Study the table and with practice you will be able to estimate wind speed. A cheap anemometer will certainly help. It is best used well away from local obstructions. Your observation could be affected by surrounding objects: buildings, hills, and so on.

You will quickly see that the 'look out the window' approach can be a very poor indicator.

> In Australia the commonly used terms for indicating very broad ranges of wind speed are *light wind* (Beaufort class 1–3), *moderate wind* (Beaufort class 4–5), *strong wind* (Beaufort class 6–7), and *gale* (Beaufort class 8–9).

Table 3.1
The Beaufort Scale: effects on land and sea, and estimated wind speed equivalents.

Beaufort number	0
Description	Calm
Effect on land	Smoke rises vertically
Effect on sea	Surface looks like a mirror
Potential wave height*	Flat (0)
Wind speed est. at 10m⁺	< 1 knot 0–0.2 m/s
	< 1 km/h < 1 mph
Weather map symbol⁺	◎
Comments	Termed *light wind* (sometimes called 'light air') by the Bureau of Meteorology

Beaufort number	1
Description	Light air
Effect on land	Smoke follows wind; wind vanes do not work
Effect on sea	Ripples can be seen
Potential wave height	1 cm
Wind speed est. at 10m	1–3 knots 0.3–1.5 m/s
	1–5 km/h 1–3 mph
Weather map symbol	—
Comments	Termed *light wind* (sometimes called 'light air') by the Bureau of Meteorology

Beaufort number	2
Description	Light breeze
Effect on land	Leaves rustle, people feel wind on faces
Effect on sea	Small wavelets with glassy appearance
Potential wave height	6 cm
Wind speed est. at 10m	4–6 knots 1.6–3.3 m/s
	6–11 km/h 4–7 mph
Weather map symbol	⌐
Comments	Termed *light wind* (sometimes called 'light air') by the Bureau of Meteorology

Beaufort number	3
Description	Gentle breeze
Effect on land	Leaves and twigs in constant motion; light flag will extend
Effect on sea	Large wavelets and crests begin to break
Potential wave height	20 cm
Wind speed est. at 10m	7–10 knots 3.4–5.4 m/s
	12–19 km/h 8–12 mph
Weather map symbol	⌐
Comments	Termed *light wind* (sometimes called 'light air') by the Bureau of Meteorology

Beaufort number	4
Description	Moderate breeze
Effect on land	Dust and loose paper are raised; small branches are moved
Effect on sea	Small waves and some whitecaps
Potential wave height	60 cm
Wind speed est. at 10m	11–16 knots 5.5–7.9 m/s
	20–28 km/h 13–18 mph
Weather map symbol	⌐ ⌐
Comments	Termed *moderate wind* by the Bureau of Meteorology

Beaufort number	5
Description	Fresh breeze
Effect on land	Small leafy trees begin to sway; inland waters form crested wavelets
Effect on sea	Moderate sized waves and many whitecaps
Potential wave height	1.3 m
Wind speed est. at 10m	17–21 knots 8.0–10.7 m/s
	29–38 km/h 19–24 mph
Weather map symbol	⌐ ⌐
Comments	Termed *moderate wind* by the Bureau of Meteorology

Beaufort number	6
Description	Strong breeze
Effect on land	Large branches in motion; powerlines whistle
Effect on sea	Large waves begin to form; extensive whitecaps and some spray
Potential wave height	2.5 m
Wind speed est. at 10m	22–27 knots 10.8–13.8 m/s
	39–49 km/h 25–31 mph
Weather map symbol	⌐ ⌐
Comments	Termed *strong wind* by the Bureau of Meteorology; a Strong Wind warning will be issued once mean winds are expected to exceed 25 knots

Beaufort number 7
Description | Near gale
Effect on land | Whole trees in motion; people have some difficulty walking into the wind
Effect on sea | Sea heaps up and white foam from breaking waves is blown in streaks
Potential wave height | 4.5 m
Wind speed est. at 10m | 28–33 knots 13.9–17.1 m/s
 | 50–61 km/h 32–38 mph
Weather map symbol |
Comments | Termed *strong wind* by the Bureau of Meteorology; a Strong Wind warning will be issued

Beaufort number 8
Description | Gale
Effect on land | Twigs break off trees; people have difficulty in walking
Effect on sea | Moderately high waves break and form spindrift; well defined streaks
Potential wave height | 7 m
Wind speed est. at 10m | 34–40 knots 17.2–20.7 m/s
 | 62–74 km/h 39–46 mph
Weather map symbol |
Comments | Termed *gale force wind* (or just *gale*) by the Bureau of Meteorology; a Gale warning will be issued

Beaufort number 9
Description | Strong gale
Effect on land | Structural damage can occur
Effect on sea | High waves; dense streaks of foam along wind lines; crests of waves begin to roll over; spray may affect visibility
Potential wave height | 11 m
Wind speed est. at 10m | 41–47 knots 20.8–24.4 m/s
 | 75–88 km/h 47–54 mph
Weather map symbol |
Comments | Termed *gale force wind* (or just *gale*) by the Bureau of Meteorology; a Gale warning will be issued

Beaufort number 10
Description | Storm
Effect on land | Trees are uprooted; severe structural damage may occur
Effect on sea | Very large waves with long overhanging crests; sea is chaotic and tumbling, and takes on a dense white appearance
Potential wave height | 16 m
Wind speed est. at 10m | 48–55 knots 24.5–28.4 m/s
 | 89–102 km/h 55–63 mph
Weather map symbol |
Comments | Storm warning will be issued

Beaufort number 11
Description | Violent storm
Effect on land | On rare occurrences, widespread damage
Effect on sea | Small to medium sized ships may be temporarily blocked from view by waves. Foam everywhere; large parts of waves blown into froth
Potential wave height | 22 m
Wind speed est. at 10m | 56–63 knots 28.5–32.6 m/s
 | 103–117 km/h 64–72 mph
Weather map symbol |
Comments | Storm warning will be issued

Beaufort number 12
Description | Hurricane
Effect on land | (Very rarely experienced over land)
Effect on sea | Air filled with foam and spray; severely impaired visibility
Potential wave height | 22 + m
Wind speed est. at 10m | 64 + knots 32.7+ m/s
 | 118 + km/h 73 + mph
Weather map symbol |
Comments | Storm warning will be issued

Notes:

* Potential wave height is *average* for a fully developed sea; considerably higher waves will occasionally occur!
+ As Beaufort classes vary in width, the likely symbols that correspond to each class are shown; one full 'feather' equals 10 knots.

The Bureau of Meteorology issues warnings for tropical cyclones when winds associated with the storm system exceed Beaufort gale force 8 (34 knots or 63 km/h). When the cyclone is near the coast, a Tropical Cyclone warning will be issued in addition to a Gale, Storm or Hurricane warning for shipping. When a tropical cyclone is located more than 800 kilometres (430 nautical miles) off the coast, a warning for gale, storm or hurricane strength winds will be issued (no Tropical Cyclone warning will be issued).

HOW TO ESTIMATE WIND SPEED AND DIRECTION FROM WEATHER MAPS

Wind speed

You can estimate wind speed from weather maps without calculations or equations. Naturally this method produces only an approximation of likely wind conditions; but once you become familiar with such an approach, you will be better equipped to understand the other important wind effects that can reduce or increase the wind near the surface (see page 59).

Daily weather maps and weather faxes[1] show the barometric pressure at mean sea level in the form of isobars, or lines of constant pressure. It is differences in the barometric pressure over large distances that cause wind. Without going into the mathematics, it is important to mention a couple of other factors that enable us to estimate the wind speed from weather maps. They include friction effects, which are increasingly important close to the ground or water surface, and latitude (explained more fully in Appendix V). For a given pressure gradient (that is, a given spacing of the isobars) the wind will be stronger near the equator than near the Poles; for a country as large as Australia (extending from 12° to nearly 44° latitude) this is an important effect.

Using Figures 3.1 a–e we can estimate wind speed and direction from normal weather maps with isobars spaced every 2 hPa (that is, 2 mb). Most maps published in newspapers and provided via fax have either this spacing or 4 hPa (4 mb) isobars, so the same method of estimation can be used, with a simple interpolation in the latter case.

The diagrams are based on equations in Appendix V. These equations should give reasonable results near land or water between 15° and 45° latitude for average surface conditions, but are not appropriate for very tightly curved isobars, such as would occur with deep depressions (lows) and tropical cyclones. Closer to the equator than 15°, or over unusually rough or smooth surfaces, they will be less reliable. Similarly, the diagrams are likely to be less accurate under stable conditions (see 'Wind and Barriers', page 63, for more detail).

These diagrams show general relationships between the horizontal pressure gradient — as determined by the spacing between isobars — and the resulting wind speed near the surface. Each shows a different, *single* pressure gradient, which is superimposed as a grid on the Australian region. (The red, yellow and circular 'grids' are different orientations of the same pressure gradient. The colours on the maps represent wind speed.) Of course, it is unlikely that you will see a real weather map with exactly the same pressure gradient over such a vast area; so, if you want to analyse a weather map for, say, Sydney and Darwin, you may have to use more than one of the diagrams to match the pressure gradients around the two areas.

The estimated wind speed has been corrected for average water and land surface friction effects and for latitude. The wind at the surface has been assumed to be 65 per cent of the geostrophic speed (see page 60) over the sea and 50 per cent over land. If the spacing on your weather map is similar to that in any of Figures 3.1 a–e, then the expected wind speed (shown as 'solid' colours ranging from blue to red) should be similar to that in the same diagram.

> Interestingly, for a given pressure gradient (spacing of the isobars), wind speeds will be much stronger near the equator than at southern latitudes. This is caused by the Coriolis effect, which depends on latitude. In the Southern Hemisphere it causes a deflection of wind (and water currents) to the left and explains why highs rotate anticlockwise and lows clockwise.

However, the main limitation of most weather maps is the small scale at which they are published; this makes it difficult to determine the actual spacing of the isobars. In Figures 3.1 a–e the isobars are shown as concentric circles, and as horizontal and vertical grids. It is important to **use only one of these** (*either* the circles, *or* the horizontal grid *or* the vertical grid) **at a time**, according to the orientation that best matches the orientations of the isobars on the weather maps you wish to analyse. In the diagrams these

---

[1] See http://www.bom.gov.au

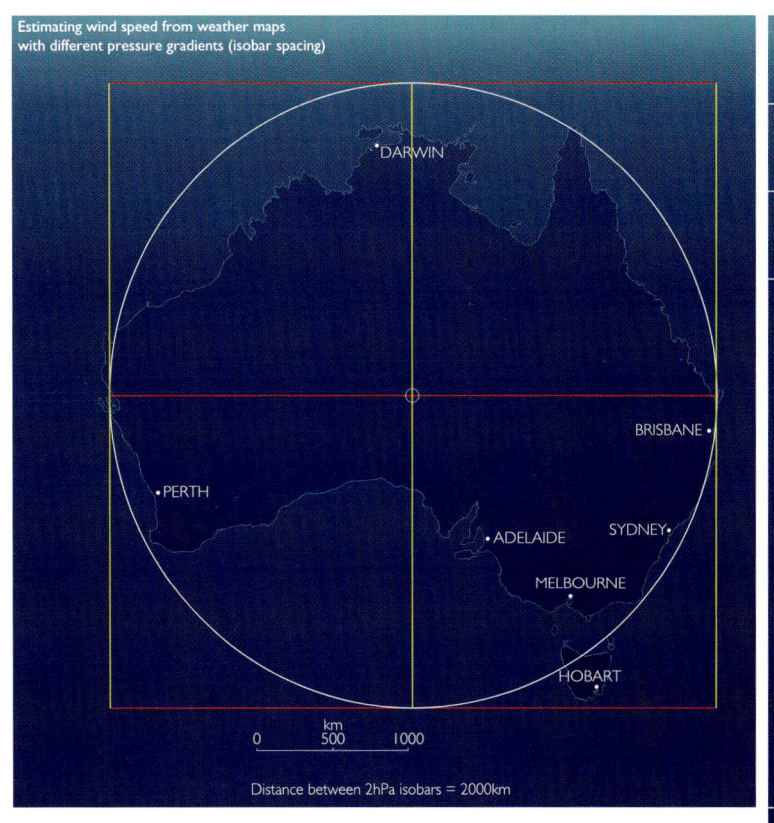

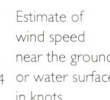

Figure 3.1a

Panel 1: 2000 km between isobars. Estimating wind speed near the ground or water surface from weather maps with 2 hPa (2 mb) isobars; speeds are corrected for latitude and water/land surface roughness.

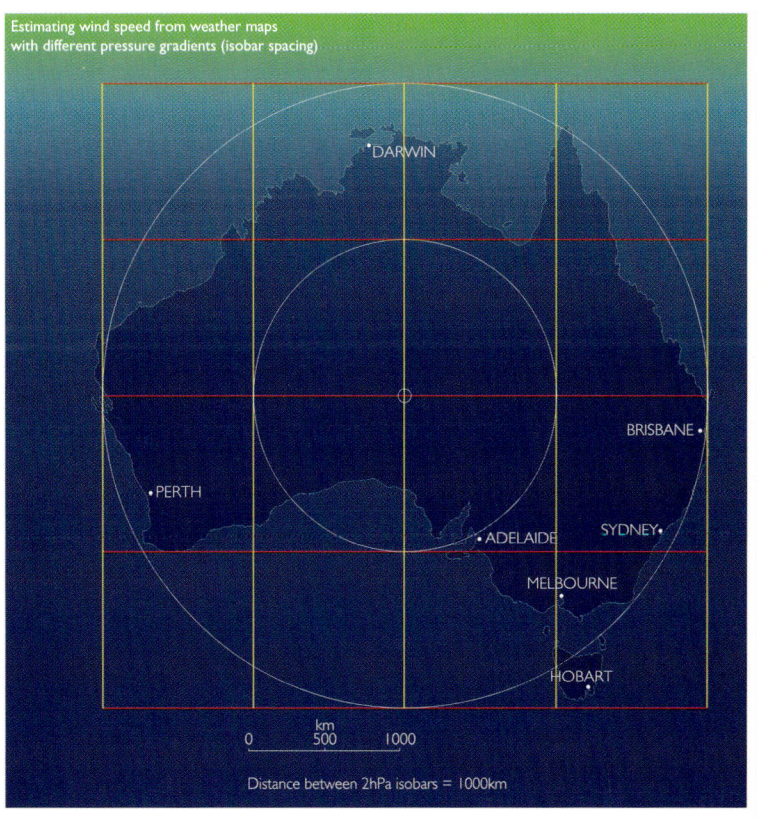

Figure 3.1b

Panel 2: 1000 km between isobars. Estimating wind speed near the ground or water surface from weather maps with 2 hPa (2 mb) isobars; speeds are corrected for latitude and water/land surface roughness.

Notes to Figures 3.1 a–e:
If your weather map has 4 hPa (4 mb) or other spacing, make sure you make an appropriate adjustment. Estimated speeds will be less reliable for tightly curved isobars; fronts, thunderstorms and other significant weather features in the vicinity may dominate or alter the local wind field.

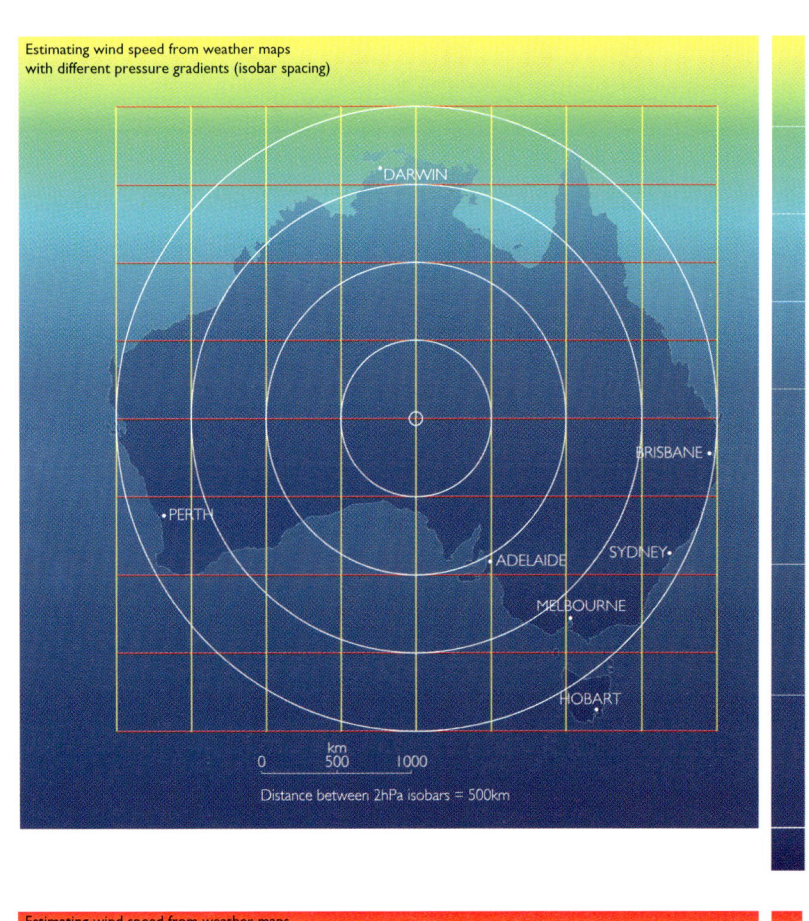
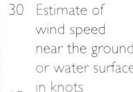

Estimating wind speed from weather maps
with different pressure gradients (isobar spacing)

Distance between 2hPa isobars = 500km

30 Estimate of
wind speed
near the ground
or water surface
in knots

Figure 3.1c
Panel 3: 500 km between isobars. Estimating wind speed near the ground or water surface from weather maps with 2 hPa (2 mb) isobars; speeds are corrected for latitude and water/land surface roughness.

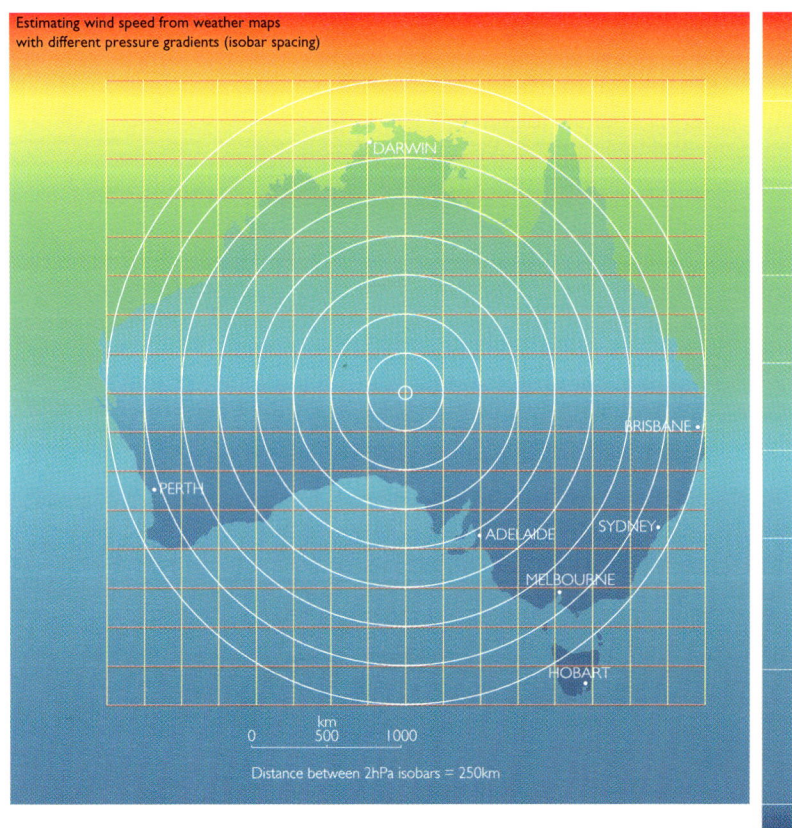
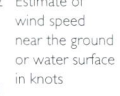

Estimating wind speed from weather maps
with different pressure gradients (isobar spacing)

Distance between 2hPa isobars = 250km

62 Estimate of
wind speed
near the ground
or water surface
in knots

Figure 3.1d
Panel 4: 250 km between isobars. Estimating wind speed near the ground or water surface from weather maps with 2 hPa (2 mb) isobars; speeds are corrected for latitude and water/land surface roughness.

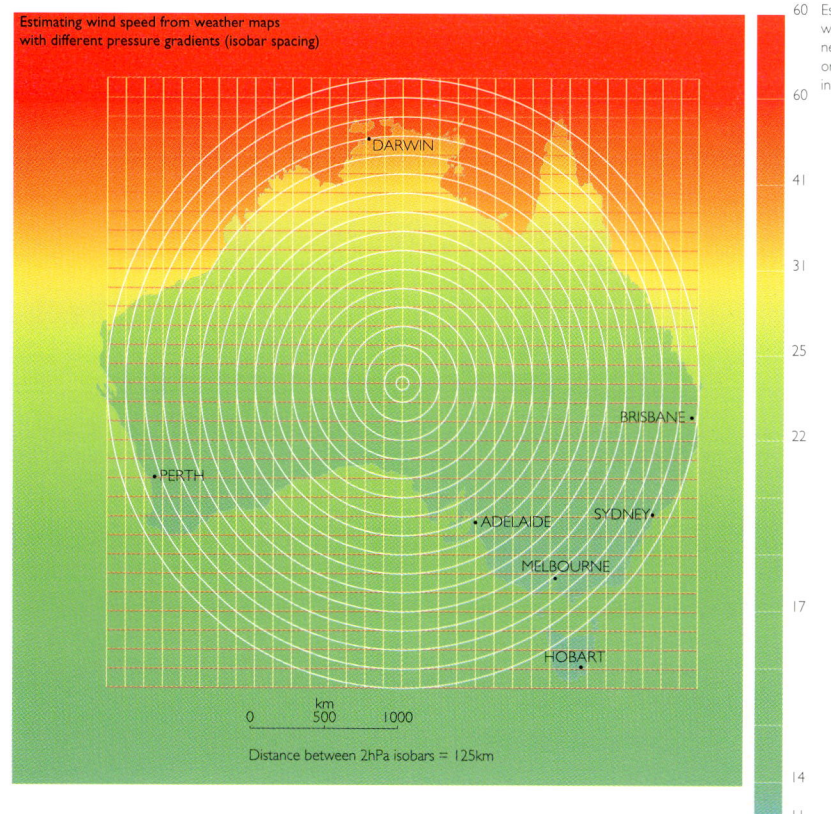

Figure 3.1e

Panel 5: 125 km between isobars. Estimating wind speed near the ground or water surface from weather maps with 2 hPa (2 mb) isobars; speeds are corrected for latitude and water/land surface roughness.

orientations are coloured red, yellow and white respectively, to make them visually distinct.

Also, as anyone who has looked at a weather map knows, isobars are not always smooth, regular circles or lines. Rather than trying to overlay your weather map onto Figures 3.1 a–e to see which of the patterns it best matches, you would be wise to start by becoming familiar with the general relationship between the spacing of the isobars and the resultant wind speed over land and over water for your latitude. With practice, you should be able to 'guesstimate' the likely wind speed for any pressure gradient on the weather map — be it from the newspaper, a weather fax or in any other form.

Weather maps may be published at different scales, so be careful when trying to match them to Figures 3.1 a–e. If your weather map has anything other than 2 hPa isobar spacing, adjust according-ly; for example, for the same pressure gradient, there will half as many 4 hPa isobars. One approach to overcoming any difficulties associat-ed with different map scales is to read off your weather map the difference in pressure between two recognisable points that are near your area and that are well separated from each other. If, for argument's sake, your weather map shows a 4 hPa difference between Sydney and Melbourne, then simply look through the diagrams until you find the one that shows a similar difference (panel number 4, with roughly a 3.5 hPa gradient). Once you get used to this approach, you should have few problems making adjustments for 4 hPa or other spacings.

Also remember that it is the pressure gradient near your region that will cause the wind. This might seem obvious, but it is important not to estimate the wind in your local area from a feature

Each diagram shows double the pressure gradient of the previous panel. Since a doubling of the pressure gradient results in a doubling of the estimated wind speed (for a given latitude) you can assume that, if the pressure gradient on your weather map is in between any two of the diagrams, so too is the wind speed. If it is much closer to one than the other, the wind speed is also likely to be closer to that diagram. In other words, you can assume that it is proportional.

on the other side of the continent. Given that each of the five diagrams has only one pressure gradient superimposed on it, it is quite likely that, in order to analyse an entire weather map, you may have to use more than one of these diagrams.

See pages 191–193 for worked examples using 'real' weather maps.

Wind direction

Figure 3.2 shows how to estimate wind direction from weather maps. It shows a simplified high pressure system (rotating anticlockwise in the Southern Hemisphere), a low pressure system, and two combinations of 'straight' flow — one in which the high is to the north of the observer and one in which it is to the south. The wind is shown in bold white above the boundary layer (say, at 1000 m), and in fine white at the surface. More wind pointers are shown on the straight flow to remind you that they can be drawn between any two isobars (this saves cluttering up the high and low with more pointers).

The left hand side of the figure — the 'brown side' — relates to land and the right hand side — the 'blue side' — to water, where the deflection is usually much less. (Deflection is the change in direction of the wind near the surface, from the geostrophic wind direction. The amount of deflection is primarily a function of surface roughness.) We have assumed an average deflection of 35° over land and 15° over water, although these can vary according to surface roughness; very rough seas could produce a 25° deflection, for example, while rough land at night could deflect the wind by up to 50°. The pressure gradient of the high pressure system is low at the centre

and increasing outward, whereas the low is the opposite.

Although Figure 3.2 is designed to show wind direction, the length of the arrows is also proportional to wind speed; this reminds us that both wind speed and direction change from the geostrophic level to the surface. Recalling that highs and lows normally travel from west to east — in Australia this usually happens every three to five days or so — Figure 3.2 can be used to give an indication of the change in wind direction with the passing of the different systems.

> Weather map symbols such as 🪶 point into the wind. But which end points into the wind? The feather points into the wind, so this one is showing a westerly wind. If you're confused, look below to the worked examples.

Once you understand this figure, it might be useful to go to Figure 3.5, which relates to surface friction and gustiness, and put the two together. Collectively, they show that, as one moves from an altitude of 1000 metres (or so) to the surface, wind speed decreases, wind direction changes (the wind 'crosses' the isobars increasingly in the

> The terms *backing* and *veering* are frequently used in meteorology; they are defined in the Glossary. In the Southern Hemisphere the wind veers (turns clockwise) from the geostrophic level to the surface; it veers more over land than over water. It follows, then, that wind veers as it passes from water to land, and backs (turns anticlockwise) as it passes from land to water. If you place yourself in any part of Figure 3.2, you will see that, as you get closer to the surface the wind direction veers.

> Figure 3.2 gives the impression that wind direction deflects smoothly from the geostrophic level to the surface; however, experienced pilots (and perhaps adventurous hang-gliders) know that most change takes place at 1000 to 100 metres; below 100 metres there is usually little change in wind direction. This process of direction change is known as backing and veering; in the Southern Hemisphere (as shown clearly in Figure 3.2) from 1000 metres the wind veers to water by about 15° and to land by about 35°. It also follows that wind blowing from water to land will veer by perhaps 20° and wind blowing from land to water will back by this same amount. Veering is usually greater under stable conditions (for example, night-time inversions, winter inversions) than in well-mixed conditions (normal day with a good breeze).

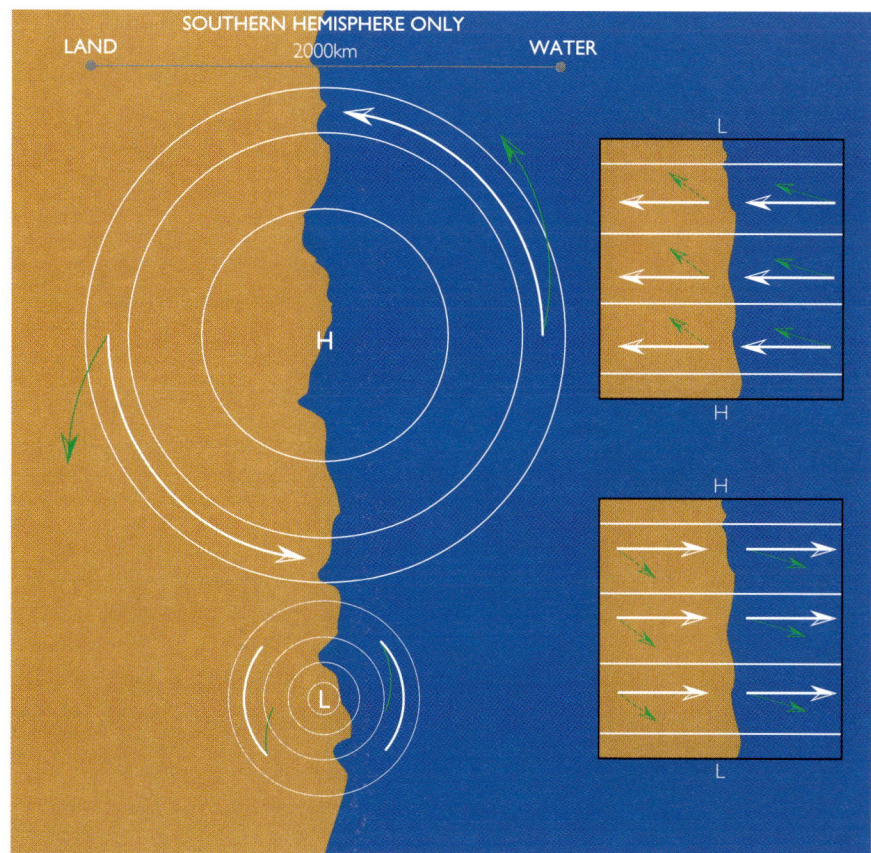
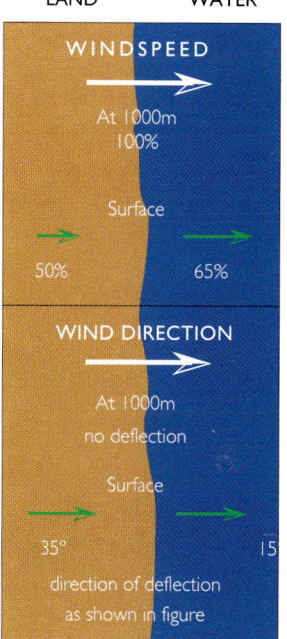

Figure 3.2
Estimating wind direction near the ground or water surface from weather maps (arrows show the direction the wind is going to, not coming from).

Notes to Figure 3.2: Fronts, thunderstorms and other significant weather features in the vicinity may dominate or alter the local wind field; these directions will be less reliable around tightly curved isobars. This figure is for the Southern Hemisphere only.

direction *from* high *to* low pressure) and gustiness and turbulence tend to increase. The rougher the surface, the greater these effects; that is, the greater the drop in wind speed, the greater the deflection and the greater the gustiness.

Worked examples

Figure 3.3 shows a real weather map, an ideal example to analyse for wind speed and direction using the method outlined above.

A quick glance at the map shows a fairly vigorous high near Tasmania and a series of weak lows over the northern part of the continent. Although the pressure gradients (spacing of isobars) are much weaker in the north, observed wind speeds as shown by the flags are not markedly different. This is because, for the same pressure gradient, stronger winds will be generated near the equator than at the Poles.

Let's see how well Figure 3.3 works in predicting wind speeds for Perth and Townsville, noting that this weather map shows wind speed in kilometres per hour rather than knots (slightly unusual, given that the flags were originally developed in knots), hence the odd speed limits on the newspaper map (each full feather is 10 knots so ⌐ is 10, ⌐ is 20, and so on).

Perth

The isobars in the Perth region are running more or less east–west and their spacing is between panel 4 and 5; that is, 250 to 125 kilometres between isobars (this newspaper weather map has 2 hPa isobars, like Figure 3.3). Panel 4 is a reasonably close match (the weather map has about 6 isobars between Albany and Carnarvon, whereas panel 4

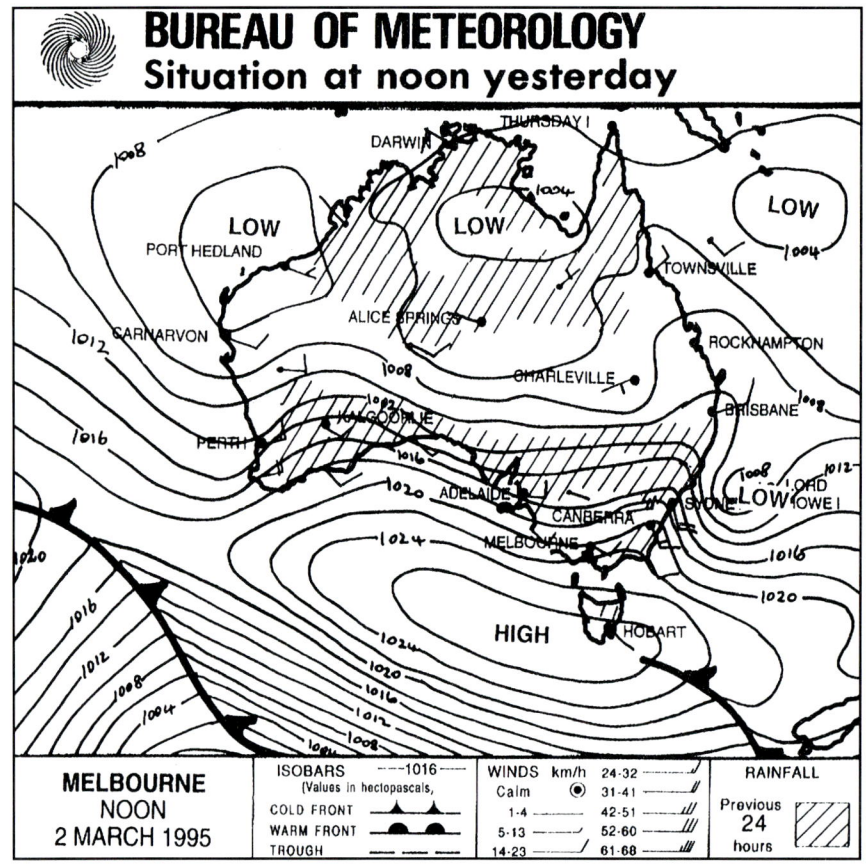

Figure 3.3
A typical small scale newspaper weather map for Friday, 2 March 1995, published in the Melbourne *Age* newspaper. (Data courtesy Commonwealth Bureau of Meteorology.)

Weather map isobar spacing and measured wind speed (over land)	Lower estimate Panel 4 estimate in knots (km/h)	Upper estimate: Panel 5 estimate in knots (km/h)
6 isobars between Albany and Carnarvon; 1–2 flags in the region indicating 14–41 km/h (see legend on map) and 1.5 flags at Perth itself, suggesting 24–32 km/h	over water — 10 (18.5) over land — 8 (15)	over water — 20 (37) over land — 16 (30)
Perth (midpoint of range) — 28 km/h	Perth (land) — 15 km/h	Perth (land) — 30 km/h

has about 4.5 and panel 5 has nearly 9). Results for the Perth region are shown in the table.

Although the agreement is reasonably good in this case, the only way to see that Perth had 1.5 flags and not 1 flag was by enlarging the weather map. A word of caution: use this method as a guide only. Reference to Figure 3.2 (wind direction) shows very good agreement, but the weather map flags point *into* the wind, whereas those in Figure 3.2 are pointing to where the wind blows *from* — a 'satellite' view, as it were.

Townsville

The isobars in the Townsville region are also running more or less east–west and their spacing is between panel 3 and 4, that is, 500 to 250 kilometres between isobars. Panel 3 is a much closer match than panel 4.

The agreement is very good in this case — panel 3 is actually quite close to the spacing on the weather map in the area. Wind direction (Figure 3.2) shows only modest agreement, which is to be

Weather map spacing and measured wind speed (over land)	Lower estimate Panel 3 estimate in knots (km/h)	Upper estimate Panel 4 estimate in knots (km/h)
3 isobars between Brisbane and the top of Cape York; 1 flag at Townsville, suggesting 14–23 km/h.	over water — 8 (15) over land — 6 (11)	over water — 16 (30) over land — 12 (23)
Townsville (midpoint) — 19 km/h	Townsville (land) — 15 km/h	Townsville (land) — 23 km/h

expected when the isobars are a long way apart and there is uncertainty as to their orientation in the local area. At any rate, if you had relied on the wind direction diagram, you should have predicted on-shore winds. No doubt you used the correct part of Figure 3.2 — you could have used just the low (that is, the low pressure system on its own) or the straight flow image, which has the low to the north and the high to the south.

SIMPLE RULES OF ORIENTATION

Remember backing and veering? Figure 3.2 shows these quite clearly. Looking from above, we can see that the wind direction changes in a clockwise direction — veers — from the geostrophic level to the surface (the opposite in the Northern Hemisphere) and it doesn't matter whether you are looking at a high pressure system or a low. It can also be seen that this veering is greater over land than over water.

Buys Ballot's law of orientation (defined in the Glossary) is a simple rule about where the high pressure system is if your back is to the wind (clearly, from Figure 3.2, it is always to your left). However, the different orientations of the observer — sometimes facing the wind, sometimes with back to the wind — can be confusing. Also potentially confusing are the likely changes in wind direction with the normal passage of highs and lows, which are usually from west to east in our part of the world. These issues are explained in Table 3.2.

> Backing and veering concepts are useful because they are independent of whether you are looking at a high or a low and also of whether you are facing the wind or have your back to it. Either way, veering means turning clockwise from the observer's point of view and backing means the opposite.

WIND FORCE

Sailors and sailboarders are usually interested in how much energy they can extract from the wind and how they can convert this energy into motion. They do this by letting the wind exert a force on their craft's sail or sails. In all but the very lightest winds (when you'd be better off going for a swim anyway), the force exerted by the wind on a sail is proportional to the square of the wind speed. We can write a simple equation for this:

Force on sail = AU^2

where A is a 'factor' and U is wind speed in metres per second.

The factor 'A' depends on the size and shape of the sail, its angle to the wind and the speed of the craft through the water. The fact that the wind speed enters the equation as a square is very significant. It means that if the wind speed doubles, the force on the sail increases four times; if it trebles, force increases by nine, and so on.

Without getting too technical, let's split A into its component parts. We do this by writing:

Table 3.2
Simple rules of orientation for wind direction, for backing and veering, and for the passage of highs and lows in the Southern Hemisphere.

Situation	Interpretation	Comments
Standing on the surface with back to the wind	The high — or higher pressure — is to your left-hand side and the low — or lower pressure — is to your right.	This is known as Buys Ballot's law (c1857).
Standing on the surface facing the wind	The high — or higher pressure — is to your right-hand side and the low — or lower pressure — is to your left.	Buys Ballot's law expressed from a different orientation.
Descending from the geostrophic level (approximately 1000 m) to the surface with back to the wind	Wind will veer (turn clockwise); you will have to turn clockwise in order to keep the wind on your back. Wind will veer more overland than water and will veer more under stable atmospheric conditions.	See Figure 3.2.
Descending from the geostrophic level (approximately 1000 m) to the surface facing the wind	Wind will veer (turn clockwise); you will have to turn clockwise in order to keep the wind on your face. Wind will veer more over land than water and will veer more under stable atmospheric conditions.	See Figure 3.2.
Wind passes from water to land	Wind veers by perhaps 15–20°.	Based on 15° deflection over water and 35° over land.
Wind passes from land to water	Wind backs by perhaps 15–20°.	Based on 15° deflection over water and 35° over land
Passage of a normal high or low pressure system from west to east	Wind direction will back or veer depending on whether you're in the top half or lower half of the system. Assuming the high or low is moving from west to east, wind will back in the top half and veer in the lower half (same for high and low).	See Figure 3.2 and imagine the high or low passing over you from west to east; in Australia most systems approach from the west.

A = (air density) × (sail area) × (force coefficient).

This tells us immediately that, at a given wind speed, doubling the air density or the sail area will double the force on the sail. But air density at sea level doesn't ever stray far from its standard value of 1.2 kg/m^3 (it changes only slightly with temperature). So the only variables within the sailor's control are the sail area and the force coefficient. Obviously one can change the sail area by changing the sail, reefing it in (or letting it out), but it is impossible to cope with the common range of wind speeds this way, since every time the wind speed doubles, a sail a quarter of the previous size is needed. Rather, the main means of controlling the force is by changing the force coefficient, and the most effective way to do this is to change the angle of the sail to the wind: sheet in — more force, slack off — less force. Fine tuning can be done by adjusting the set of the sail. For example, on a boat tightening the leech to make the sail fuller will increase the force coefficient and therefore the force on the sail.

Another factor affecting the magnitude of the force coefficient is how the area of the sail is distributed in space (the 'aspect ratio' of the sail). Most sailors know that, given two sails of the same area, a tall one will be more efficient — especially in strong winds — than a short fat one. This is because wind speed increases with height, so that a tall thin sail has more of its area up where the wind is stronger, and the effect is exaggerated by the dependence on wind speed squared. The price paid for the greater efficiency

of high aspect ratio sails is obviously a greater 'heeling moment' (or the tendency to pull the boat over sideways).

Not all the force that the wind exerts on a sail is useful. By convention the force at right angles to the apparent wind is called lift, and that in the same direction is called drag. (The apparent wind is the wind measured on the craft and thus is a combination of true wind speed — measured on the water — and boat speed.) As a result we can split the force coefficient into the sum of a drag coefficient and a lift coefficient; both this sum and the ratio of the two coefficients will change with sail angle and other factors.

All of which goes to illustrate how local variations in wind speed are exaggerated when translated into force on a sail, because the sail force is dependent on wind speed squared and on where this force operates on the sail.

The variations in wind speed that we want to consider are of two kinds: variation in the vertical and variation in the horizontal. The first depends primarily on the roughness of the underlying surface; the second on local topographic features.

THE ATMOSPHERIC BOUNDARY LAYER AND GUSTINESS

Wind speed usually increases with height above land or water surfaces. Changes in wind speed with height are also called 'wind shear', a term derived from the shearing or sliding effect of fast-moving air (above) slipping over slower moving air (below). Friction between moving air and the ground leads to much lower wind speeds near the surface.

The term *boundary layer* is used to describe a variable region between the surface, where the wind speed is low, and the unrestricted air well above; right at the surface the wind speed is zero.

Appendix V contains a simple mathematical model for estimating wind speeds above land and water surfaces.

To understand the connection between the geostrophic wind and the winds experienced close to the ground we need to know a little more about the boundary layer. It is the layer of the atmos-

phere that is influenced directly by the ground surface. One of the chief characteristics of the boundary layer is turbulence — the unpredictable gustiness that is so much a part of near-surface winds. Turbulence has two causes: wind shear due to the friction of the wind against the surface, and buoyancy, which occurs when the air in contact with the surface becomes hotter than the air above, and expands and rises, to be replaced by cooler, sinking air.

On all but the stillest, hottest days, friction dominates the patterns of wind and turbulence in the lowest fifty metres or so of the boundary layer, the surface layer. At higher levels, however, buoyancy dominates.

The sun's heating of the ground generates buoyant 'plumes' which, by mid-afternoon in temperate latitudes, have produced a boundary layer one to three kilometres deep. The rising plumes and the sinking air masses that take their place mix the atmosphere very effectively, except in the surface layer. Figure 3.4 shows the main features of the boundary layer over land.

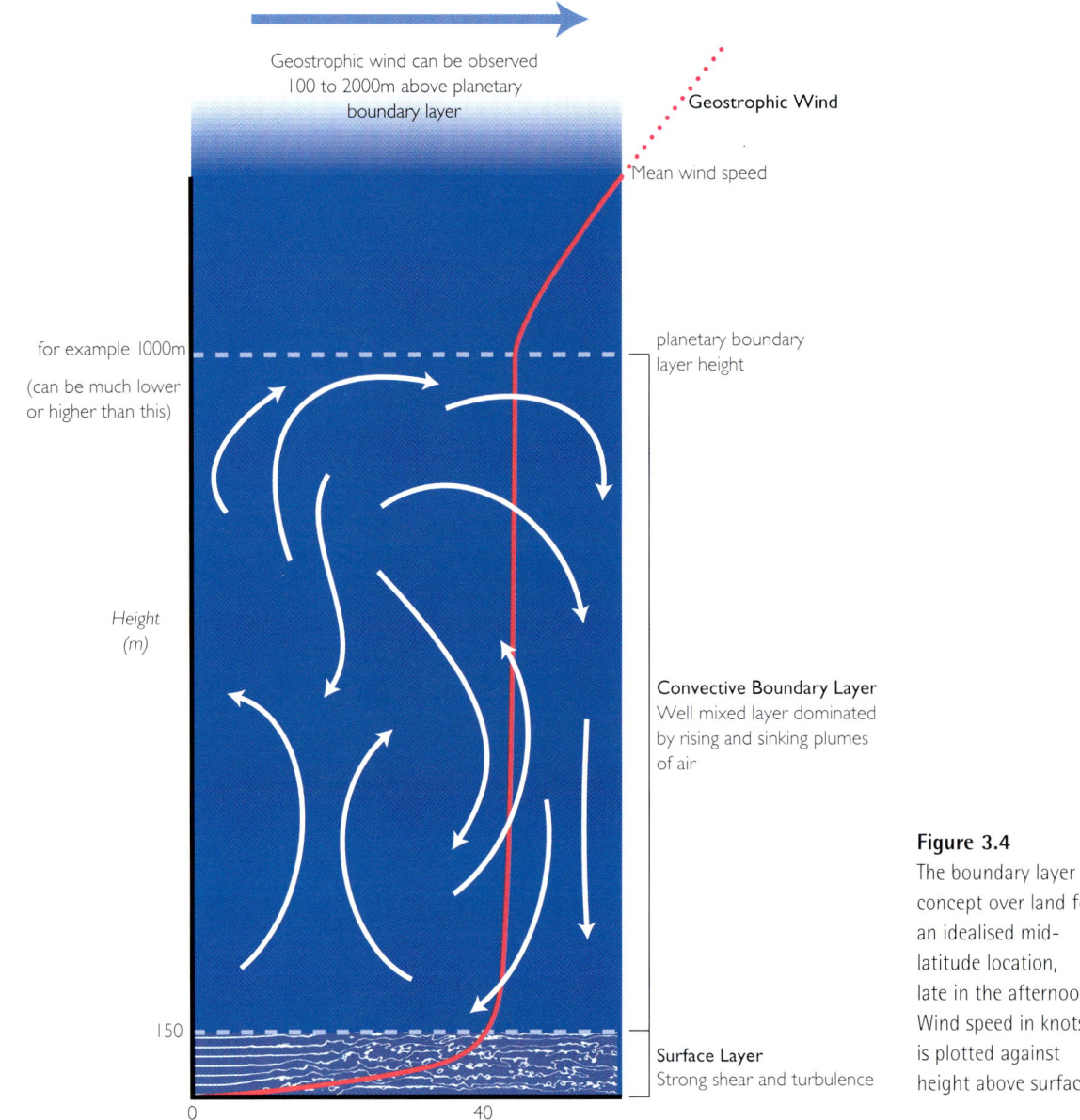

Geostrophic wind can be observed 100 to 2000m above planetary boundary layer

Geostrophic Wind

Mean wind speed

for example 1000m

(can be much lower or higher than this)

planetary boundary layer height

Height (m)

Convective Boundary Layer
Well mixed layer dominated by rising and sinking plumes of air

150

Surface Layer
Strong shear and turbulence

0   40

Wind speed

Figure 3.4
The boundary layer concept over land for an idealised mid-latitude location, late in the afternoon. Wind speed in knots is plotted against height above surface.

Some variations in this simple picture have very important consequences for coastal winds. First, the production of buoyant plumes requires the ground to be warmer than the air above and this occurs because of solar heating. As a result, the process switches off when the sun goes down. From then on, the ground loses its heat by radiating to the sky and cools the air in contact with it. The buoyant plumes rapidly run out of energy (usually within 30 minutes of sunset) and the turbulent boundary layer 'collapses', reducing from 1–3 kilometres to 100 metres or so in depth. When the sun rises the next morning, the process starts again, with the boundary layer deepening through the day to reach its maximum height by early afternoon.

When the geostrophic wind (that's the wind above the boundary layer) is of reasonable strength (more than 10 knots, say) the resulting turbulence, which is caused by friction, adds to the mixing caused by buoyancy during the day. At night the stabilising effect of the cool ground opposes the wind shear; air near the ground is cooler and denser and hence is reluctant to move upwards. Only on windy, overcast days do surface frictional effects account for all the turbulent mixing in the boundary layer.

The second important variation is the difference between sea and land. Unlike land, the sea is slow to warm and cool. Furthermore, evaporation helps to keep the sea surface relatively cool, so that buoyant plumes are weak or absent over the sea. This pronounced difference in the heating of the boundary layer by land and sea leads to the sea breeze described later in this chapter.

Wind speed increases in the surface layer as one gets higher. How fast the wind speed increases depends on whether the ground surface is rugged or smooth. A value called the wind exponent (α) can be used to work out how fast the speed will increase (see Appendix IV for more detailed information).

The wind exponent α applies only to wind in the surface layer — from the surface to fifty metres or so — and usually when the wind is moderate to strong. When winds are light, α cannot be easily linked to surface roughness alone, as it also responds to the buoyancy effect. With this proviso, the value of α varies from about 0.1 for smooth water and flat hard ground up to about 0.4 for dense urban areas with tall buildings and mountainous terrain. The practical consequences are as follows:

- The smaller the value of α, the faster the wind speed increases above the ground or water surface. This results in what might be called a 'fat' wind profile; that is, there will be more wind energy close to the ground — and hence more energy between the ground or water surface and any given height, such as the top of a mast. Conversely, at higher values of α (over rough surfaces), there is a 'thin' wind profile, with less wind energy close to the ground — and hence less energy between the surface and any given height.
- Small values of α also mean less turbulence; and large values, more. In thin wind profiles resulting from very rough surfaces you can expect much higher turbulence or gustiness than over smooth surfaces.
- The smaller the value of α, the shallower the surface layer. For example, if α is 0.1 (smooth surface) the surface layer might be only 50 metres deep. However, if the surface is aerodynamically rough, the surface layer might be as deep as 150 metres.

It must also be pointed out that, over rough surfaces, the formula doesn't work below the height of the roughness; don't, for example, try to predict wind speeds between buildings or trees.

The α formula describes the shape of the wind profile immediately above the ground or water surface. Sailors, sailboarders and hang-gliders should consider this carefully. For example, assume that the wind is measured at 2 metres above the surface and you are a sailboarder used to rigging a 5m^2 sail to handle 25–30 knots in Sydney's Pittwater — which is surrounded by hills and buildings. You will be in for a big surprise if you try to use the same sail to handle 25–30 knots in Sandy Point at Wilsons Promontory in Victoria. There, the prevailing winds blow over very smooth sand dunes, directly off Bass Strait, and for the same wind speed at 2 metres above the surface, the locals are using 3.8 and 4m^2 sails, if not smaller.

The following points summarise the essentials of wind speed and gustiness for a given wind speed above the surface layer:

- Smooth surfaces usually produce a shallow surface layer with relatively little turbulence (gustiness) and relatively high wind speeds near the ground (or water). Above the surface layer, the wind speed will vary little to the top of the boundary layer. On the ground or water surface, the average wind will be relatively fast and smooth.
- Rough surfaces usually produce a deep surface layer with high turbulence (gustiness) and relatively low (but gusty) wind speeds near the ground. On the ground or water, the average wind speeds near the surface will be slow and rough (turbulent).
- Intermediate surfaces usually produce conditions between those described above.

Figure 3.5 shows hypothetical surface layer wind profiles for different values of α. A word of caution: these do not take into account barriers such as buildings, hills or trees in the immediate area of the observer — that is, effects that might occur if you were sailing in the lee of an individual hill or building. Such local effects are dealt with later (see pages 63–70). In this sense, the curves in Figure 3.5 should be considered examples of what the wind profile might look like if averaged over a considerable distance (and time).

The roughness of the sea surface is caused by waves and it increases with wave height. Since wave height increases with wind speed, so does α,

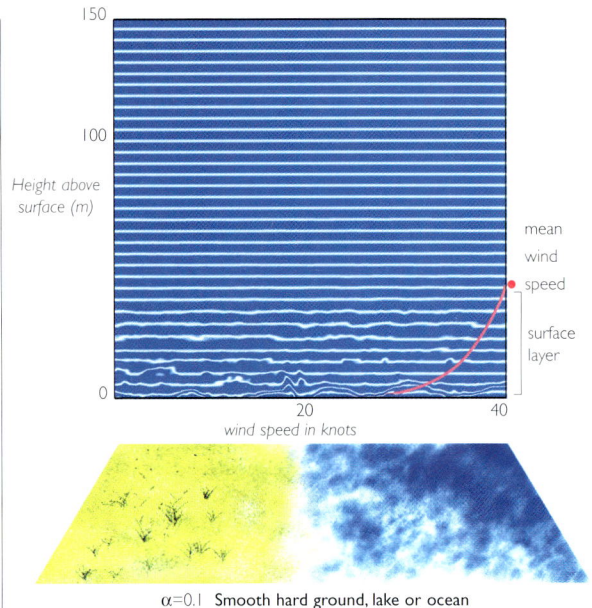

α=0.1 Smooth hard ground, lake or ocean

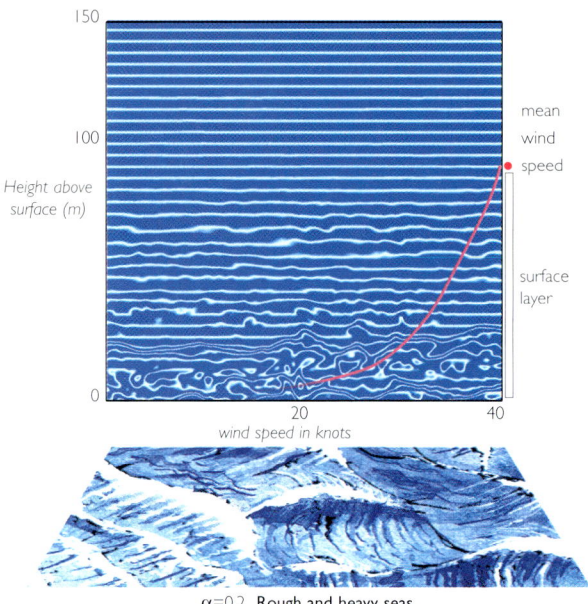

α=0.2 Rough and heavy seas

This is schematic in the sense that it represents the average conditions over a significant horizontal distance

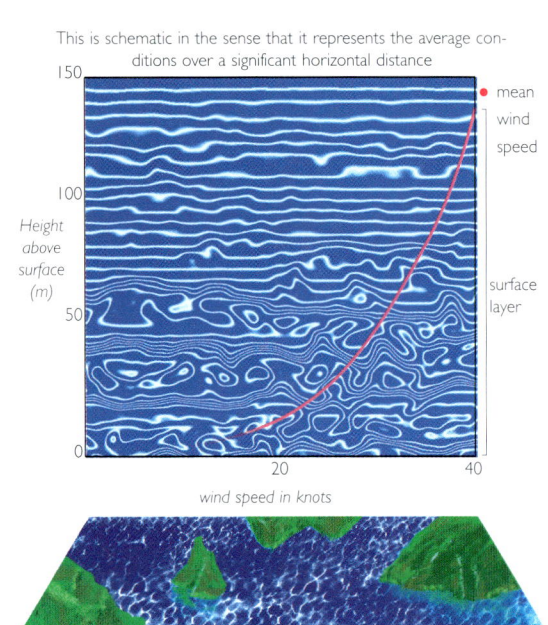

α > 0.3 Inside harbour or lake surrounded by hills, buildings etc

In the top left profile, which is over a smooth surface, there is a lot of wind energy close to the surface and the surface layer is about 50 metres deep. At 10 metres above the surface, the wind speed is about 35 knots. At 20 m, which might be close to the top of a mast on a yacht, the wind speed has not increased much.

The profile above right, over a very heavy sea, shows less wind energy close to the surface and a surface layer about 100 metres deep. At 10 metres above the surface the wind speed is about 25 knots, increasing to nearly 30 knots at 20 metres.

The profile at left, over the very rough lake/estuary is rather different. The surface layer is 150 metres deep; at 10 metres above the surface, the average wind speed is less than 20 knots and at 20 metres above the surface the wind is not much stronger than at 10 metres. You can imagine that the wind would also be rather gusty.

Figure 3.5
The surface layer concept, highlighting the different shape of the wind profile over different surfaces with different aerodynamic surface roughnesses.

but for the range of wind speeds that relate to sailing rather than survival, the value of α given for water surfaces in Appendix IV will suffice. If a lake or a bay is surrounded by rough ground with a much larger α than the water, then for a long way from shore (up to 1 kilometre) the wind profile will retain the shape of that over the rough land surface, except in the lowest few metres above the water.

Figure 3.5 shows a 40 knot wind at the top of the respective surface layers.

We have seen that the rougher the surface —

aerodynamically speaking — the greater its influence on the wind profile. Rough surfaces also produce gustier winds than smooth surfaces, a point of real significance for those who depend on the wind for motive power. (People who learned to sail on inland lakes are pleasantly surprised by the smoothness of the on-shore winds on the coast.)

Gustiness is very hard to quantify in a practical way. However, some useful points can be made. If we have two sites, one a 'smooth' lake surrounded by tall crops (α = 0.2) and the other a 'rough' lake surrounded by hills ($\alpha > 0.3$), and the region has an annual average wind speed of 16 knots, we can expect the smooth lake to experience, annually, 15 gusts to 20 knots, 12 gusts to 40 knots, 4 gusts to 60 knots and 1 gust to 80 knots. In sharp contrast, at the rough lake we can expect 1400 gusts to 20 knots, 1100 gusts to 40 knots, 400 gusts to 60 knots, 60 gusts to 80 knots and 5 gusts to 100 knots. These figures are based on probability equations and should be used with great caution (think of the probability when playing Two-up or when trying to toss 3 heads in a row). But the numbers show a very sharp increase in the gustiness of the wind for a relatively modest increase in surface roughness. Although lakes are used in this example, the principle holds true for near off-shore winds in much the same way.

To take an example over a much shorter period — say, 5 minutes: if the wind speed at 100 metres above the surface is 40 knots, then one can expect the peak 1–3 second gust to be 56 knots. Closer to the ground, at 30 metres, the peak gust could be 60 knots and at 2 metres 70 knots. Of course, the rougher the ground, the greater this effect.

WIND AND BARRIERS

The behaviour of wind blowing around (and over) three-dimensional objects such as islands, headlands, ridges and even mountain ranges is important for users of the Australian coast, particularly those who either 'harness' the wind for motive power (for example, hang-gliders and sailors) or those who can be directly influenced by it (powerboaters and surfers).

As a rule, the stronger the wind, the stronger the effect of topography on it. Many sailors will have experienced fluky and gusty airflows close downwind of steep islands or near cliffs in an off-shore wind. In landlocked estuaries or where mountains come close to the shore, the steady wind that blows out at sea may be translated to an unpredictable horror, with strong and seemingly random shifts in direction and strength. This makes it important to know about the kind of local wind and turbulence effects you may encounter and the reasons behind them. As the patterns of wind flow around a topographic feature 100 metres high by 1 kilometre long are quite different from those around one of the same shape but ten times bigger, the subject is best dealt with by referring to small scale effects and large scale effects.

Effects at small scales

'Small scale' means landscape features no bigger than tens of metres in height and up to some hundreds of metres across. Together with the wind strength, this scale determines what wind patterns to expect. In the main, the winds described in the following scenarios are fresh breezes or, if the winds are light, it is daytime.

The small islet

Imagine sailing downwind towards a small steep islet that is roughly circular. Its height is h and diameter d. The islet is quite steep, so we can expect to see all the possible wind patterns demonstrated, especially separation or reversal of wind direction behind the islet — a phenomenon that occurs behind steep obstacles only. The breeze is fresh, around 20 knots and not too gusty. As we approach and round the islet we encounter three areas with quite different wind regimes. These are shown in Figure 3.6, where the wind's streamlines close to the water are shown as blue, those about halfway up the islet as green and those a small distance above the islet as yellow. The grey 'cloud', particularly noticeable downwind of the islet, is the area where most wind disturbance occurs.

Immediately upstream of the islet, in what we might call region A, the average wind speed reduces and turbulence or gustiness increases. The winds start to swerve to the right or left of the islet. Region A does not extend very far upwind (the upper limit is about equal to the height of the islet, h) — its extent being shown by where the three sets of streamlines start to become distorted upwind of the islet: this is well upwind of the grey region of maximum disturbance.

As we round the islet to the right or left (region B), the average wind speeds up and the turbulence levels, both in absolute terms and relative to the average wind, decrease. This effect also extends barely one islet height out from the shoreline and is most pronounced close in — the extent is shown by where the three sets of streamlines are no longer

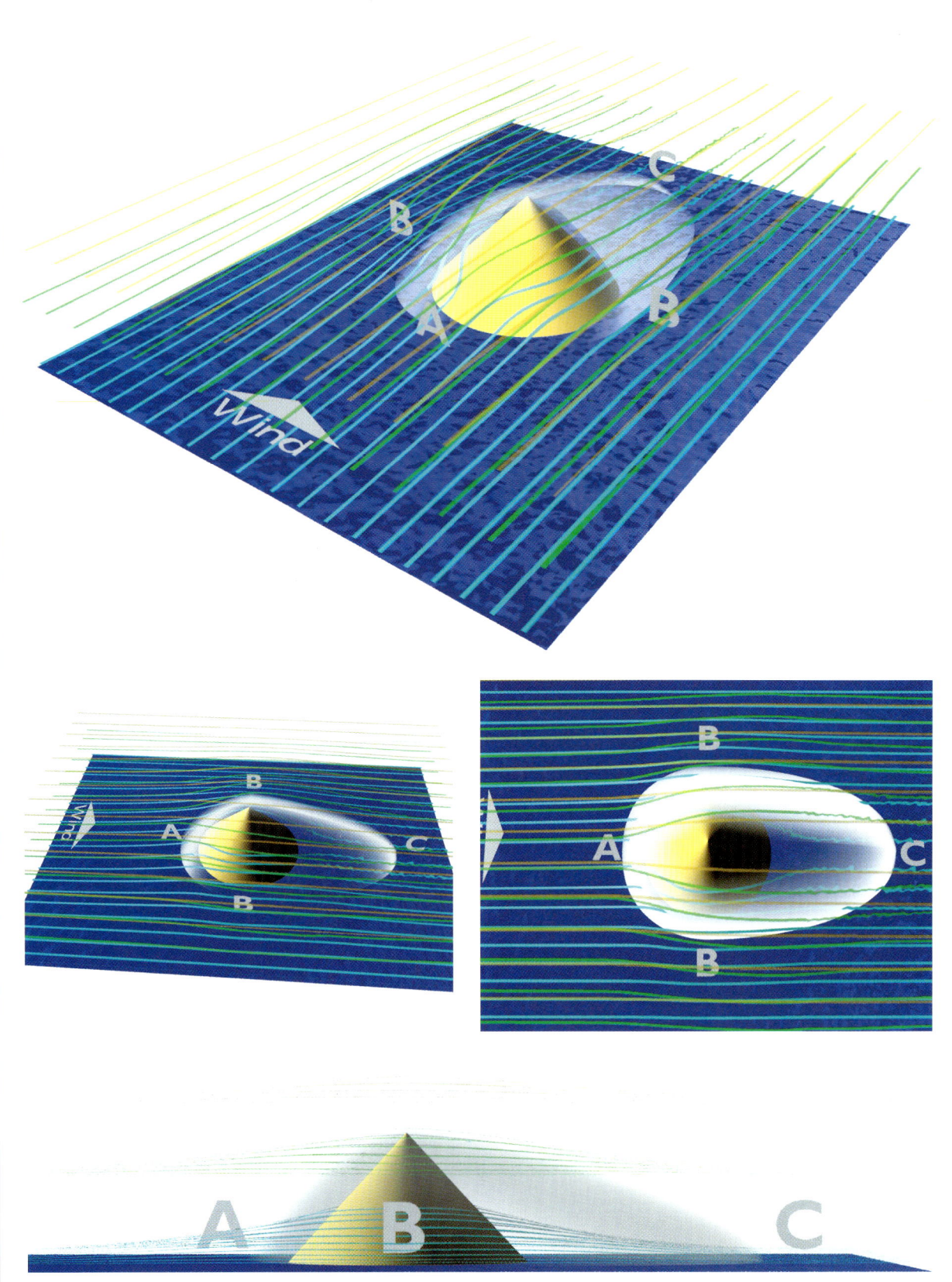

Figure 3.6
Idealised wind flow patterns around a small, circular islet in a fresh breeze.

distorted by the islet and, again, this is well outside the grey region of maximum disturbance.

Continuing to sail into the lee of the islet we encounter the most pronounced changes. Immediately behind the islet, if it is steep enough, the average wind reverses direction (region C). However, this will not be very apparent to the sailor, as it is masked by a large increase in turbulence. The steady breeze upwind of the islet is replaced by strong wind gusts, which have a higher probability of being in the direction of the islet than in the prevailing downwind direction. These gusts are at least as strong as the prevailing breeze. If the islet is not steep enough to cause this separation and reversal in wind direction, then conditions in its lee will still be highly turbulent and gusty, but with a higher probability of being in the prevailing wind direction than back towards the islet. The extent of region C is where the three sets of streamlines are no longer distorted by the islet and, yet again, this is well downwind of the grey region of maximum disturbance.

How steep is 'steep enough' depends on the roughness of the islet's surface. Steepness is measured as the average slope of the islet from its highest point to water level on its lee side. If the islet is covered with trees or bushes, a downslope angle of about 20° is enough to induce reversed or 'separated' flow in the lee; if the surface is smooth or grassy, then a slope of 30° is required (see Figure 3.7). This figure applies to a roughly circular islet. If it is of long aspect ratio instead of circular, with its long axis across the wind, then separation occurs at downslope angles as low as 10° for a rough surface and 20° for a smooth (Figure 3.8).

The downwind extent of the region of intense

gustiness and possibly reversed flow depends on the shape of the islet. Usually it extends no more than 3 or 4 h (h = height of islet) downwind, but if the islet has a steep rocky crest, this can be increased to as much as 10 h. Even outside the region of possible reversed flow, increased turbulence will be detected many islet diameters downwind. In this 'wake' region we also find reduced

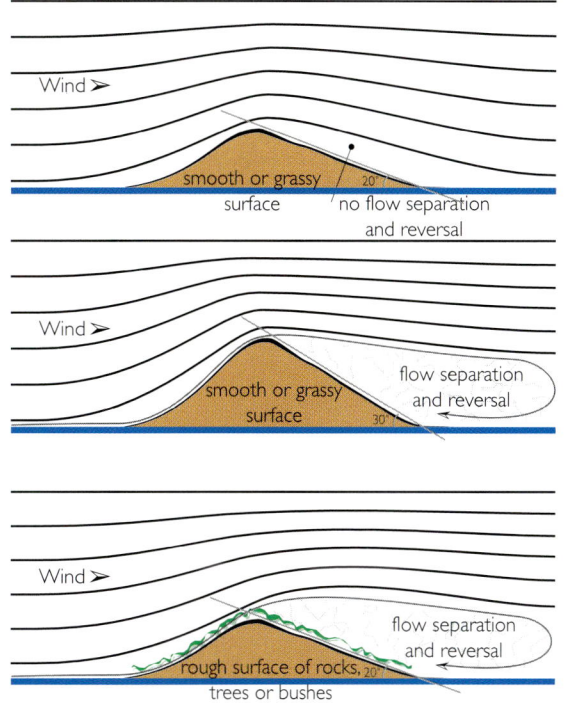

Figure 3.7
Idealised wind flow patterns around a small, circular-shaped islet in a fresh breeze – the effect of slope and aerodynamic roughness. (Based on data provided by J. Finnigan.)

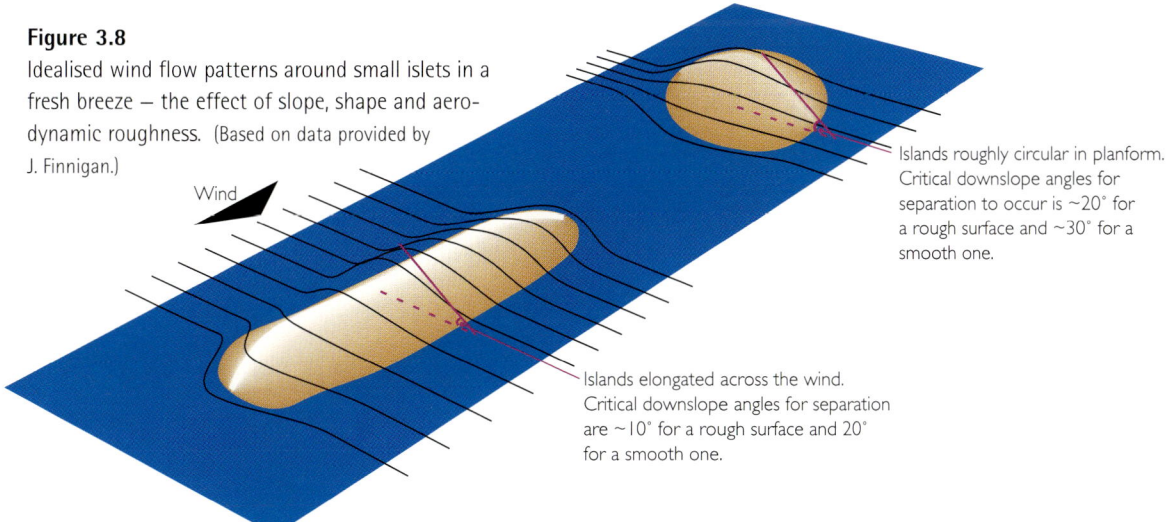

Figure 3.8
Idealised wind flow patterns around small islets in a fresh breeze – the effect of slope, shape and aerodynamic roughness. (Based on data provided by J. Finnigan.)

Islands roughly circular in planform. Critical downslope angles for separation to occur is ~20° for a rough surface and ~30° for a smooth one.

Islands elongated across the wind. Critical downslope angles for separation are ~10° for a rough surface and 20° for a smooth one.

average wind speeds. In strong winds, very choppy wave conditions will also often be found in the lee of the islet.

All these effects will be found around any small islet, but their strength depends directly on wind strength and islet steepness. Obviously, if the islet hardly breaks surface, its impact on winds flowing over it will be almost imperceptible.

> The magnitude of changes to wind speed and the strength of turbulent gusts in the lee are proportional to the strength of the oncoming wind and to the ratio of islet height to diameter – double the oncoming wind and wind speed doubles; double h/d and the wind speed changes double.

Estimating gust strength using a simple formula
The strength of the strongest gust that might be encountered close behind the islet can be estimated by the following rule of thumb which is based on multiplying the wind speed at a specified height by a shape factor for the island.
Step 1: Divide the height of the islet by its diameter, multiply by 7 and add 1
Step 2: Estimate the wind strength approaching the islet at a height of h/5 above sea level using Figure 3.9.
Step 3: Multiply the results of step 1 and step 2 together to get the strength of the strongest gust.

Expressed as a formula, the rule of thumb is:

Maximum gust strength = $U_{h/5} \times \left(1 + 7\dfrac{h}{d}\right)$

where $U_{h/5}$ denotes the undisturbed wind strength at a height h/5 metres.

Here is an example. Say the wind as estimated on board your boat is 30 knots and you are approaching a circular island 1 kilometre wide and 80 metres high.
Step 1: Enter height (80 m) and diameter (1000 m) as follows: (1+7 × 80/1000) = 1.6.
Step 2: From Figure 3.9 establish that, if the wind is 30 knots at boat height (assumed to be 2 metres above the water), then at 16 metres (h/5) it would be about 37 knots.
Step 3: Multiply the results of steps 1 and 2 to arrive at a 58 knot gust speed.

Another example: the wind as estimated on board your boat is a 'serious' 50 knots and you

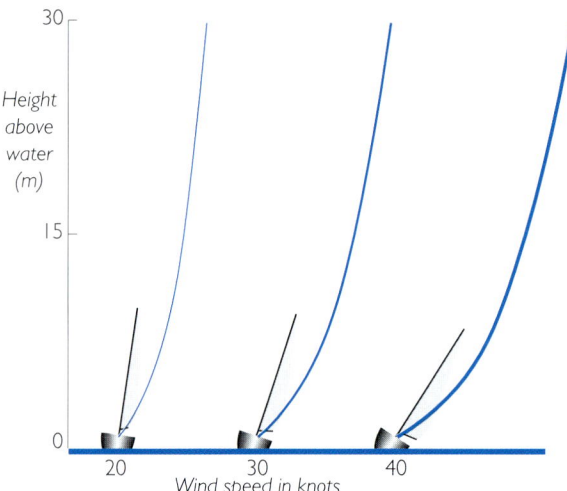

Figure 3.9
Wind speed profiles for a 20, 30 and 40 knot wind speed, measured at boat height (assuming a wind coefficient of 0.1).

are approaching a much smaller island which is 100 metres wide and 10 metres high.
Step 1: Enter height (10 m) and diameter (100 m) as follows: (1+7 × 10/100) = 1.7.
Step 2: Note that h/5 is boat height (2 m), so wind speed remains at 50 knots.
Step 3: Multiply the results of steps 1 and 2 to arrive at a scary 85 knot gust speed.

> The shape factor (1+7h/d) can vary between close to zero (for a low and wide island) to about 2.0 (for a steep-sided island). The shape factor cannot be used meaningfully outside certain limits: the height of the island must be 100 metres or less and its diameter less than 1000 metres. Although useful, estimated gust speeds should not be treated as absolute – at best, they are estimations.

All the wind patterns described above are general and will alter depending on the detailed shape of the islet. The effects can be strong. For example, the wind patterns behind Ailsa Craig, a steep rock in the middle of the Clyde estuary in Scotland, are notorious because they form a strong vortex that has resulted in some close calls for helicopters servicing the nearby Royal Navy anchorages.

The wake of the islet

The region of reduced average wind speed and increased turbulence extends downwind in a growing zone called the 'wake'. The wake rapidly spreads so that its crosswind extent exceeds the diameter of the islet. At the same time the strength of the changes to mean wind and turbulence also quickly decrease. Although meteorologists with sensitive anemometers can detect the influence of the islet fifty diameters downwind, it is usually imperceptible to the sailor by ten diameters. The average wind speed recovers fastest, usually reaching 90 per cent of its upwind value by ten diameters downwind. The turbulence persists a little longer (see Figure 3.10).

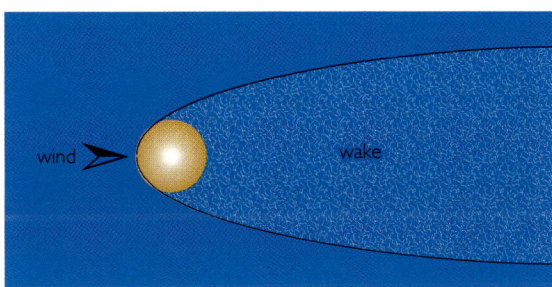

The wake is an area of decreased average windspeed and increased turbulence

Figure 3.10
Idealised wind flow patterns around a small islet in a fresh breeze — the concept of the wake. (Based on data provided by J. Finnigan.)

The headland

Much of what we have said about an islet can be applied to a headland as well, given that it is effectively an island attached to the shore on one side. The simplest situation to analyse is that of a headland extending a considerable distance at right angles to the coast, with a strong wind blowing parallel to the coastline. As Figure 3.11 shows, three distinct regions develop: upwind, around the point of the headland, and in the lee.

Upwind

There is a slight deceleration of the average wind speed just upwind of the headland and an accompanying increase in gustiness or turbulence.

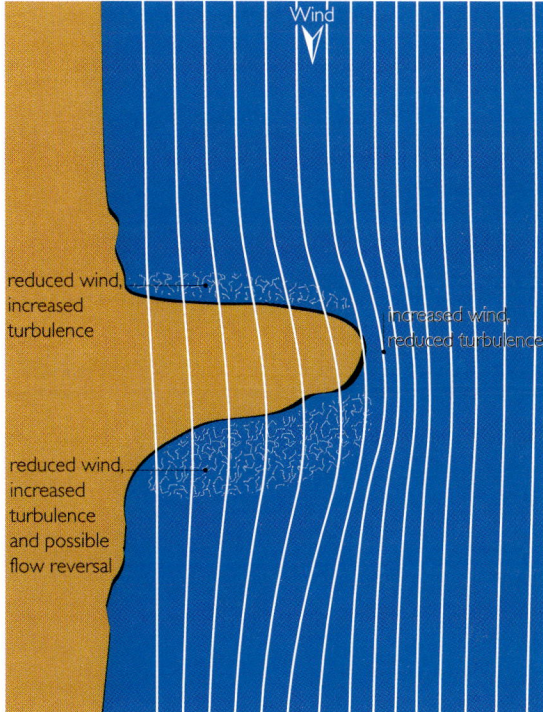

Figure 3.11
Idealised wind flow patterns around a headland in a strong breeze. (Based on data provided by J. Finnigan.)

However, just as for an islet, this will be imperceptible unless the headland is high or is flanked by cliffs.

Off the point

There will be an increase in average wind speed and a decrease in turbulence as the wind rounds the point. Again, this will be imperceptible unless the headland is large or is flanked by cliffs.

In the lee

Downwind the wind speeds will be reduced in a region that increases with the height of the headland. Just as for the islet, the extent of changes to the wind and turbulence depends on the steepness and roughness of the headland.

Other wind directions

When the wind is not crossing the headland at right angles — either because it isn't blowing parallel to the shore or the headland is not at right angles to the shore — then a sufficiently long headland will have the effect of 'steering', or deflecting, the wind along its length. This is illustrated in Figure 3.12. As for all these phenomena, the effect may be imperceptible to the sailor unless

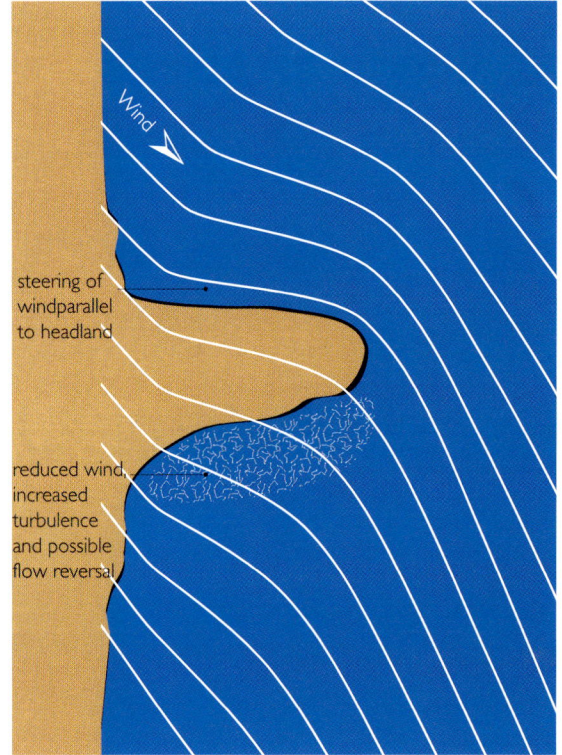

Figure 3.12

Idealised wind flow patterns around a headland in a strong breeze — the concept of wind steering, when the wind approaches the headland at an oblique angle. (Based on data provided by J. Finnigan.)

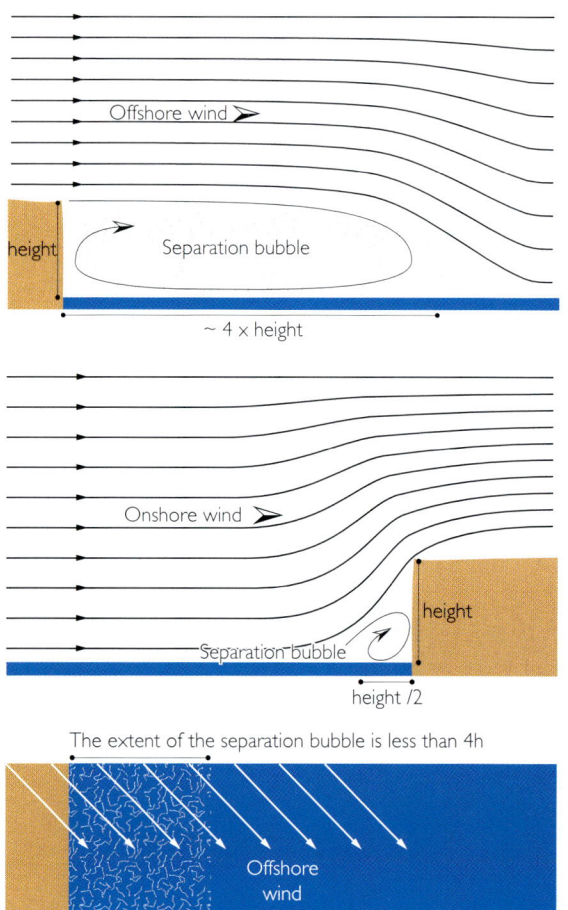

Figures 3.13 a–c

Idealised wind flow patterns around a cliff in a strong breeze — the effect of the angle of approach of the wind. (Based on data provided by J. Finnigan.)

the headland is sufficiently large and steep. Of course, the same effect will be observed with winds at angles to the long axes of elongated islands such as those in Figure 3.8.

A second effect of a wind not at right angles to the headland is that it reduces the extent of the region of separated flow or enhanced turbulence in the lee of the headland (see Figures 3.13 a–c for the analogous effect behind cliffs).

Cliffs

As far as the wind is concerned, cliffs are features which are steep enough to cause separation and reversal of the wind direction close to them. When the wind is offshore, the extent of the separation 'bubble' is three to four cliff heights (see Figure 3.13a). Remember, as we saw in the context of the islet, that the 'reversed' average wind direction really means that the probability of a strong gust in the direction opposite to the prevailing wind is somewhat higher than one in the downwind

direction; it is not a wind one can set a sail to!

An onshore wind also leads to a separation bubble at the foot of the cliff, as in Figure 3.13b, but this extends a shorter distance than when the wind is offshore. Although the average flow reverses, once again this really means that the probability of gust direction changes. Don't rely on this wind reversal to keep your boat off the rocks!

When the wind is at less than a right angle to the cliff, the extent of the separation bubble decreases but the intensity of the turbulent gusts may increase (Figure 3.13c). As in the case of the headland, in onshore winds there is a deflection or steering of both the average wind and the gusts along the line of the cliff.

Combinations of effects

It should be obvious by now that there is a certain 'stock' set of effects when the wind blows over an obstacle, whether it be an islet, a headland or a cliff. These are:

- upwind deceleration accompanied by an increase in turbulence
- upwind (and downwind) separation if the obstacle is steep enough
- acceleration around the side of the object accompanied by a decrease in turbulence
- downwind deceleration accompanied by an increase in turbulence extending a considerable distance downwind in a wake
- steering of the wind parallel to the shore, if the coastline is steep enough. This effect is purely local and not to be confused with shifts in the direction of large scale winds as we move from sea to land.

Any real situation is likely to be more complex than the idealised cases above but with common sense the examples may be combined. The average slope of an islet or headland may be too shallow to produce separation; but, if the islet/headland is surrounded by cliffs, separation will certainly occur. Then, the extent of the reversed flow region would be determined by the cliff height, not the overall height of the islet or headland (see Figure 3.14).

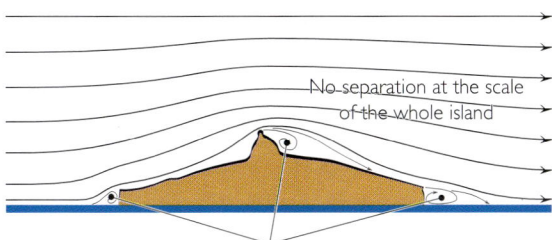

Separation caused by local features such as cliffs and steep hills

Figure 3.14
Idealised wind flow patterns around an island of complex shape in a strong breeze — separation occurs around local features but not around the island as a whole. (Based on data provided by J. Finnigan.)

One combination effect not yet discussed is 'channelling', or increases in speed caused as the wind squeezes between two islets or an islet and the shore. A typical situation is illustrated in Figure 3.15. Once again this may be thought of as simply exacerbating the speed-up experienced as the wind rounds one islet by superimposing the same

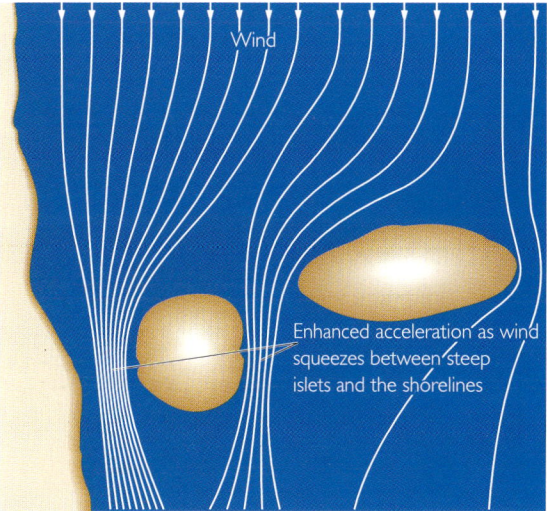

Wind

Enhanced acceleration as wind squeezes between steep islets and the shorelines

Figure 3.15
Idealised wind flow patterns around a series of islands and a nearby shoreline in a strong breeze — the effect of channelling of the approaching wind. (Based on data provided by J. Finnigan.)

phenomenon from its neighbour. As before, the increase in speed is accompanied by a decrease in turbulence.

The possible number of combinations is too large to explore with examples. Suffice to say that, for our purposes, all these effects may be considered additive. For example, a wind that has had its speed reduced and turbulence increased in the lee of a headland may then speed up and its turbulence may reduce as it squeezes between two downwind islets.

Medium–sized low islands

Let us increase the scale now to larger islands some kilometres in horizontal extent but not much higher than a hundred metres or so at their peaks. The average slope (much less than 20°) will not cause separation and reversal of flow in its lee. Nevertheless, local features such as cliffs or steep hills may cause local separation and turbulence but the scale of such effects will be set by the scale of the *local feature*, not by the island as a whole (see Figure 3.14).

The island as a whole has a different effect. We can think of it as a region of increased surface roughness (as characterised by the wind exponent α — see page 61). An 'internal boundary layer' will develop over the island, with increased turbulence and reduced average winds near the surface

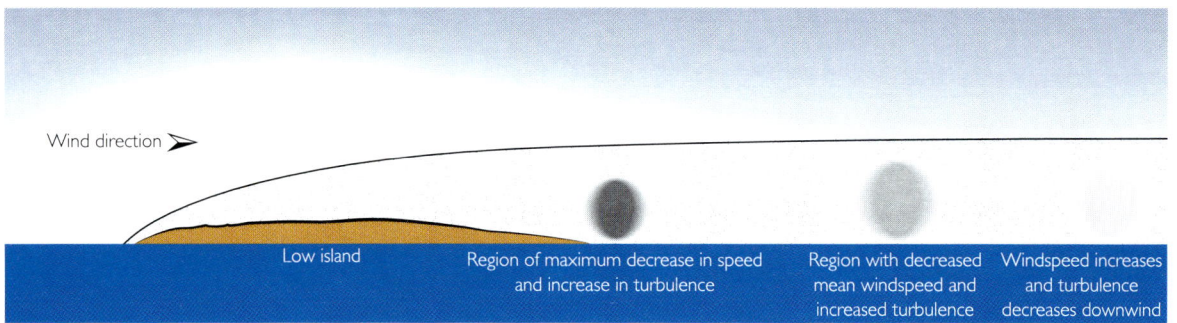

Wind direction

Low island | Region of maximum decrease in speed and increase in turbulence | Region with decreased mean windspeed and increased turbulence | Windspeed increases and turbulence decreases downwind

Figure 3.16

Idealised wind flow patterns around a medium sized, low island in a fresh breeze – the creation of an internal boundary layer (wake) downwind. (Based on data provided by J. Finnigan.)

(see Figure 3.16). Downwind, this boundary layer persists as a wake, which behaves in much the same way as the wake behind a steep islet. A sailor will notice reduced average wind and increased gustiness for about five diameters behind the island. Up to recent times, the land smell carried in this wake was an important source of navigational information for indigenous Pacific Ocean voyagers.

Effects at large scales

Large scale in the present context means features sufficiently high that, as the air flows over them, the whole boundary layer is deflected upwards. Since the depth of the boundary layer grows throughout the day, reaching a couple of kilometres on summer afternoons over land (less over the sea), the concept of large scale grows with it.

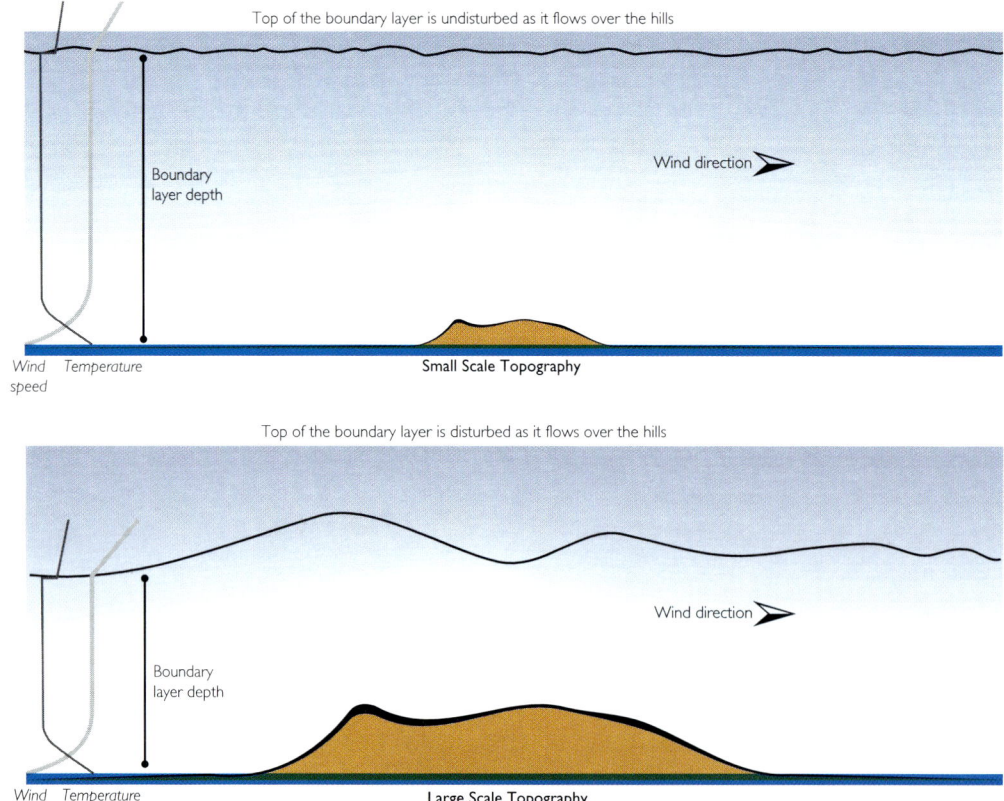

Top of the boundary layer is undisturbed as it flows over the hills

Boundary layer depth

Wind direction

Wind speed | Temperature | Small Scale Topography

Top of the boundary layer is disturbed as it flows over the hills

Boundary layer depth

Wind direction

Wind speed | Temperature | Large Scale Topography

Figure 3.17

A comparison between airflow patterns over a small scale feature (for example, an islet) and a large scale feature (for example, a large island or mountain range). Note the effect on the boundary layer. (Based on data provided by J. Finnigan.)

A good rule of thumb is that any offshore or near shore feature taller than 500 metres is a candidate for large scale effects. Relatively few landfalls — be they islands, cliffs or peninsulas — of this scale are encountered around Australia. The small scale features already discussed certainly deflect airflows over themselves but don't disturb the boundary layer top, except in very stable conditions when winds are light anyway (see Figure 3.17).

The first surprising departure from small scale effects is that, in the lee of large scale features, winds may *increase* while those upwind *decrease*. Figure 3.18 illustrates three possible airflow patterns that may form over a ridgeline as the wind blows at right angles to it. The ridge may be a coastal mountain range, an island or a peninsula of 500 to 1000 metres or more in height (much of the Great Dividing Range, where it is close to the coast, would qualify, for example).

What is shown in this figure is the change in wind speed close to the surface — at about 10 metres, say — as we move over a ridge. Three lines are plotted, corresponding to three different situations: neutral stability, moderate stability and substantial stability. Let's look at these more closely.

The airflow around topography of any scale is actually determined by the balance between *inertial forces* (the tendency of the air to keep on moving in the direction it is already going) and *buoyancy forces*. All of the effects discussed so far are caused by inertial forces. The air's tendency would have been to keep moving in the direction it was already going. But there was an island, a cliff or a headland in the way, and the wind had to find a way round this. Complicated trade-offs between pressure and turbulent mixing then produced the kind of airflow patterns already described.

Buoyancy effects add another important element to the balance of forces. They occur when air layers of different density interact. Air is a gas and its density changes as it warms up or cools down, or as its pressure changes. Warm air is lighter, or less dense, than cool air. If nothing else were disturbing the atmosphere, the coolest, densest layers of air would be nearest the ground, with warmer, less dense, layers above. This isn't what we experience, of course — it is generally cooler on top of mountains than at the bottom — but, as it turns out, we can easily explain this seeming contradiction.

As air expands it cools, so its density depends not only on its temperature but on its pressure too. Fortunately we can correct for this effect fairly simply, at least for most cases that are relevant here. At increasing distances above the ground, atmospheric pressure falls, air expands and becomes less dense, and air temperature falls in proportion. For dry air this background rate of temperature decrease with height is almost exactly

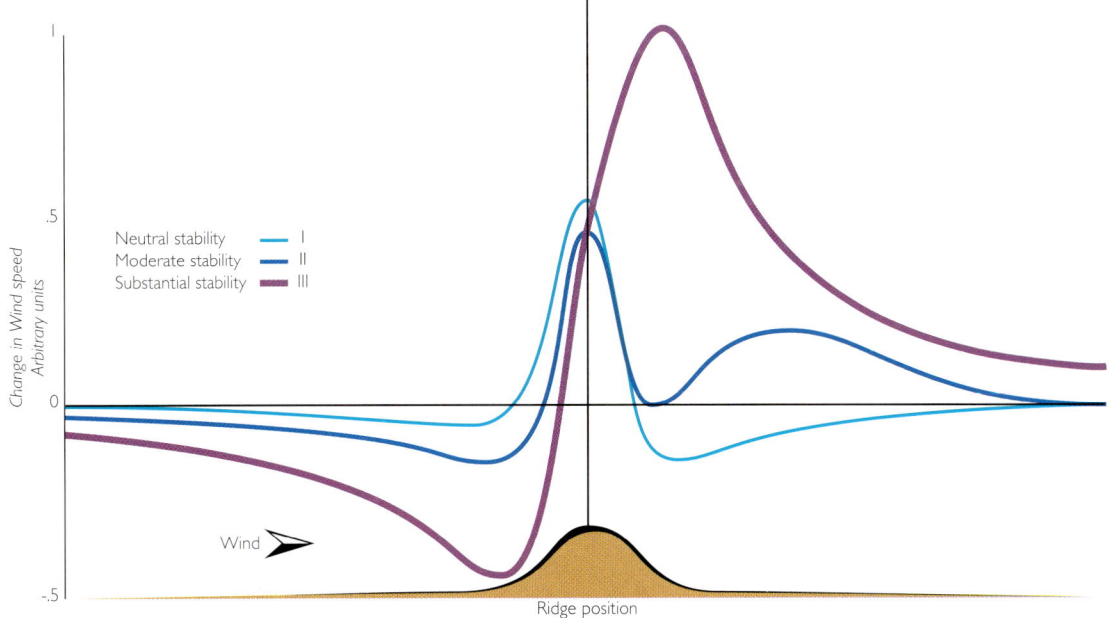

Figure 3.18

Idealised wind speed patterns over a large scale feature in a fresh to strong breeze — the effect of atmospheric stability.

(Based on data provided by J. Finnigan.)

1° Celsius per hundred metres of height gain (in mid-latitudes) and is called the (dry) adiabatic lapse rate. Now, if we take the simple step of correcting all our temperatures by adding 1°C for every 100 metres we rise above ground level, then we can take the corrected temperature as an indication of the density. Temperatures corrected in this way are called 'potential temperatures'.

We are now in a position to understand buoyancy effects. Most of the atmosphere most of the time is 'stably stratified'; that is, the less dense layers of air (layers with higher potential temperature) are lying on top of more dense layers (with lower potential temperature). If we disturb this situation — for example, if we lift a dense air layer by making it flow over a hill — it will end up amongst less dense layers, and gravity will pull it back towards the surface. Sometimes it may develop sufficient momentum on its way down for it to overshoot its proper level and end up amongst more dense layers, thus being forced back up again to higher levels. This can happen more than once, resulting in an oscillatory or wavelike motion.

The one region of the atmosphere that is not usually stably stratified is the daytime boundary layer. If conditions are very windy, then the turbulence caused as the air interacts with the ground will mix the lower layers of the atmosphere to the same density and potential temperature. (Remember: the actual temperature will still fall by 1°C for every 100 metres because of the adiabatic lapse rate.) The lifting of one air layer above another as the wind blows over a hill won't result in any buoyancy forces in this case. In these circumstances we say the atmosphere is 'neutrally stratified'.

More often, the ground surface heated by the sun warms the air layers in contact with it. These are now less dense than the air layers above and will rise spontaneously until they find a level where their new density matches their surroundings. The atmosphere is now 'unstably stratified' — there are less dense layers below dense layers and this leads to energetic turbulent mixing as buoyant plumes of air heated by the warm ground rise to find the level where their density matches their surroundings. By mid-afternoon on an Australian summer's day, this height, which defines the top of the boundary layer, can be 2 to 3 kilometres. Over the ocean, except in the tropics near the equator, the boundary layer is usually neutrally stratified and not as deep as the unstably stratified boundary layer over land. This is because the large capacity of the ocean to absorb heat means ocean surface temperatures don't respond as strongly as the land surface to daily solar heating.

The actual wind flow and turbulence patterns encountered around any given topography depend on the balance between inertial forces and buoyancy forces, and this is where the concept of large scale and small scale topography becomes important. Large hills and mountains will always distort the stably stratified atmosphere above the boundary layer (Figure 3.17), so buoyancy forces will always affect flow patterns around them. Smaller features will experience buoyancy forces if the boundary layer is stably stratified or shallow — that is, at night time or early morning, so long as the wind is not too strong (strong winds will mix the boundary layer to a neutrally stratified state, whatever the time of day or night). Hence the distinction between large and small scale topography. In the strong winds that give rise to dangerous conditions when they interact with topography, flow patterns around small scale features are always dominated by inertial forces and we can ignore buoyancy effects.

Finally, the buoyancy force on an air layer depends on the difference between its density and the density of the air layers it ends up amongst as it is lifted in its passage over a hill. For a given hill height, this difference will be larger if the stratification (that is, the rate of increase in potential temperature with height) is larger. Conversely, for the same stratification, the density difference will be larger if the air layer is lifted over a larger hill.

We have seen throughout that the flow patterns are a result of the balance struck between buoyancy and inertial forces, and we know that, for a given hill height, the inertial force increases with wind speed. So, in deciding what flow patterns we will encounter, we have to account for three variables: hill height, stratification and wind speed. Appendix VI explains how scientists do this using a concept called the Froude Number. Here, it will be more helpful to present the same information in a pictorial way.

Predicting wind patterns around large scale features

Figure 3.19 combines the effect of stratification, hill height and wind speed in a single diagram to demonstrate the practical meaning of the neutral, moderate and substantial stability curves in Figure 3.18.

The figure has three 'dimensions': height, width and depth. The height of the figure (or the y-axis on a normal graph) represents the height of the hill or mountain, taking into account that this height must be large enough to affect the top of the boundary layer. The width (or the x-axis on a normal graph) is wind speed. The depth is atmospheric stability — the slice representing the most stable atmosphere is 'furthest' from the reader and that closest to neutral stability (little or no change in potential temperature with height) is 'closest'.

The shades of blue on each slice represent neutral, moderate and substantial stability, as in Figure 3.18. Remember that atmospheric stability is not independent of wind speed. If the wind is strong enough, the atmosphere will become mixed to some extent; and when this happens, atmospheric stability will tend towards neutral. To deal with this, the colour/shading is designed to leave the impression that strong wind combined with strong stability is less likely than strong wind combined with weaker stability. It is

clear, for example, that substantial stability (which can result in a dramatic reduction of wind speed in front of a hill and an even more dramatic increase behind it, as shown in Figure 3.18) is most likely when the hill is high, the wind speed light to moderate and the atmosphere very strongly stratified, while neutral conditions are approached over lower hills in strong winds.

Here are some concrete numerical examples:
- Neutral stability, together with moderate to strong winds, would occur with a hill at the lower end of the height range (say around 250 metres, more or less).
- Moderate stability would correspond to a temperature *increase* with height of about 5°C per 100 metres over a hill about 500 metres high and to a wind speed of about 30 knots.
- Substantial stability could correspond to a wind speed of 10 knots, and a rate of temperature *increase* with height of about 5°C per 100 metres over the same 500 m hill.

Since this is the situation in which the most dramatic changes occur, it is worth looking at

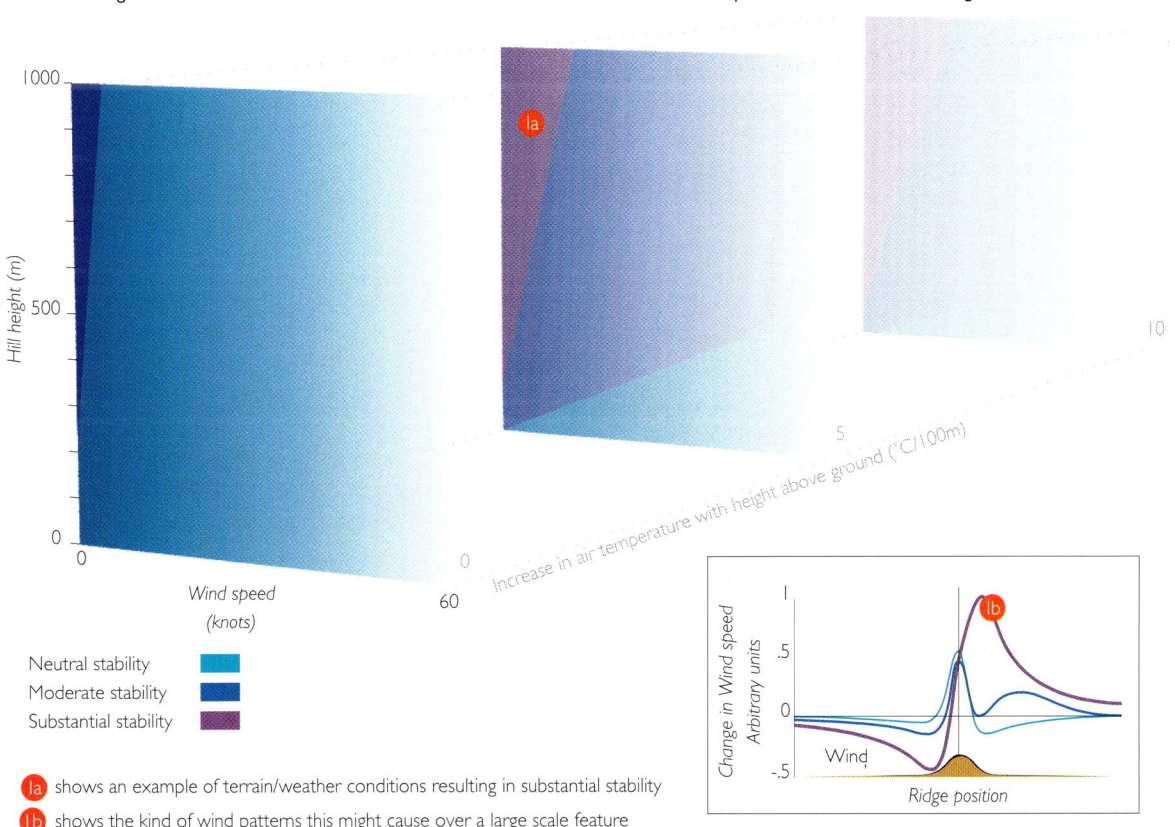

This figure shows the terrain/weather conditions which result in the different wind patterns shown in the inset figure

Neutral stability
Moderate stability
Substantial stability

Ia shows an example of terrain/weather conditions resulting in substantial stability

Ib shows the kind of wind patterns this might cause over a large scale feature

Figure 3.19
A graphical interpretation of the conditions resulting in different wind flow characteristics around large scale features.

Lenticular clouds often form above the wave crests revealing the presence of lee waves

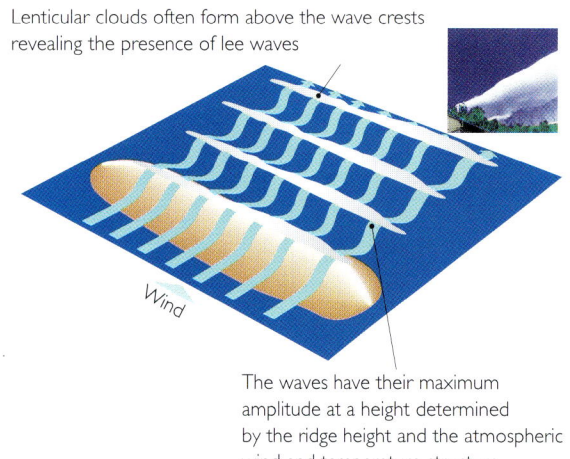

The waves have their maximum amplitude at a height determined by the ridge height and the atmospheric wind and temperature structure

Figure 3.20
Idealised wind flow patterns over a large scale feature in a fresh to strong breeze — the formation of lee waves (sometimes accompanied by lenticular clouds). (Based on data provided by J. Finnigan. Photograph R. Badham.)

some other combinations that also correspond to *substantial stability*. If the wind speed was much stronger — say, 50 knots — and the rate of temperature increase with height was the same, the hill would have to be well over 1000 metres high for conditions to be those of substantial stability (and a candidate for extreme gustiness). If, however, the hill was only 100 metres high (that is, small scale), the corresponding wind speed for the same atmosphere would be a mere 4 knots and gustiness would hardly be a problem.

Let's return to Figure 3.18 and look at curve 1 (neutral stability). Here we see the pattern we have already described for low ridges, headlands and islets. The wind drops slightly in front of the ridge, speeds up over the crest and drops to a greater degree in a sheltered region in the lee. The situation has been drawn with a hill or ridge of low slope, so there is no separation. (With a separation region, the speed reduction in the lee would be much more pronounced.)

Moving next to curve 2, the moderate stability case, we see quite a different pattern. The speed reduction upwind is much larger; compared to upwind, there is no shelter immediately behind the hill; and, surprisingly, a region of secondary speed-up has appeared well behind the ridge.

Curve 3, the substantial stability situation, shows even greater changes. Now the wind speed reduction in front of the ridge is greater than the leeside shelter was in neutral conditions. Even more surprising is the speed-up in the lee; it is twice as large as the hill top speed-up in neutral conditions. This strong downslope wind is a potentially destructive dynamic weather phenomenon caused by the interaction of the topography with strong wind shear and stability in the atmosphere above.

If the atmosphere is highly stable close to the ground and less strong at higher levels, and if the wind shear doesn't change too much with height, then another phenomenon may be encountered: lee waves, which are sometimes accompanied by lenticular clouds (see Figure 6.12). Illustrated in Figure 3.19, these consist of a pattern of bands of increased and decreased wind speeds downwind of, and parallel to, the ridge.

The most extreme form of dangerous downslope wind occurs when the atmosphere is even

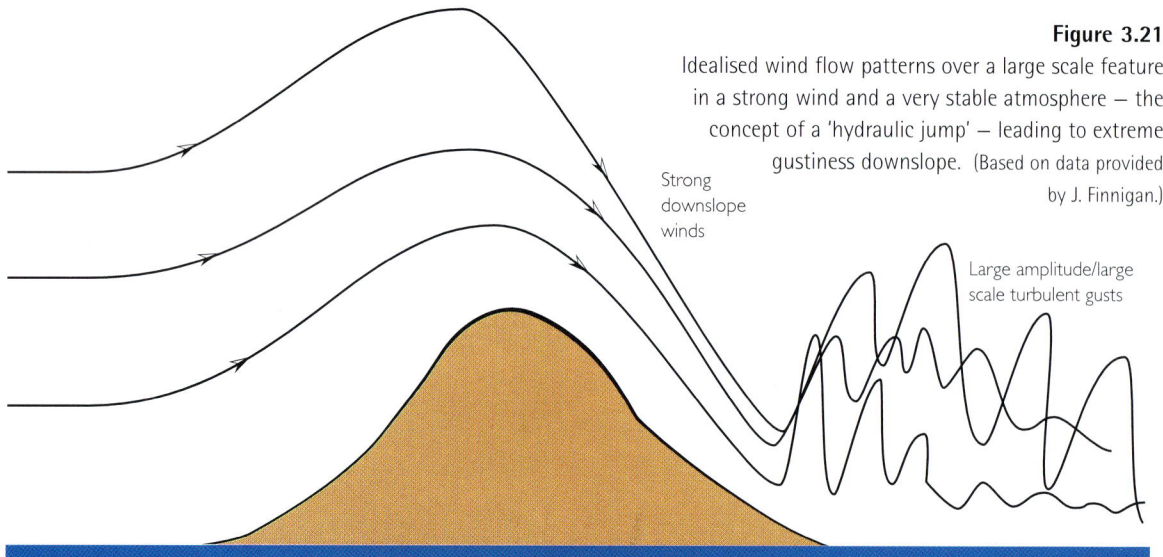

Figure 3.21
Idealised wind flow patterns over a large scale feature in a strong wind and a very stable atmosphere — the concept of a 'hydraulic jump' — leading to extreme gustiness downslope. (Based on data provided by J. Finnigan.)

Strong downslope winds

Large amplitude/large scale turbulent gusts

more strongly stratified than in our substantial stability case and when winds at high levels in the atmosphere are strong. Then the strong down-slope speed-up ends in an abrupt deceleration, where the energy of the mean wind is taken up by large amplitude turbulent fluctuations, as shown in Figure 3.21. Since these turbulent gusts are on the scale of the whole mountain, they can last for many minutes at a time and reach speeds of 50 or 60 knots. Scientists believe that this phenomenon is akin to what we observe when water pours over a smooth boulder in a stream, ending in a break-ing wave on the downstream side of the boulder. The technical term for this is a 'hydraulic jump'.

The clearest examples of these effects are seen when the wind blows at right angles to a long ridgeline — be it a coastal mountain range, a peninsula, a large headland or an elongated island. If we consider, instead, an isolated moun-tain or a large island of more circular cross sec-tion, then similar patterns to those in Figures 3.18 and 3.20 may be observed on the centreline, but the overall flow patterns are three dimensional and often include well formed vortices behind the obstacle (see Figure 3.22). If the approach flow is very stable indeed, then the wind doesn't flow over the mountain or island at all, but goes round it and the eddying patterns are confined to the lowest layers of the atmosphere, with little turbulence being generated.

If the same very stable conditions are found in the flow approaching a mountain wall, then the wind can't get around the obstacle and the airflow below a certain level is 'blocked' — that is, it just comes to a dead stop. This is by no means unusual in large inland mountain ranges, but the special conditions required would be unusual near the coast.

Finally, it is worth noting that distinct counter-rotating vortices sometimes form behind large isolated islands in steady airstreams, even if the islands are not particularly steep. These vor-tices sometimes form 'vortex streets' — beautiful trains of counter-rotating whorls which, in cases

Figure 3.23
Idealised wind flow patterns over a large scale feature in a moderate to strong breeze — the formation of vortices.

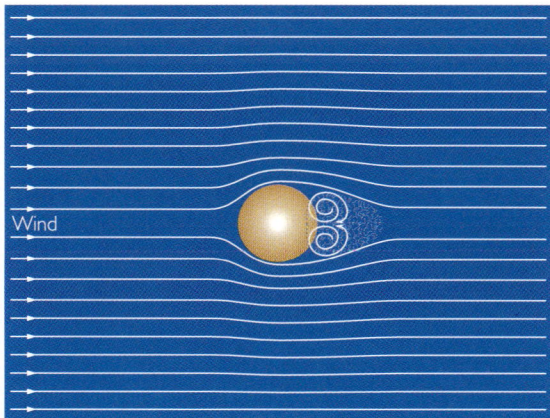

Figure 3.22
Idealised wind flow patterns over a large island in a stable airflow — the formation of two-dimensional vortices behind the island. (Based on data provided by J. Finnigan.)

where they have been made visible by clouds, have been observed from space. Figure 3.23 shows an artist's impression of this phenomenon behind Jan Meyen land in the Arctic Ocean. The diameters of the individual vortices are similar to, or larger than, the diameter of the island causing them. This phenomenon could lead to baffling wind shifts for a sailor downwind of the island.

Just as with the smaller features we discussed earlier, these phenomena may be encountered in various combinations, given the complexities of a real atmosphere blowing over a real coastline. Unfortunately, the vital importance of stability in determining what will occur, the fact that the depth of the atmosphere that is stably stratified varies through the day, and the relationship between the obstacle height and this depth (expressed in the Froude Number — see Appendix VI) all make it very difficult to offer relatively simple rules of thumb, as could be done for small features. If we add to that the critical importance of upper level wind shear, which usually accompanies stable windy conditions, then it is clear that we can rarely make firm predictions about the local winds.

Some practical generalisations

What *can* users of the coast draw from this discussion, then, that is of practical use? First, a realisation that shelter is not necessarily to be found on the lee side of these large features. Instead, winds may be much stronger than upwind. Second, a few signs to look for that may give a clue that these effects are occurring. A reliable indicator is the appearance of lenticular clouds above a mountain or an island (see Figure 6.12). These are dense, cigar shaped clouds (sometimes called 'hogsbacks') that form and remain stationary relative to the feature rather than being blown downwind. They form when moist air condenses as it is carried up over the mountain. If lee waves are present, they may be marked by a succession of such clouds, each one marking the crest of a wave.

Stable, strongly sheared airflows — Kelvin–Helmoltz billows (named after two famous nineteenth century scientists) — often reveal themselves well away from any high topography by another kind of wave cloud. These are ranks of regular, elongated clouds, often with as many as twenty in a set of billows. Sometimes the sky is full of these features. They are a warning that mountain wave patterns and downslope winds may be active around tall features of the coast scape

Finally, there is no substitute for local knowledge. Areas subject to downslope winds are often well known to locals. Tasmania's steep coastline may be a likely candidate, as would be some of Australia's Antarctic islands, although local knowledge there may be confined to the penguins and elephant seals. As some kind of yardstick for comparison, cruising yachtsmen report these effects in Patagonia.

SEA BREEZES

Sea breezes are local winds whose effects are commonly felt over an area of about 100 to 200 kilometres inland *and* offshore. Sea breezes can be characterised as 'distortions' of larger scale wind patterns and are caused by strong temperature gradients, which are common at the coast because the land and water are usually at different temperatures. Generally, the larger the temperature gradient between land and water, the stronger the sea breeze.

Even though the sea breeze is a common phenomenon, it is difficult to quantify. The offshore part of the breeze, in particular, has not been studied to any great extent. Sailors of ancient times used the sea breeze for their fishing trips: after a night at sea, they would harness it to bring them back to shore. However, if they sailed too far out and overestimated the reach of the sea breeze, they would have to wait for another day or resort to oars!

The idealised sea breeze system

The circulation of sea breezes occurs on most coasts and in large lakes in summer. This is the time of year when the land surface warms up after sunrise, heating the air above it. The same thing happens over the sea, but to a much lesser extent because the sea absorbs a large part of the heat energy and currents 'take' the heat away from the

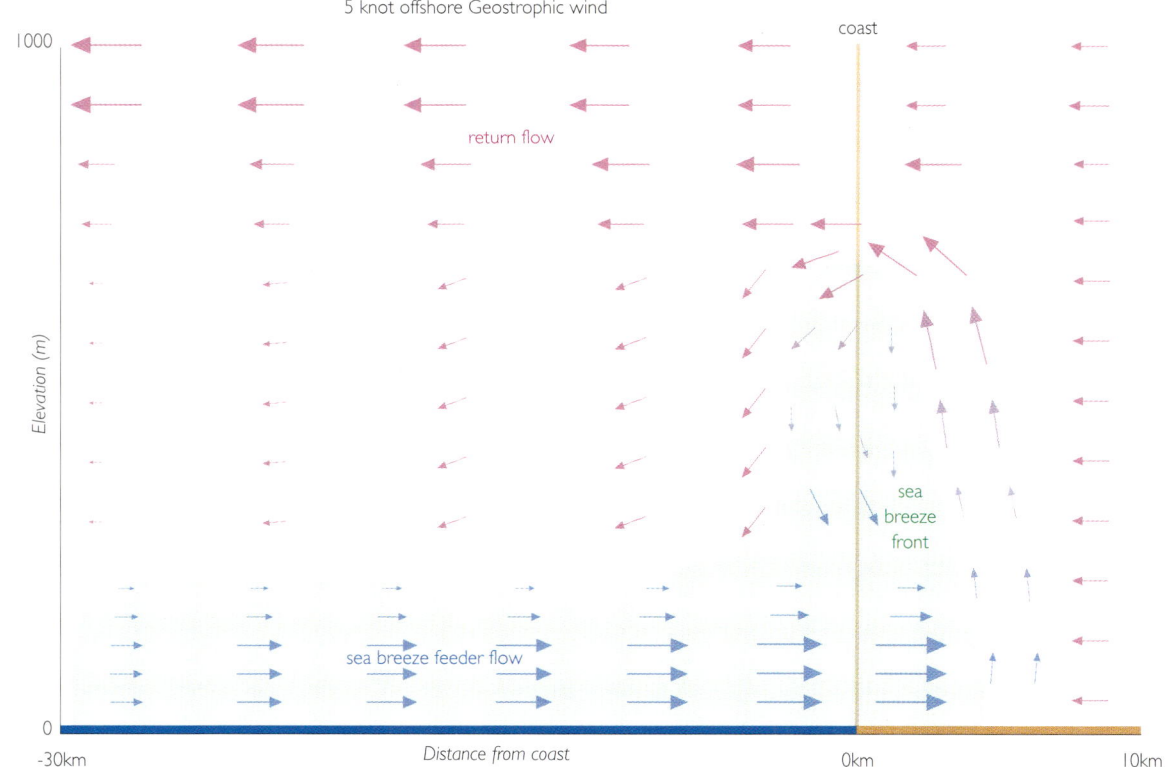

Figure 3.24

Idealised structure of a sea breeze occurring during light offshore wind conditions in the late morning. (Based on analysis by K. Finkele.)

surface. The warm air over the land rises, expands and creates a low pressure relative to the air over the cool sea. The pressure difference is more pronounced near the surface, where the air is warmer.

As a result of the pressure difference, a wind will blow from sea to land. This wind is called the 'sea breeze feeder flow' and is part of the sea breeze circulation or system (see Figure 3.24). It is this feeder flow that most people call a sea breeze.

Figure 3.24 is based on aircraft observations over a long straight coast near Coorong in South Australia, during light (2.5 m/s, or 5 knots) offshore wind conditions. The green area is the sea breeze feeder flow, defined as where the onshore wind component is at least 1 m/s (2 knots). The wind speed in the sea breeze feeder flow increases steadily over the water, as shown by the increasing length of the blue arrows from left to right.

The 'sea breeze front' is characterised by large gradients of temperature, humidity and wind direction. At the front, the feeder flow 'collides' with the geostrophic wind. The collision of air is referred to as 'convergence' (the opposite is divergence) and is shown by the opposing blue and red arrows near the sea breeze front. Wind speeds

tend to increase in the vicinity of the sea breeze front. For this sea breeze, the offshore extent of the circulation is considerably greater than the landward extent.

The height of the sea breeze feeder flow is relatively shallow (200 to 400 metres) and depends on atmospheric stability and mixing depth over land and over water. The mixing depth is the depth to which turbulence and buoyant plumes are effective in mixing the near-surface air upward. A higher water temperature usually results in a deeper maritime boundary layer, which in turn increases the depth of the feeder flow (a higher water temperature also implies a reduced temperature difference between land and water and hence a weaker sea breeze). As a general rule, the larger the depth of the feeder flow, the smaller the average wind speeds within it, and this is why warm water temperatures usually result in weak sea breezes.

The convergence of wind at the sea breeze front creates a strong and narrow updraft — as shown by the near-vertical blue arrows in Figure 3.24. The height that this updraft can reach is limited by the damping effect of the stable atmosphere aloft. The 'return flow' at higher levels is shown as pink and purple arrows in Figure 3.24 and is in the opposite direction to the sea breeze feeder flow.

The return flow is less frequently observed than the sea breeze feeder flow, but can be of significant strength. It transports warm and dry air out to sea. As shown by the decreasing size of the pink arrows, the return flow slows down and sinks. This completes the sea breeze circulation.

The time of onset of the sea breeze is primarily dependent on the extent of heating of the land. The circulation of the sea breeze is very important in moving the air pollution of cities near coasts. Sometimes a brown haze can be seen some distance offshore in the mornings. This is the previous day's pollution moved out to sea by the return flow of the sea breeze the day before, or it can be pollution moved offshore by the land breeze during the night.

Development of the sea breeze system

The effect of geostrophic wind

Figure 3.25 shows the development of a sea breeze system for a range of offshore geostrophic wind speeds — no onshore geostrophic wind speeds are shown because sea breezes will not occur with onshore winds unless they are very weak. This figure and Figure 3.26 use the same colours as a reminder that the likelihood and strength of the sea breeze (and its location) are dependent on the strength of the offshore geostrophic wind. The extent of the sea breeze system is defined as the area where the feeder flow is at least two knots (onshore).

The figure indicates that a sea breeze system usually begins offshore, and that the sea breeze system in strong offshore winds is displaced offshore and remains there. Most people would say that, under those conditions, there is no sea breeze, but it would be more accurate to say that the sea breeze does not reach the coast.

For light to moderate offshore winds (5 to 10 knots) the offshore extent grows twice as fast as the inland extent. The sea breeze grows offshore until late afternoon, then retracts as the heating over land decreases. At the seaward end of the sea breeze feeder flow is a generally broad region of calm winds. This is the region where the

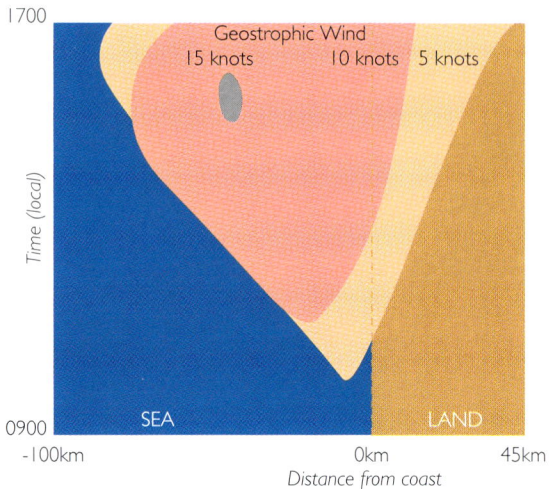

Figure 3.25

An idealised plan view of the horizontal development of a sea breeze for various offshore geostrophic wind speeds.

(Based on analysis by K. Finkele.)

large scale wind is less affected by the pressure difference due to the land–water temperature gradient. The wind will change its direction where the onshore component of the wind in the sea breeze feeder flow disappears and the wind will again tend towards the large scale wind speed and direction.

For strong offshore winds, the sea breeze system remains offshore or may not form at all.

Development in time

The sea breeze system moves inland as the land heats up and this continues as long as a temperature difference is maintained (and it requires a large enough water body to supply the cool air for the sea breeze feeder flow). The rate at which the sea breeze front moves inland is known as the 'sea breeze penetration speed' (which is usually 1–3 m/s, or 2–6 knots). However, the wind speed within the sea breeze feeder flow is much larger than the speed at which the sea breeze system itself moves inland. For example, the feeder flow can well exceed 10 m/s, or 20 knots, near the sea breeze front (as many Western Australians know). The inland penetration speed slows down as a result of friction and the backing (direction changing anticlockwise) due to the Coriolis force, until the sea breeze comes to a standstill after sunset — although sometimes the sea breeze continues well into the night (this depends on latitude and other factors).

Many users of the coast will also be interested in the rate at which the sea breeze feeder flow expands seaward (we might call this the rate the 'offshore' penetration speed of the sea breeze). At the seaward end of the circulation, there is no defined front as there is on land. A convenient definition of the sea breeze feeder flow is that we can call it a 'sea breeze' when the wind speed perpendicular to the coast is at least two knots (in cases when the large scale wind is not in that direction). There is a large (5 to 10 kilometre) region at the seaward end of the feeder flow where the wind speed is less than two knots. The offshore penetration speed is not slowed by friction as much as over land and the turning (backing) of the Coriolis force seems to play a small role. The horizontal extent of the sea breeze increases until late afternoon, then diminishes (this is clearly shown in Figure 3.25).

Predicting sea breezes

The factors influencing sea breeze strength are (in order of significance): the geostrophic wind (direction and speed), the temperature gradient between land and water, the coastline geometry (bays, gulfs, islands, peninsulas, and so on) and the coastal terrain.

The major influence — the geostrophic wind (U_g)

As the sea breeze can be characterised as a 'distortion' of the geostrophic or large scale wind, the likelihood of a sea breeze depends firstly on the strength and direction of the large scale wind. The sea breeze attempts to balance the pressure differences which occur near the surface, and these pressure differences are sensitive to any change in the land–sea temperature difference.

Wind can be broken into vectors — in this case, onshore and offshore components. (Appendix III explains, by means of simple graphics, the concept of wind according to its onshore and offshore components as well as coast-parallel components.) The likelihood of a sea breeze occurring for the different wind regimes is summarised in Table 3.3 and shown graphically in Figure 3.26.

Figure 3.26 shows a coastline with sea to the left and land to the right, and arrows superimposed on it representing different strengths and directions of the geostrophic wind. The top panel shows a 5 knot geostrophic wind, the middle one 10 knots and the lowest one 15 knots. The colour of the arrows shows the likelihood of a sea breeze.

> It is the orientation of the geostrophic wind in relation to the coastline that is important, not whether the coastline is running north–south or east–west.

In order to use Figure 3.26, all you need is an estimate of the geostrophic wind on any particular day and you will be able to plot this relative to the orientation of the coast in your particular area. Estimates of the geostrophic wind can be obtained from weather maps or forecasts. (Strictly speaking, forecasts are usually predictions of surface winds rather than geostrophic winds. The surface wind can be assumed to be 50 per cent of the geostrophic wind.)

The onshore geostrophic wind
An onshore geostrophic wind brings cool and generally moist air inland from offshore. Inland areas that have not been affected by the cool airflow will heat up more strongly than areas near the coast. With an onshore wind there may still be a slight horizontal temperature gradient near the coast and this may induce a speeding up of the wind, but such a wind is generally not referred to as a sea breeze. In these conditions a sea breeze

Table 3.3
Predicting the likelihood of a sea breeze from geostrophic (large scale) wind speed and direction.

Geostrophic wind speed in knots	Geostrophic wind direction (relative to the coast)		
	onshore	*coast-parallel*	*offshore*
0–4	likely	likely	very likely
4–12	no	likely	strong sea breeze
>12	no	likely	unlikely and displaced offshore

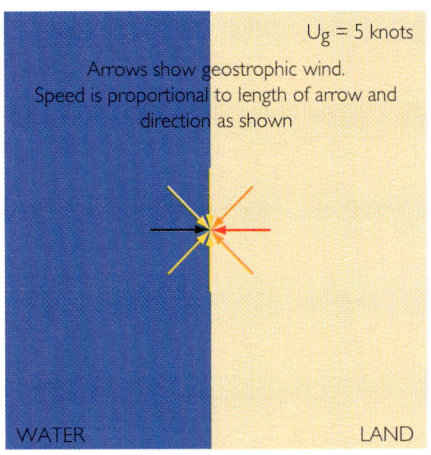

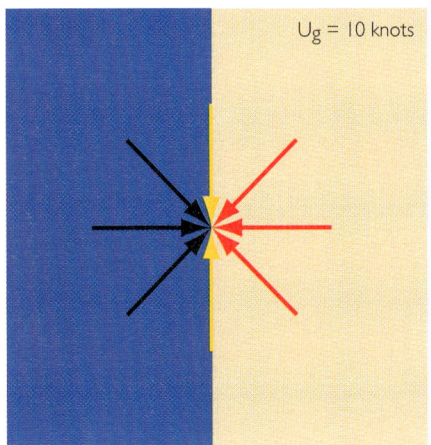

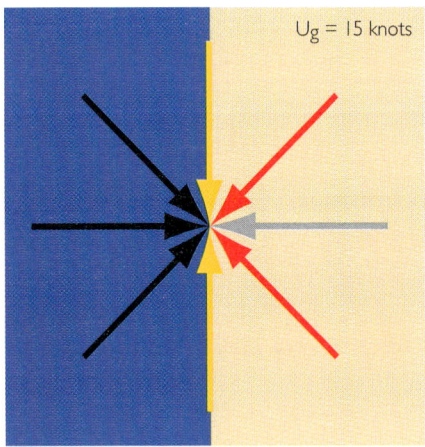

Likelihood of sea breeze
is shown by colours

→ Strong sea breeze

→ Very likely

→ Likely

→ Unlikely (displaced offshore)

→ No sea breeze

Figure 3.26

Predicting the likelihood of a sea breeze from consideration of the geostrophic (large scale) wind. (Based on analysis by K. Finkele.)

front is not normally felt, as the temperature and humidity gradients are small and span large distances. With small horizontal gradients of wind speed over large distances, there is only a very weak updraft and some slight distortion of the large scale wind aloft associated.

At 5 knots

Looking at the left hand or water side of the figure, you can see that a 5 knot onshore geostrophic wind will effectively cancel out any chance of a sea breeze, unless it strikes the coast at a fairly oblique angle — say, less than 30°.

At 10 knots or above

A 10 knot or stronger onshore geostrophic wind will cancel out any chance of a sea breeze.

The coast-parallel geostrophic wind
(and no geostrophic wind)
The coast-parallel geostrophic wind, or calm geostrophic wind, brings very little or no cool moist air inland from offshore or warm dry air offshore. As the sun rises and the land heats up, the temperature gradient between land and water will not really be altered by the large scale wind. The onset of the sea breeze in such conditions can be early in the day (depending on the temperature difference) and can occur close to the coast, providing the wind is not too strong — say, less than 15 knots. The sea breeze frontal gradients in such wind conditions are moderate. The updraft at the front is moderate to strong and a return flow develops aloft. The offshore extent of the sea breeze is moderate as well.

The offshore geostrophic wind
The offshore geostrophic wind is favourable for sea breezes, as long as it is not too strong (as shown in Figure 3.26, the threshold seems to be about 12 knots). For light offshore winds, the sea breeze initially forms offshore. The sea breeze then expands and moves inland, reaching the coast by mid-morning. For moderate offshore wind speeds (4 to 12 knots), the sea breeze, again, forms offshore but will have a slower inland penetration speed due to the opposing large scale wind. Large gradients form at the sea breeze front and the strong convergence at the front creates a strong updraft. The sea breeze might reach the coast by late morning or early afternoon, or

become stagnant or even be pushed back out to sea by the large scale wind. The return flow can be significant due to the strong updraft. The return flow is in the direction of the large scale wind and results in increased wind speeds aloft. The return flow also is very effective in transporting warm and dry air offshore and the horizontal extent of the sea breeze can be significant. Frontal gradients of a sea breeze formed under moderate offshore wind conditions are the strongest; therefore the sea breeze has a large vertical extent.

Strong offshore geostrophic wind speeds (> 12 knots) will move warm air offshore rapidly. If the land heats sufficiently, a horizontal temperature gradient will be created offshore. If the gradient is strong enough, the distortion of the large scale offshore wind will produce an 'onshore' wind component, which will occur far offshore. This distortion might be only small, just as it was in the case of the onshore large scale wind, but it will nevertheless be felt offshore.

At 5 knots

Looking at the right hand side of the figure, you can see that a 5 knot offshore geostrophic wind will be conducive to a sea breeze, particularly if it is blowing directly offshore or nearly so.

At 10 knots

Looking at the right hand side of the figure, you can see that a 10 knot offshore geostrophic wind will be conducive to a strong sea breeze. It is interesting that, as the wind direction approaches coast-parallel, the sea breeze strength is likely to decrease.

At 15 knots

Looking at the right hand side of the figure, you can see that a 15 knot offshore geostrophic wind may produce a sea breeze, but if one occurs, it may be displaced offshore. As the angle approaches 45° (offshore), the chances of a strong sea breeze increase and then decrease again as the wind approaches coast-parallel.

The changing geostrophic wind
The above discussion assumes that the geostrophic wind stays constant, which is not always the case! If a sea breeze has developed and the geostrophic wind swings around to less

favourable conditions, say from a moderate offshore to light onshore, the sea breeze inland penetration speed will be enhanced and the sea breeze front will move very rapidly inland. The wind speed within the sea breeze feeder flow might be 20 knots and the wind speed ahead of the front might be only 10 knots but in the same direction. This will still maintain the sea breeze front with a moderate updraft and the return flow will alter the large scale wind aloft somewhat, but may not be strong enough to reverse the direction of the large scale wind.

If the offshore component of geostrophic wind varies in speed, which can be the case when the direction changes, the sea breeze is affected in onset, strength and inland movement. For example, if the offshore component of the geostrophic wind is small early in the day, the sea breeze reaches the coast early and has a moderate sea breeze front. Should the winds then pick up, the gradients at the front are strengthened and the return flow increases. In the event that the wind speed continues to increase, the sea breeze front can be pushed out to sea and the temperature gradient at the front can weaken due to the lack of heating from underneath.

Should the geostrophic wind swing around from an onshore to offshore component during the day, we might get a late onset of a sea breeze near the coast. The onshore wind has transported cool air from offshore over land, so the heating is not as effective and the air temperatures are generally lower due to the cooling. This will produce a weaker sea breeze which starts later in the day.

The secondary influence — the temperature gradient between the land and water surfaces

For a sea breeze to form under favourable wind conditions (light winds), the temperature gradient between land and water can be as little as 5° Celsius. Of course, during the day the temperature gradient increases as the air warms over land, and the maximum temperature gradient might be as high as 20°C between the inland air and the air offshore. The important thing to remember is that, although the ground can reach very high temperatures (40–50°C) or the sea surface temperature might be rather low (10–15°C), it is the temperature of the air above the surface that is important in determining the sea breeze strength.

The land surface temperature

The land heats up after the sun rises. The hotter the land becomes, the stronger the sea breeze. The amount of heating of the land and near-surface air is related positively to the amount of solar radiation reaching the ground and negatively to the amount of evaporation from the ground and the plants.

As cloud cover increases, solar radiation usually decreases (the exception may be high and wispy clouds, but this effect is usually short-lived). Latitude and time of year influence the amount of solar radiation. Near the equator, solar radiation is usually stronger than in temperate regions, which results in a more rapid heating of the land and therefore an earlier onset of the sea breeze. However, day length tends to increase away from the equator, which results in a slower, but longer-lasting sea breeze (and greater inland penetration).

The extent to which the land heats up will also be influenced by soil type: a sandy soil loses its water more quickly by evaporation and drainage than a clayey soil. Sands are therefore much drier than loams or clays and will heat up more rapidly.

The sea surface temperature (SST)

The temperature of the water will strongly influence the mixing of air above it. The lower the air temperature over water, the stronger the sea breeze. As long as the water temperature is lower than the air above it, the sea breeze is not likely to be greatly influenced by water temperature.

For very cold water (less than 10°C), the air above the water is likely to be stable and have a shallow mixing depth, perhaps only 10 metres. Under these conditions, the mixing will be weak. Hence, the air temperature up to the top of the sea breeze will not be strongly influenced by the water temperature. The sea breeze will then be a function of the difference between the air temperature over the water and land, and largely independent of water temperature.

For very high water temperatures (hotter than 25°C), which are likely to be associated with an unstable atmosphere, the mixing depth will be very great and may approach those over the land (~500–1000 metres). The mixing will increase the air temperature above the water to approximate that of the water. The sea breeze in these conditions will be weaker the higher the water temperature, due to the reduced temperature gradient between the air above the water and the land.

Stability considerations aside, if the wind is blowing over extensive areas of water, it will eventually become influenced by the temperature of that water.

> A temperature gradient between land and water of about 5°C is necessary to induce a sea breeze under favourable conditions (light to calm large scale winds). A larger temperature gradient will increase the strength of the sea breeze. A large temperature gradient (> 10°C) is necessary for a sea breeze to form with a light to moderate offshore geostrophic wind.

The combined sea breeze index

The combined effect of large scale wind and temperature gradients can be expressed in a 'sea breeze index' (Simpson, 1994). It represents the ratio of the two major controlling forces: the inertial force (proportional to the square of the *offshore* component of the geostrophic wind speed, U_g, in m/s) and the buoyancy force (proportional to the difference between the temperature of sea and land, where ∂ means difference).

$$\text{Sea breeze index} = \frac{U^2}{\partial T}$$

A sea breeze will occur if the sea breeze index is lower than a critical value. Unfortunately, the critical value is not universal and will depend on location. Measurements from the north-east of Lake Erie gave a critical value for the lake breeze of approximately 3.0 (the units are $m^2 s^{-2} {}^\circ C^{-1}$). Measurements from Thorny Island produced a critical value of 7.0. The higher critical value of Thorny Island is partly due to using the winds at a height of 1000 metres, whereas at Lake Erie surface winds were used.

The index reflects what was stated earlier for strong offshore geostrophic winds: if these are larger than about 6 m/s (12 knots), the sea breeze is unlikely to form, no matter how large the temperature gradient between land and water. On the other hand, if the winds are calm to light, a small temperature gradient is sufficient to produce a sea breeze.

Now, let's relate this information to a real situation — the sea breeze in the Carnarvon region of Western Australia, which is arguably the strongest and most reliable in Australia. Look back at Figure 2.43 — the 3 pm 'light wind and above' map for October. This shows a high (80 per cent) probability of wind and a predominant south-west direction. We can't say that the 80 per cent relates only to sea breezes, but certainly sea breezes are a strong influence. Now look at Figure 2.27, the October maximum air temperature map; Figure 5.15, the October sea surface temperature map; and finally Figure 2.13, the inferred October geostrophic wind direction. Collectively these show very favourable long-term conditions for sea breezes: the monthly average maximum air temperature is quite hot at around 30°C; the average SST is coolish at about 20°C; and the average geostrophic wind direction is offshore (south-east). Although we do not have a sea breeze index for this area (and these are average values, not daily ones), if the geostrophic wind on any particular day was 10 knots (5.2 m/s, recalling that the index requires wind in m/s), the value of the index would be favourable (that is, small), as follows:

$$\text{Sea breeze index} = \frac{U^2}{\partial T}$$
$$= \frac{5.2^2}{10}$$
$$= 2.7 m^2 s^{-2} {}^\circ C^{-1}$$

If you compare the corresponding temperature and wind data for many other parts of the Australian coast, you will see that they are less favourable, partly explaining the reliable and strong sea breezes in Western Australia. You should now be able to assess the long-term sea breeze likelihood for other parts of the coast. You will see that, in some areas, the SSTs are quite high relative to the air temperatures (that is, a small temperature gradient) and in other areas the geostrophic winds tend to be onshore.

Other influences

Coastline geometry can be straight, concave (a bay or gulf) or convex (peninsula or island). Figure 3.27 shows how a convex coast can create converging (colliding) sea breezes, which would be stronger than those on a straight coast, which, in turn, would be stronger than those on a concave coast (where the sea breezes are diverging). Either the land or the water can be limited in horizontal extent, so these effects can also be limited. For a curved coastline, keep in mind that the large scale wind component perpendicular to the coast is most important. Generally, the sea breeze front moves inland parallel to the coastline, provided the geostrophic wind does not change significantly along the coast.

Assuming the geostrophic wind does not change significantly over a limited land body (such as a peninsula or island), one coast might experience onshore geostrophic winds whereas the opposite coast will experience offshore conditions. For a land body forming an acute angle, such as Cape York Peninsula, and with large scale winds favourable to sea breezes on both coasts, the sea breezes collide some distance inland and continue to move inland in a common direction. The spectacular roll cloud which is associated with the colliding sea breeze on Cape York is called the 'Morning Glory'. Other colliding sea breezes form between a straight coast and a bay, such as Port Phillip Bay. Here this colliding sea breeze interacting with the Dandenong Ranges forms a wind pattern called the 'Melbourne eddy'.

For a limited water body (a bay or gulf) a sea breeze will form under favourable conditions. This bay breeze is a localised effect of the coastline geometry. The larger scale coastline might be quite different, which will induce a 'continental'

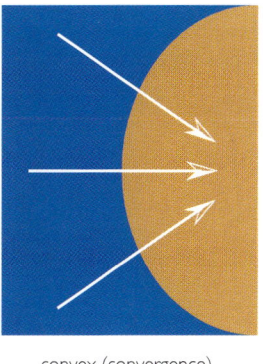
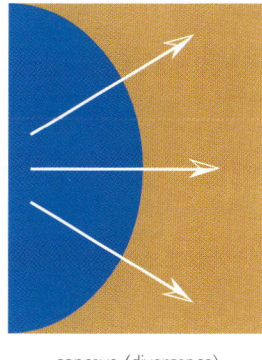
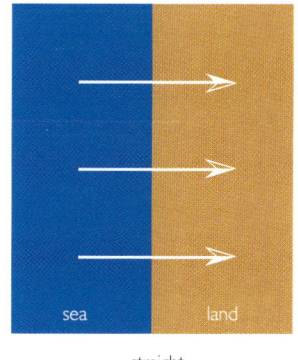

convex (convergence)   concave (divergence)   straight

Figure 3.27
The influence of coastal geometry on sea breezes.

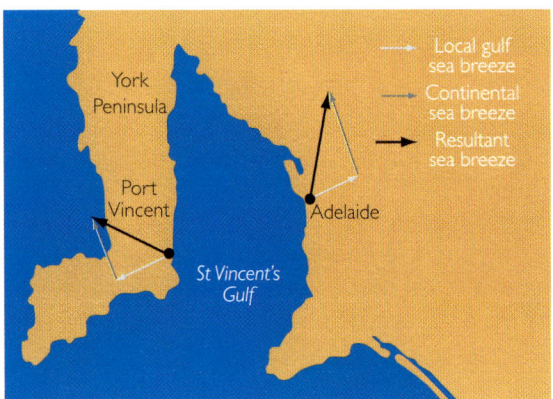

Figure 3.28

A schematic representation of how a sea breeze and a gulf breeze can combine in St Vincent's Gulf. (After Physick & Byron Scott, 1994.)

sea breeze. For example, St Vincent's Gulf might induce sea breezes moving perpendicular to either coast (Adelaide and east coast of the York Peninsula). However, in the late afternoon the continental sea breeze, which is parallel to either coast, adds its wind component to the local sea breeze, as shown in Figure 3.28.

In addition, large portions of Australia's coastline are near hilly or complex terrain. Hills generally enhance the strength of the sea breeze by enhancing the frontal gradients. However, the sea breeze front does not move as far inland as it would if there was no terrain. The terrain presents an obstacle for the sea breeze to 'climb over' before it can move further inland.

The lake and gulf breeze

The lake and gulf breeze is a 'sea breeze' formed by the temperature difference between land and a limited water body. The water temperature of lakes and gulfs is generally higher in summer than sea surface temperature and increases by a few degrees during summer. Smaller and shallower lakes warm up faster during summer than larger or deeper lakes. Depending on its circulation, the water temperature can be significantly different (5–10°C) from one shore to the other. For a lake or gulf breeze to develop, the limited water body has to be big enough to create a significant pressure difference between land and water; that is, the temperature difference between the air above the land and water has to be significant to create a feeder flow.

Lakes smaller than 50–100 kilometres in diameter may have lake breezes for only a short time before the return flow transports enough warm air out to the centre and warms the feeder flow, and the lake breeze 'runs out of steam'. Larger lakes or limited water bodies are less affected by the warming of feeder flow since their size ensures a bigger supply of cool air. There are few lakes in Australia that are large enough to develop definite lake breezes but there are a number of gulfs where breezes occur (for example, St Vincent's Gulf).

The land breeze

The land breeze is generally weaker than the sea breeze. The basic mechanisms are the same as for the sea breeze, except that the temperature difference between land and water is reversed. At night, the land can cool to well below the water temperature. The water temperatures of lakes and gulfs in summer can reach 20–25°C and the land temperature near the coast can be as low as 5–10°C during a still clear night (even in summer). This night temperature gradient causes a 'reversed sea breeze' — that is, a land breeze. The land breeze (like the sea breeze) seems to be most prevalent in Western Australia.

OTHER IMPORTANT WIND EFFECTS

Day–night (diurnal) variation

Except in unusual circumstances — such as where very tightly spaced isobars bring to an area very strong winds that can blow consistently for twenty-four hours a day — the wind speed usually falls off at night to half daytime levels, or it can die completely. Wind direction may also change dramatically. This diurnal variation is due to two main factors: the formation of a stable atmosphere near the surface — which results in a decoupling of the surface and geostrophic winds — and the removal of strong daytime horizontal temperature gradients, which can locally enhance or reduce the geostrophic wind.

On land at night, local cooling can be sufficiently intense to create a very stable atmosphere near the ground surface. When this happens, local winds can effectively become decoupled from those aloft (that is, from the geostrophic wind). When it is very clear and cloudless the surface 'wind' can consist of a thin and cold layer of air sliding down the sides of mountains at about 1 metre per second into low lying areas, seemingly 'oblivious' to the winds aloft — which, of course, are being caused by the large scale pressure patterns. The greater the inversion, the more stable the atmosphere becomes and the greater this decoupling effect.

If the large scale winds are sufficiently strong, they will overcome local surface cooling and strong winds will be felt down to the ground. If they are moderate, surface cooling can 'overcome' or mask the macro-scale wind and local conditions on the ground can vary — from no wind at all, to cold air drainage into the valleys, up to a gentle breeze. Wind direction, similarly, can vary.

Over water at night, the effect just described is far less pronounced, due mainly to the reluctance of large water bodies to change temperature, especially over 'short' periods of time (such as days or weeks). Therefore, a much smaller drop in wind speed and a smaller change in wind direction can be expected at night on the sea.

Surface friction effects near the coast

As was signalled by Figures 3.1 a–e and 3.2 (estimating wind speed and direction from weather maps), there is a subtle — and little known — wind effect that takes place quite close to the coast. The effect — of particular relevance to sailors, hang-gliders and sailboarders — results in either higher or lower wind speeds than expected in a fairly narrow strip near the coast, up to five kilometres from land. Figure 3.29 shows this surface wind as arrows: both wind speed (as shown by the length of the arrows) and wind direction (as shown by the orientation of the arrows) are different over land and water.

Where the wind is blowing parallel to the coast: for argument's sake, the coast will run 'north–south', with the land to the 'west' and the water to the 'east' (top part of the figure) and land to the 'east' and the water to the 'west' (lower part of the figure). But we need to remember that it is the orientation of the coast relative to the geostrophic wind that counts.

Imagine now that the wind is blowing from the 'north' (ideally you have just imagined a high to the 'east' of the imaginary figure and a low somewhere to the west). Over the water the wind will be blowing perhaps 15° from the north, but over the land it will be more like 35° or so (as was also shown in Figure 3.2). By extending the arrows over the water and over the land, it is easy to see that they are growing further apart, in the phenomenon called divergence. The diverging streamlines will result in a *decrease* in wind speed for perhaps 1 kilometre out to sea under unstable conditions and up to 5 kilometres out to sea under stable conditions. This is shown as a dark blue strip in the diagram.

The lower part of Figure 3.29 shows a case where the land is to the 'east' and the water to the 'west' in this imaginary area (there are many places in Western Australia with this orientation). The streamlines will converge and wind speeds will be *higher* over similar distances. This is shown as a red strip. It is difficult to put a precise figure on the amount of increase or decrease from this effect, but it may be around 20 per cent. Depending on your situation, you may wish either to seek or to avoid this area.

In the figure the geostrophic wind is shown coming only from the 'north'; to picture what would happen if it were approaching from the 'south', you need only turn the page upside-down!

> Now generalising for the Southern Hemisphere: if the wind is blowing parallel to the coast and you are facing it, and if the land is on your left, the wind will decrease in an area close to the coast. If the land is on your right, the wind will be enhanced. If you would rather 'have the wind on your back', reverse these generalisations.

Horizontal temperature gradients

It is not hard to accept that wind is a function of horizontal pressure gradient — as shown by the spacing of the isobars on weather maps — as well as atmospheric stability. However, there is an effect that is not as intuitive but which helps explain both the day/night differences in wind and why the observed wind on a given day (or night) might be stronger or weaker than that expected

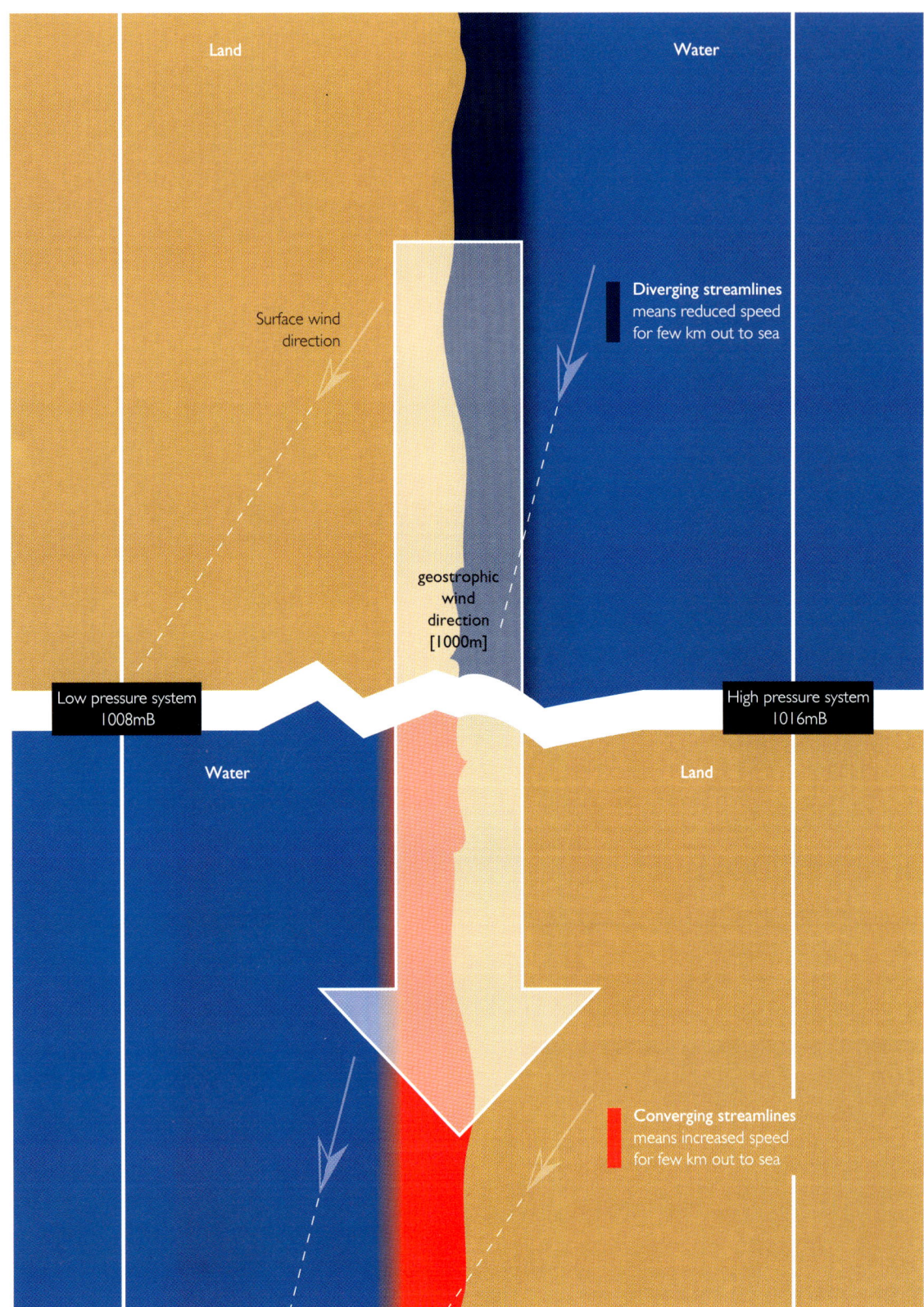

Figure 3.29

A schematic representation of how friction effects can either enhance or reduce the wind speed in a narrow zone near the coast; this effect is strongest for winds blowing parallel to the coast.

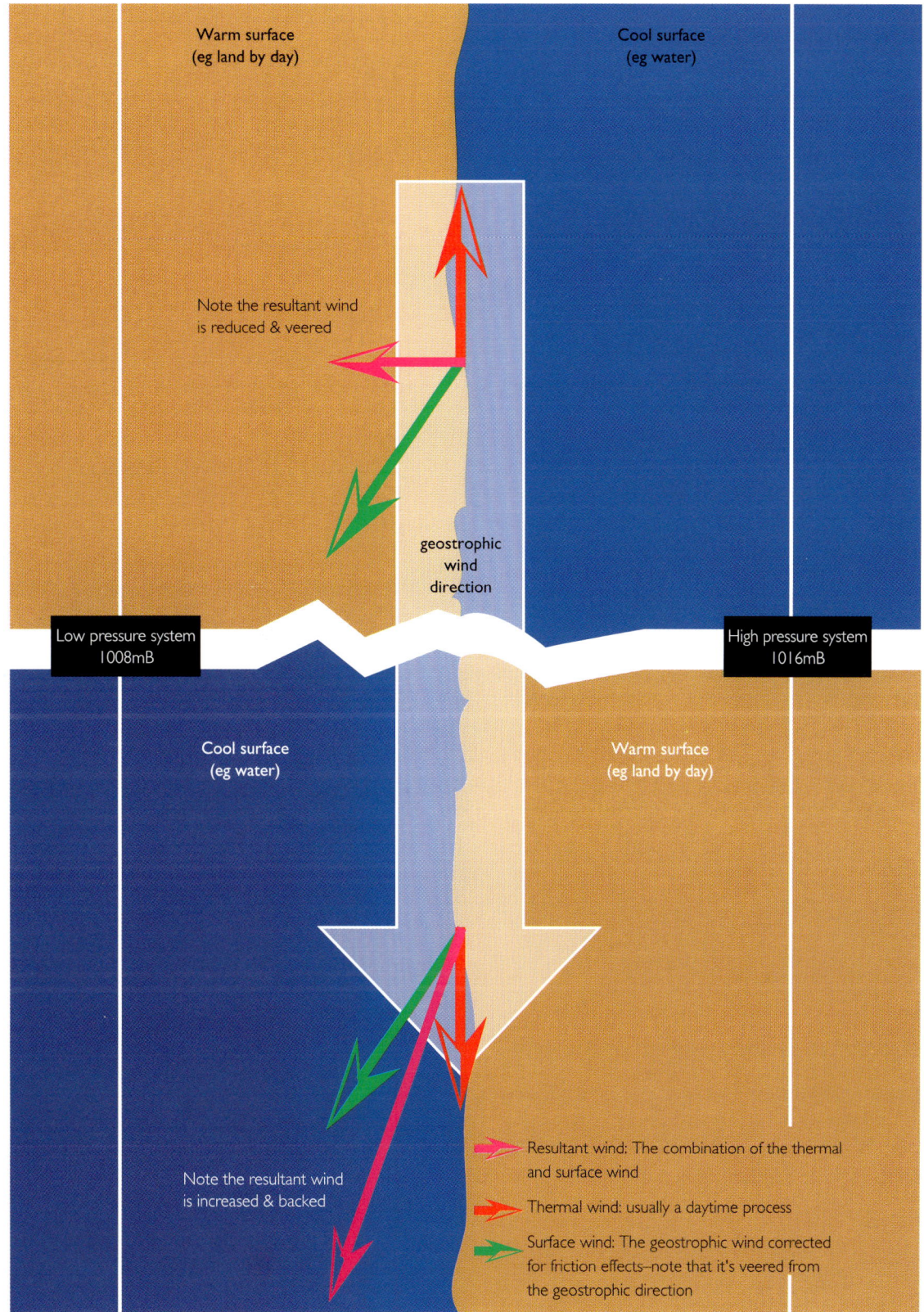

Figure 3.30

How thermal effects can enhance or reduce the wind speed in a narrow zone near the coast. This effect is strongest for winds blowing parallel to the coast.

from the pressure patterns on weather maps. In a sense, this is a more general — and usually larger scale — effect than the sea breeze we discussed earlier: the situation in which a regional or large scale horizontal temperature gradient is superimposed on a pressure gradient. This effect exists over much larger distances, both horizontally and vertically, than the local sea breeze.

Closely analogous to the way in which horizontal pressure induces wind, horizontal temperature gradients, which are common in areas where land meets water, also induce wind that blows parallel to the isotherms (lines joining areas with equal temperature) rather than to the isobars (lines joining areas with equal pressure). In the Southern Hemisphere, this 'thermal wind' blows along the isotherms with the low temperature to the right (that is, if you face the 'thermal wind', the higher temperature will be on your right). The stronger the temperature gradient, the stronger the thermal wind.

Without going into detail, it follows that, depending on the orientation of the isobars on a particular weather map in relation to the isotherms, the thermal wind can enhance or reduce wind speed on the ground. For example, if the isobars and isotherms follow a similar orientation — that is, both pressure and temperature decrease or increase 'together' — then the thermal wind will enhance the wind felt at the surface. However, if they oppose each other — the pressure gradient decreases at the same time as the thermal gradient increases — the thermal wind will decrease the wind felt at the surface. This effect is shown schematically in Figure 3.30. As in the previous figure, the geostrophic wind is from the 'north'. In this figure the resultant wind is derived by adding the surface and thermal wind vectors — if you are not sure how to do this, see vector addition in Appendix III.

In the upper part of the figure the cooler surface is to the 'east' and hence the thermal wind is from the 'south'. The surface wind, as expected, is from the north-east. (For simplicity's sake, the surface wind is shown veered by 35°, which is 'standard' over land.) The resultant wind — achieved by 'adding' the thermal and surface wind — is considerably slower and veered over what might have been expected as a result of predicting the surface wind by considering the pressure gra-

dient alone (that is, by estimating the surface wind from the geostrophic wind).

In the lower part of the diagram the only thing that has changed is the orientation of warmer and cooler surfaces; here the cooler surface is to the 'west' and hence the thermal wind is from the 'north'. It is clear from the diagram that the surface and thermal winds are blowing in similar directions, and the resultant wind is considerably faster and backed over what might have been expected as a result of predicting the surface wind by considering the pressure gradient alone.

The friction and thermal effects described above could be very useful for seeking stronger winds on a day when the winds are just below what you might want, or for seeking lighter winds on a day when the winds are above what you might want. Remember: it is the orientation of the geostrophic wind in relation to the coast that is important in the 'friction effect', and in the case of the 'thermal effect' you also have to consider the orientation of the thermal gradient.

Now, generalising for the Southern Hemisphere: if the isobars on the weather map align with isotherms in the area — that is, both 'highs' and 'lows' are more or less coincident (the two sets of lines 'slope' in the same direction) — then winds will be enhanced, and sometimes, this can be a strong effect. If, however, the high pressure corresponds to the low temperature (or the converse), or the two sets of lines are 90° to each other, the wind could be reduced or even eliminated.

Local and regional effects notwithstanding, the method described earlier in this chapter for estimating wind speed and direction from weather maps will generally yield quite good results. You can apply these principles to fine tune your predictions for areas where the land and sea temperatures are distinctly different.

CHAPTER FOUR

Wave Essentials

Waves can be beautiful or terrifying, depending on your vantage point. Sitting on a headland watching a perfect break can be truly uplifting; clinging to a yacht or powerboat in the face of rising winds and a rapidly growing swell can be truly terrifying. Like a good understanding of wind behaviour, a fairly detailed knowledge of wave behaviour is essential to safe active enjoyment of Australia's coastline.

WAVE HEIGHTS IN THE AUSTRALIAN REGION

Virtually everywhere in the Australian region experiences waves of at least one metre more than half of the time. Put another way, if you leave the coast there is at least a 1 in 2 chance that the average significant height of the waves you encounter will exceed one metre. The great majority of waters off Western Australia, South Australia and Tasmania have probabilities at or near 100 per cent, meaning that, on average, waves are *always* higher than one metre. Probabilities are slightly less off the New South Wales coast, perhaps averaging 90 per cent, and the Queensland coast slightly less than this, especially inside the shelf/ reef systems. Wave probabilities are lower across the Top End, between Weipa (Queensland) and Port Hedland, varying from 40 per cent to 50 per cent. The lowest probabilities are found in a large region centred on Broome.

The pattern of two metre waves is quite different. There is a very large area to the north where probabilities are 10 per cent or less. Here it may still be possible to experience large waves — certainly during cyclone season — but not commonly. However, virtually anywhere south of Carnarvon — through Albany, Adelaide, Hobart and up the east coast to Brisbane — experiences 50 per cent probability or more. Given that much larger waves will occur than the significant wave height, and that the statistics refer to *average* height, all users of coastal waters need to ask themselves whether they and their craft are fully prepared for at least 2 metre high waves. A 50 per cent probability really means that waves two metres and above are very common in these waters.

For some people, the thought of waves three metres and larger is not a pleasant one. Although relatively few areas experience a 50 per cent or higher probability of such an event, anywhere to the south of Carnarvon or to the south of Sydney has at least a 20 per cent probability.

The broad scale patterns

Figures 4.1 to 4.3 are wave maps based on data recorded more or less continuously from the 'Geosat' satellite during 1986–89 (the satellite altimeter data and analysis was provided by Lawson and Treloar Pty Ltd, Sydney.) From the satellite data it was possible to calculate a data point for every 2° of latitude (about 120 nautical miles, or 220 kilometres) and use normal mapping techniques to produce a reasonably smooth and reliable surface of the average significant wave height.

A few restrictions apply to interpreting the maps:
- They may not be as reliable around the Great Barrier Reef or between the Reef and the mainland as they are in more exposed/deeper water areas. This also applies to Bass Strait and other areas where the depth of water is less than 100 metres (see Figure Int.1). In such areas it has been assumed that wave heights are similar to those of adjoining areas with deeper water but in Bass Strait, in particular, the waves may be steep sided and close together.
- Less data are available close to the coast and in 'narrow' areas, due to the satellite altimeter taking time to lock onto the sea surface,

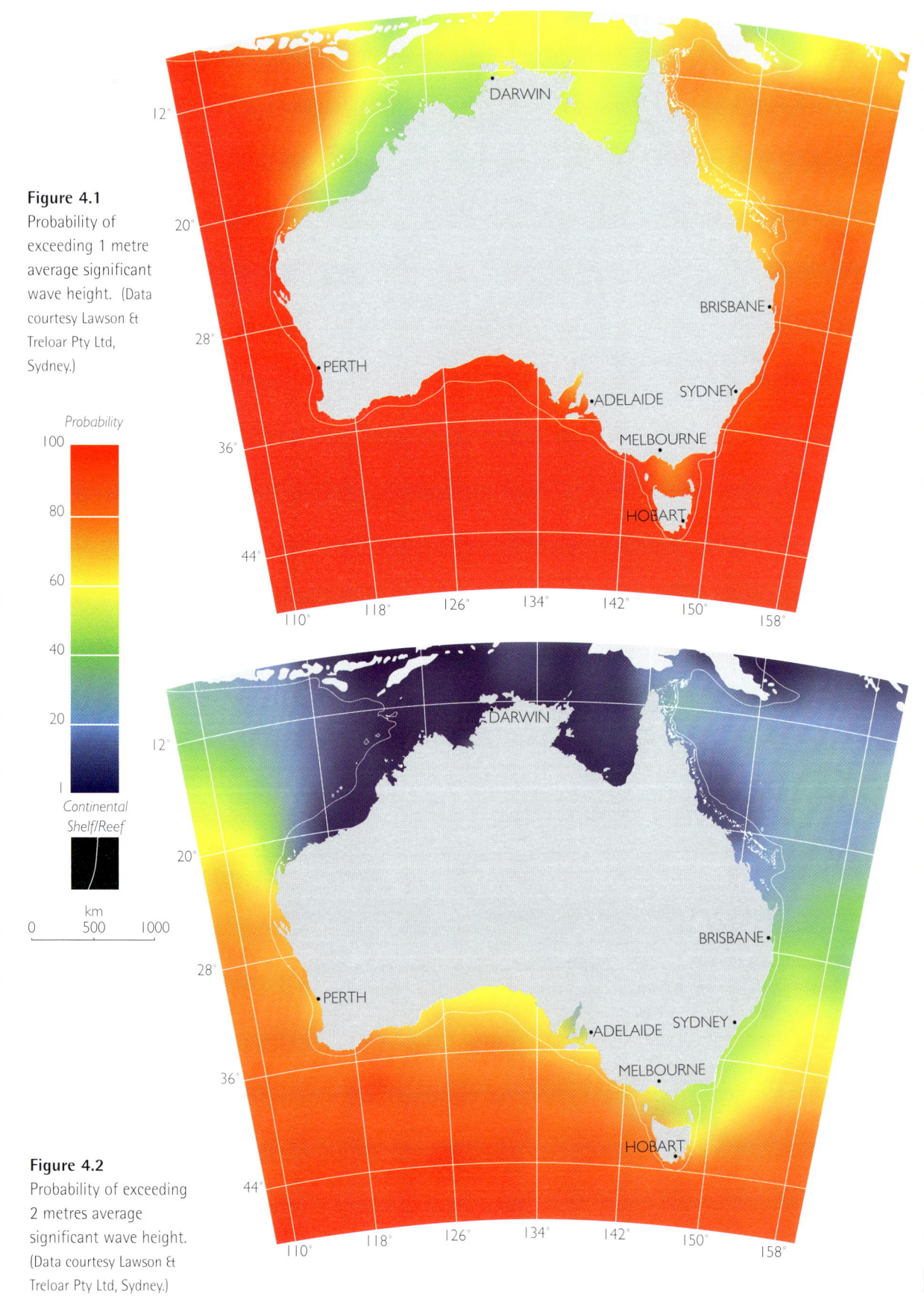

Figure 4.1
Probability of exceeding 1 metre average significant wave height. (Data courtesy Lawson & Treloar Pty Ltd, Sydney.)

Figure 4.2
Probability of exceeding 2 metres average significant wave height. (Data courtesy Lawson & Treloar Pty Ltd, Sydney.)

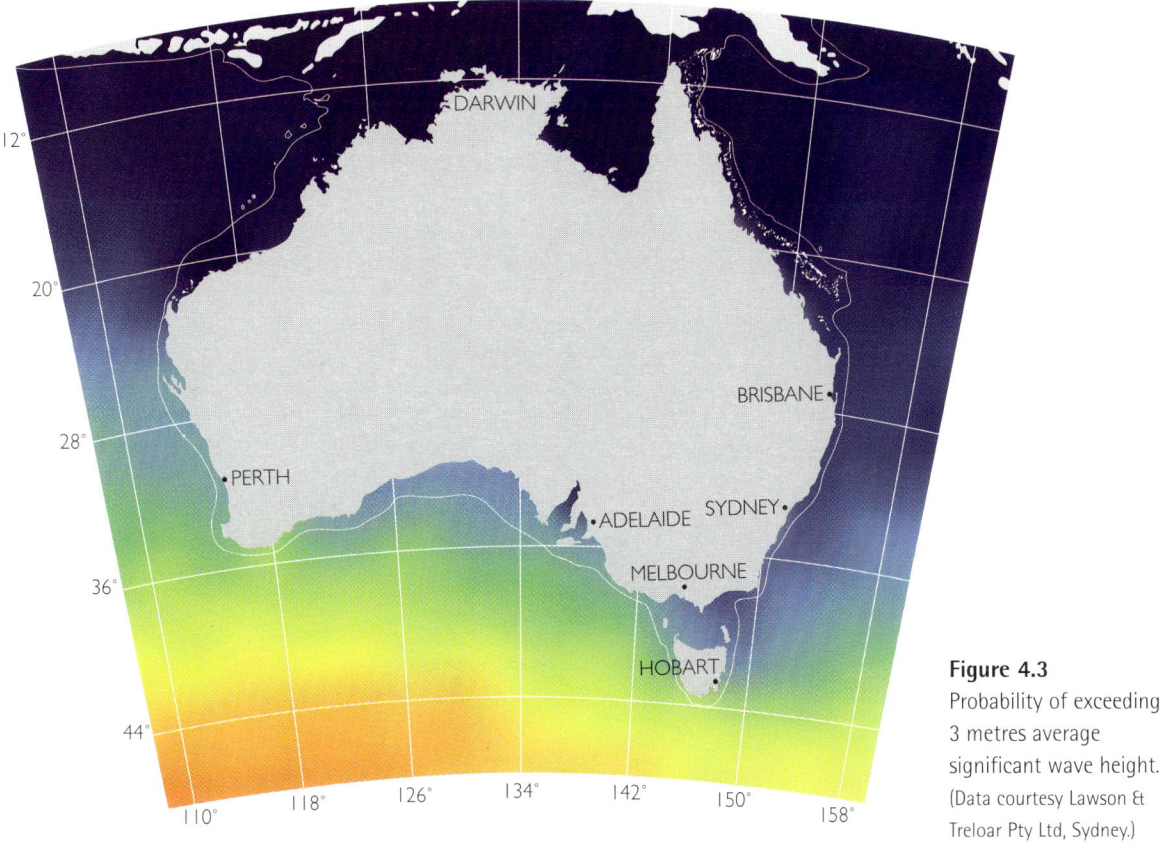

Figure 4.3
Probability of exceeding 3 metres average significant wave height. (Data courtesy Lawson & Treloar Pty Ltd, Sydney.)

resulting in a slightly less reliable wave height estimate.

- The maps show the probability of exceeding the average significant wave heights — that is, H_s — of 1, 2 and 3 metres. The data have been averaged, over approximately 3 years and over areas about 100 nautical miles square; therefore the maps show only broad scale patterns rather than detailed local ones.

- Very much larger (and smaller) waves will occur. On average a wave almost twice H_s will occur every 1000 waves (and this might be in only 3 hours if there is a wave every 10 or 11 seconds).

- The maps show the probability of exceeding a particular wave height at any time. A value of 70 per cent means that, based on three years' data, there is a 70 per cent chance of observing larger waves than the height shown on the maps (that is, 1, 2 or 3 metres). In the case of the 3 metre map, a figure of 70 per cent would indicate that you are in an area that frequently experiences large waves. On the other hand, a figure of 5 per cent would suggest that waves 3 metres and above are quite rare (although they may occur).

The wave maps show the probability of exceeding average significant wave heights — they do not show how big the waves can become under any particular conditions (daily predictions are the subject of the next section). For example, if an area has a probability of 70 per cent for two metre waves, there is no telling how big the biggest waves might be. All that can be said is that two metre waves and larger are experienced quite frequently (70 per cent of the time).

To summarise and interpret the patterns shown in the wave maps:

- Consistently large waves are found almost anywhere off the coast of the southern states of the mainland and Tasmania; for any given latitude, the probabilities are usually higher on the west coast, reflecting higher wind conditions and the general direction of west to east synoptic systems (especially cold fronts from the Southern Ocean).

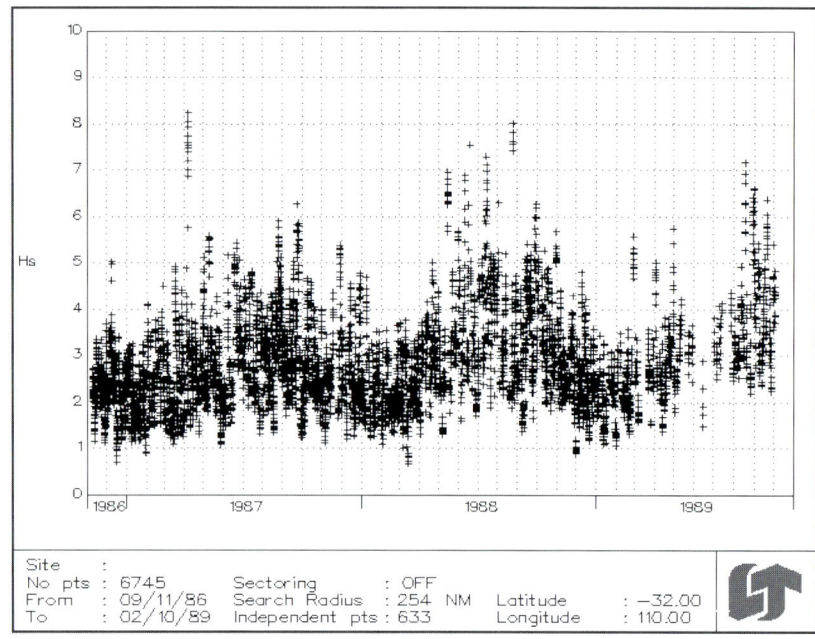

Site :
No pts : 6745   Sectoring   : OFF
From   : 09/11/86 Search Radius : 254 NM   Latitude : −32.00
To   : 02/10/89 Independent pts : 633   Longitude : 110.00

Figure 4.4
Time series record for a site to the west of Perth. (Data courtesy Lawson & Treloar Pty Ltd, Sydney.)

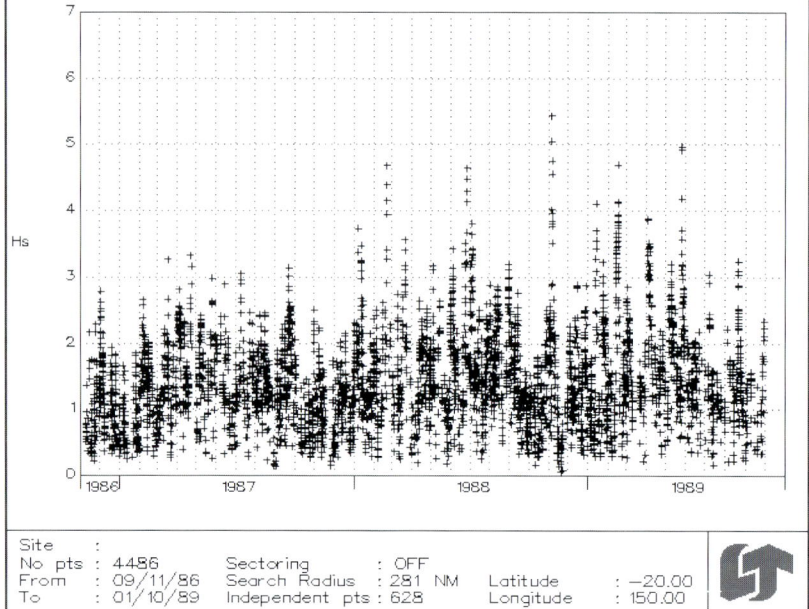

Site :
No pts : 4486   Sectoring   : OFF
From   : 09/11/86 Search Radius : 281 NM   Latitude : −20.00
To   : 01/10/89 Independent pts : 628   Longitude : 150.00

Figure 4.5
Time series record for a site to the east of Townsville. (Data courtesy Lawson & Treloar Pty Ltd, Sydney.)

- Areas with the lowest probability of large waves are found to the north of the mainland, especially in the Indian Ocean, Timor Sea, Arafura Sea and Gulf of Carpentaria regions.
- Wave heights in the Southern Ocean are frequently large; anyone who doubts this should read Kay Cottee's book *First Lady*.

What about seasonal effects?

Experienced sailors know that average wave heights tend to vary with the seasons. In some parts of the world the largest waves are associated with winter storms, but is this a reasonable generalisation around Australia?

Figure 4.4 shows a three year series of records of H_s from a site well to the west of Perth. Although there is a great deal of scatter in the record, as you would expect (reflecting the passing of synoptic pressure systems), there is a definite *tendency* for the largest waves to occur in winter. This is not to say that large waves do not occur in the other months; merely that they are more common in winter. Very large waves can and do occur in most exposed sea/ocean areas. There are many records above seven metres, even though the probability of exceeding three metres may be 'only' 35 per cent.

Figure 4.5 is for a site to the east of Townsville, well away from the influence of winter storms associated with the 'roaring forties'. The wave data are for the same three years and do not show the winter trend of the previous example. Large waves can occur throughout the year, although some of the largest are in the summer months, probably associated with tropical storms (most prevalent in the warmer months) and the occasional cyclone. By comparison, the probability of exceeding three metres is less than 5 per cent, very much less than to the west.

Given that the wave climate maps are based on only three years, and this is unlikely to be sufficiently reliable to discern seasonal patterns, it would be safer to assume that large waves can occur anywhere at any time of the year. Some areas regularly experience large waves and some do not. As one moves into southern waters, especially toward the Great Southern Ocean, larger waves become more frequent, especially in winter.

ESTIMATING WAVE HEIGHTS ON ANY PARTICULAR DAY

On any particular day, local winds can produce waves that are very much larger or smaller than the averages that we have just seen. Before looking into this, we need to define some terms and to make some generalisations about what causes waves.

When boating or sailing, one usually experiences waves that have originated from many different areas; each contributing area may be experiencing different tidal and current influences, will have different bathymetry (the shape of the sea floor) and will certainly have different weather conditions. It is not surprising that, when these combine, they produce an amazing spectrum of wave heights, wavelengths and periods (time between successive crests).

Some of the most common attributes of waves are:
- **wave height** — the distance from the trough to the crest
- **significant wave height** — the height of the highest one-third of the waves
- **wavelength** — the distance between successive waves measured from crest to crest or trough to trough
- **wave period** — the time between successive waves
- **wave velocity** — the speed at which the wave is travelling (not to be confused with wind velocity)
- **fetch** — the distance the wind has been blowing over water, normally for a given wind speed the greater this distance, the larger the waves become, until a saturation state is reached and the waves cannot grow any more.
- **duration** — the length of time for which the wind has been blowing. Normally for a given wind speed the longer this duration, the larger the waves become, until saturation.

Almost everyone has observed waves of some kind. A stone thrown into a pond or wind acting on water can produce waves. Waves can also be produced by undersea earthquakes (tsunami) or by the gravitational attraction of the moon and sun (tides). When two or more sets of swell arrive at a location (most obvious on a beach), they may combine to *reinforce* each other — producing larger waves, or to *extinguish* each other — producing smaller waves. The net result is called 'surf beat' and is a common sight at beaches when the surf is running (that is, worth paddling out for, if you are a surfer). Figure 5.24 shows a beat pattern, in this case for tides rather than waves.

Surf beat is characterised by a series of a dozen or so low waves, followed by several high waves (usually 3 or 4) and then by a quiet period, before repeating itself. Surf beat's periodicity is about 70 seconds, but experienced people — especially surfers — will know that surf beat can have much longer or shorter periods than this. In the case where two or more long swells are approaching the coast, the periodicity could be several minutes.

General factors controlling wave height

The main factors affecting wave height are:
- **wind speed** — the higher the wind speed, the larger the waves
- **wind duration** — the longer the wind has been blowing, the higher the waves will become
- **fetch** — the longer the distance for which the wind has been acting (over water), the larger the waves will become
- **water depth** — the deeper the water, the larger the waves (excluding cases where deep water abruptly becomes shallow)
- **the presence of ocean currents** — an opposing current tends to reduce wavelength and to increase wave height; a current travelling in the same direction as the waves tends to increase wavelength and decrease wave height.

'Rogue', 'freak' or 'king' waves

The phenomenon of significant wave height (H_s) can help to explain what we call 'rogue', 'freak' or 'king' waves. On average, approximately 1 in 1000 waves reaches 1.85 times the significant wave height. Given that, at sea, a typical wave period might be 10 to 11 seconds (and correspond to an average wavelength of about 200 metres), then a 'rogue' wave will occur on average once every 3 hours (that is, 1000 waves with period 10 to 11 seconds will pass the observer in about 3 hours). All boaters should consider this carefully: if the waves are averaging 2 metres and the boat is handling them with little trouble, don't become complacent — sooner or later (usually in less than 3 hours!) a wave of 4 metres or larger will pass. The 'one-in-a-thousand' concept is useful not only for those at sea; experienced rock fishers will always tell you never to turn your back on the ocean. Unexpectedly large waves have washed many of their fellow anglers off the rocks.

> It is not a matter of *whether* a large wave will occur, but *when*! Frequently — and all too frequently in the media — one hears reports of 'rogue' or 'freak' waves. Such reports actually contribute to a belief that these large waves cannot be predicted and therefore accidents and fatalities are an act of God. Large waves *will* occur; what we are less sure of is precisely *when*.

Although very useful, it is important to bear in mind that the 1 to 1000 relation is a statistical generalisation, much the same as expecting an equal number of heads and tails in coin tossing (which will certainly happen in the long term, as all gamblers know).

All wave height graphics in this section refer to significant wave height, as do most forecasts and wave height reportings.

Wave heights in open water

Seas and swell

Most people who use coastal, sea and oceanic waters (that is, non-inland waters) use two terms to describe the state of the water at any given time. These are the *seas* and the *swell*.

Seas (or sea waves) are waves generated more or less locally by wind and still under the influence of this wind. Sea waves usually do not move very far away from the area in which they were generated.

Swell (or swell waves) are derived from sea waves that have travelled well away from the area where they were generated — which may be hundreds, if not thousands, of kilometres out to sea. They arrive as regular, organised 'sets'. The distance between successive swell waves is usually longer (that is, swells have longer wavelengths than seas) and have a smooth, rounded appearance. They can be formed by tides, coastal landslides, storms at sea, undersea earthquakes, and so on; but most commonly they are caused by wind sustained over a long period of time and over a long fetch (recalling that fetch is the distance the wind has travelled over water).

> Because seas are still under the influence of the wind that caused them, they generally travel in the same direction as the prevailing wind. Sea waves tend to be more irregular (choppy, disorganised) than swell waves.

As wind is always gusty, it is not hard to understand that it will push down harder on some parts of the water than others, causing distortions or ripples. These rapidly grow to the point where they are big enough to affect the flow of the wind. Once this happens, the wind will exert a stronger force on the windward side than the leeward side of the chop, causing it to grow rapidly in size (and produce what are called 'newly formed seas'). Provided that the wind continues to blow, waves will continue to grow so long as they receive more energy from the wind than they dissipate (usually by breaking). As will be shown in more detail later, wave height depends on sea surface roughness (the rougher the surface, the more 'grip' the wind has), wind speed (the stronger the wind, the larger the waves), wind duration (the longer the wind blows, the larger the waves will become) and fetch (the longer the fetch, the bigger the waves).

Newly formed seas are very chaotic and disorganised. The largest waves are formed by storms — often a series of storms — at sea. As the waves continue to grow, the windward wave surface becomes steeper and steeper until the wave height approaches about one-seventh of the wavelength. Once the steepness reaches 1:7, the wave usually breaks, forming whitecaps and spray. Waves merge, overtake each other, cancel each other out and generally interact randomly until, after a certain amount of time, the energy received approximately equals that lost; this is called a 'fully developed sea'. Both the time taken and the fetch required for a fully developed sea increases with increasing wind.

> Sea waves (seas) generally move more *slowly* than the wind that generated them due to 'slippage' between the wind and the waves it causes.

As the waves continue to develop they tend to fan out from the storm area, gradually forming lower, longer and more rounded swell waves. Since the waves have fanned out (this is called 'angular dispersion') their energy is dissipated over a greater area, hence the wave height is reduced. Swell waves lose little energy as they move and can thus travel vast distances.

> Swell waves (swell) generally move *faster* than the wind that generated them due to complex interactions (for example, 'leapfrogging') which take place over several days and hundreds to thousands of kilometres of ocean.

Figure 4.6 shows a model of the development of seas and swell in open ocean under the influence of a storm. This is shown in terms of both fetch (distance) and time, since waves take a finite time to travel such distances.

The transition of seas into swell does not happen over short distances or in short periods of time. We have already seen that seas tend to be higher and steeper than swell, but by how much? And how fast do seas and swell travel in relation to the wind that formed them? In order to answer these questions clearly, we need to look more closely at wave behaviour.

Initially, the wind starts to blow on calm water, forming ripples that quickly become chop. Chop tends to be fairly steep (5 per cent, or 1:20) and moves much more slowly than the wind causing it; a rule of thumb is that chop moves at about 20 per cent of the wind speed.

However, as the wind continues, the waves become not only bigger, but steeper as well; in a fully developed sea, where energy lost equals energy gained, the waves are very steep (10 per cent, or 1:10) and move at about 50 per cent of the wind speed. The wave height will depend on wind speed.

As the waves move well away from where they were generated, they become very much flatter (2 per cent, or 1:50) and may end up travelling as fast as, or faster than, the wind that generated them. When fully mature, swell wave velocity is proportional to wavelength. Despite the theoretical distinction between seas and swell, it is unlikely, after fifty or so hours of constant wind — particularly if it's strong, that there will be any real difference between a well-developed sea and a swell; some waves could have travelled 500 kilometres away from where they first formed.

Relationships between swell-wave velocity, wavelength and wave period

In mid-latitude waters, the following simple relationships apply to swell-wave velocity, periodicity and wavelength (where velocity is in metres per second, wavelength in metres, and periodicity, T, in seconds).

$$\text{swell wave velocity} = 1.25\sqrt{\text{wavelength}}$$

$$\text{swell wave velocity} = 1.56 \times \text{waveperiod}$$

$$\text{wavelength} = 1.56T^2$$

These equations are practical and useful for

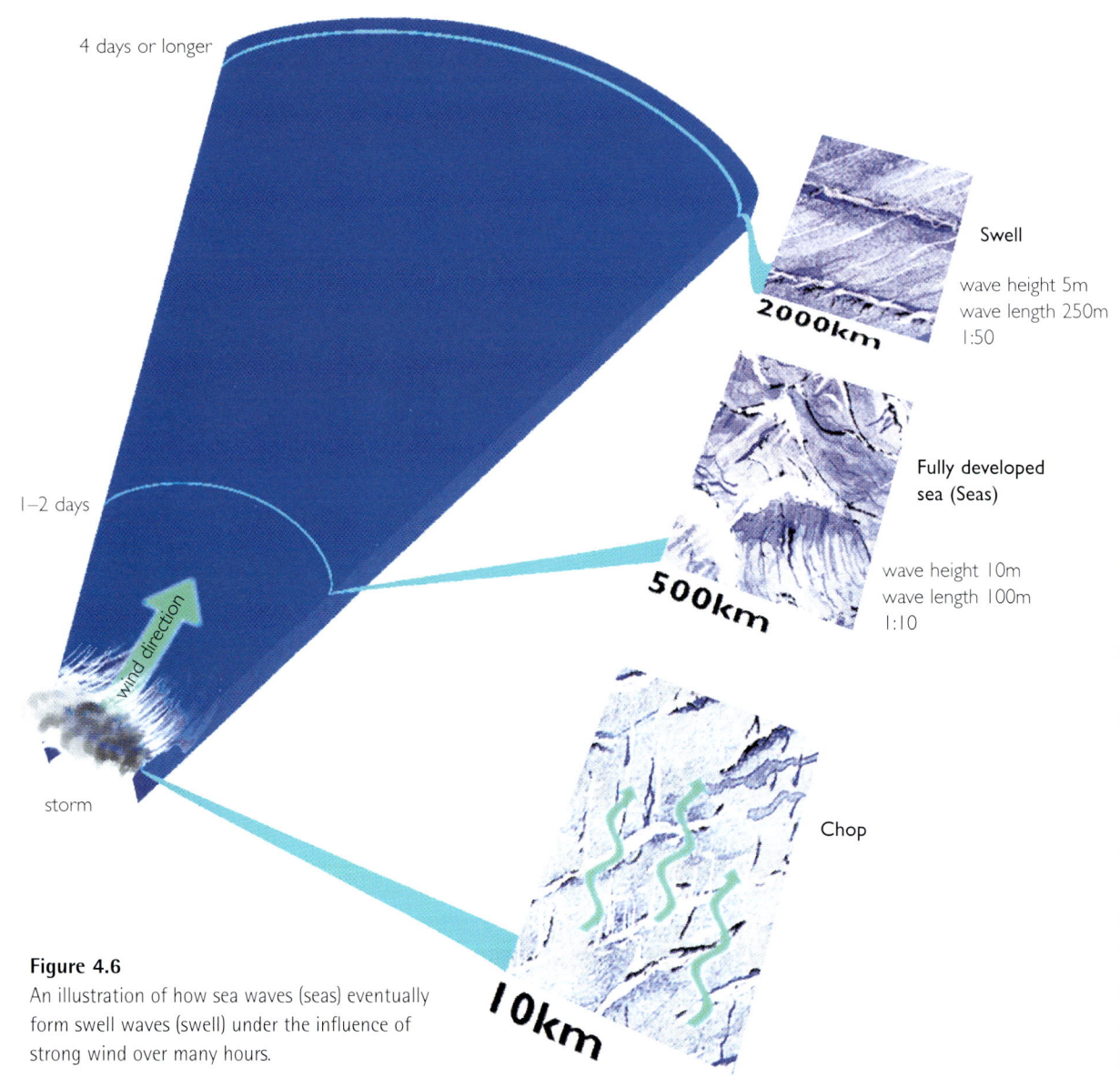

Figure 4.6
An illustration of how sea waves (seas) eventually form swell waves (swell) under the influence of strong wind over many hours.

people engaged in coastal activity. For example, if you are at sea and measure the time between successive dominant swell waves at 15 seconds, then you know the wavelength is about 350 metres (a very long swell), and its velocity a speedy 23 m/s (45 knots).

Classification of seas and swell

Tables 4.1 to 4.3 show how seas and swell are commonly classified and referred to in forecasts.

The Beaufort Scale and wave characteristics

Table 4.4 shows, for each Beaufort class, the approximate size of a fully developed sea, a characteristic wave period and wavelength, and (importantly) the fetch and duration required to reach the fully developed sea.

> Coastal users should study this table carefully; it contains a great deal of useful information for estimating the ultimate wave conditions for combinations of wind speed, wind duration and fetch.

You can substitute values for either wavelength or wave period from this table into the equations

Table 4.1

A classification of sea waves (seas) by wave height and appearance.

Classification	Likely max. ht (m)	Effect/appearance
Calm	0	No waves
Rippled	0.1	No waves breaking on beach
Smooth	0.5	Small breaking waves
Slight	1.3	Waves rock buoys and small craft
Moderate	2.5	Sea becoming furrowed
Rough	4.0	Sea deeply furrowed
Very rough	6.0	Disturbed sea, steep faced rollers
High	9.0	Very disturbed sea, steep faced rollers
Very high	14.0	Towering seas
Phenomenal	14.0+	Hurricane seas

Table 4.2

A classification of swell waves (swell) by wave height.

Classification	Wave height (m)
1 Low swell	0–2
2 Moderate swell	2–4
3 Heavy swell	> 4

Table 4.3

A classification of swell waves (swell) by wavelength.

Classification	Wavelength (m)	Typical wave period (s)
1 Short	0–100	5
2 Average	100–200	10
3 Long	> 200	> 15

Table 4.4

Characteristics of fully developed seas for different Beaufort classes.

Beaufort	Wind speed (kt) U	Potential wave height (m) H	Wave period (s) T	Wavelength (m) L	Fetch (km) –	Duration (hr) –
0	0	0	0	0	0	0
1	1–3	0	0.3	0.1	0.06	0.1
2	4–6	0.1	1.4	2	0.56	0.7
3	7–10	0.2	2.4	6	5.9	2.3
4	11–16	0.6	3.9	16	24	4.8
5	17–21	1.3	5.4	31	65	9.2
6	22–27	2.5	7	51	140	15
7	28–33	4.5	9	80	290	24
8	34–40	7	11	115	520	37
9	41–47	11	13	165	960	52
10	48–55	16	15	225	1570	73
11	56–63	22	17	300	2500	100
12	> 63	> 22	> 17	> 340	> 2500	> 100

Note: These numbers should be treated as indicative rather than as absolute; for example, there is no guarantee that 4.8 hours and Beaufort 4 will always result in a 0.6 metre wave. The numbers have been derived from long-term observations over a range of conditions and are generalised (averaged) to some extent. Note also that wavelength and wave periods differ from values predicted by the formulas.

above to estimate wave velocity — but if you do, then don't expect exact answers, since Table 4.4 shows averaged quantities, whereas the equations are theoretical relationships.

From this table, we can make the following statements about fully developed seas:

- light wind speeds produce small short-period waves
- light winds produce fully developed *small* seas in a relatively short time and require a relatively short fetch to do so
- strong winds produce large long-period waves
- strong winds produce fully developed *large* seas in a relatively long time and require a long fetch (that is, in a strong wind, waves continue to build with time and distance).

What does this mean in practical terms for a person at sea with rising winds? If the wind is rapidly rising, do not fool yourself into thinking that the waves won't get much bigger than they already are. Depending on just how strong the wind is (or becomes), the waves may continue to build for days. They will increase in size with increasing fetch and duration up to the fully developed state.

Predicting open ocean wave heights from wind speed and wind duration

Figure 4.7 is derived from data mainly from the east coast of America. It shows potential wave heights in deep water, for a wide range of wind speeds (x-axis) and wind durations. The colours represent wave heights in metres. The errors associated with this figure are typically less than one metre.

Don't use the figure as an absolute reference — at best it is a guide, since local influences can never be accounted for in a generalised figure such as this. Another important warning: Figure 4.7 is most accurate when the water is relatively flat to start with — that is, there is little or no swell or seas running. If there is a significant 'background' swell, or (to a much lesser extent) seas, the effect of local wind may be strongly influenced by whether it is opposing, or in the same direction as, the swell.

Figure 4.7 is designed so that wave height is shown intuitively by the 'height' of the figure; this is reinforced by the colours, showing low waves as a 'cool' blue, grading through 'warm' yellows and oranges and finally up to 'hot' shades of red. The figure can be read in a number of ways; if you

The general rule is that an opposing wind tends to flatten a swell and place a sea on top of it, whereas one in the same direction will tend to enlarge the swell. A very interesting feature of the interaction between wind and existing swell is that, regardless of whether the wind is in the same or the opposite direction as the pre-existing swell, it does not usually alter the dominant periodicity of the pre-existing swell. Put another way, the energy of the wind tends to be concentrated into the same periodicities as the existing swell.

want to look up a combination of 30 knots and 10 hours duration, then follow the line from 30 knots and the line from 10 hours to the point where they meet. If you want to see how a 40 knot wind will build waves in time, then start at the 40 knot line (where it begins is zero duration) and follow it 'up the hill'; climbing up the 40 knot line indicates increasing duration up to 50 hours. You can also estimate other combinations of wind speed and duration by interpolating (judging) between the lines on the figure.

To understand the figure better, let's look at several examples of different wind speeds (although it is possible to estimate wave conditions for any combination of wind speed and duration within 0–60 knots and 0–50 hours). In the examples it is assumed that the wind speed

The waves described here are waves generated in deep water; when these strike shallow waters — where depth is less than half the wavelength of the wave — friction against the sea bottom can cause them to grow in height and become steep sided (so-called short seas). More worrying for boaters is where the water depth is less than about 80 per cent of the *height* of the wave — at this point the wave will tend to break. Bass Strait is generally less than 60 metres deep and some of the waves generated by storms in the Southern Ocean can grow to a menacing height as they encounter these waters. See the section on shallow water waves for more information.

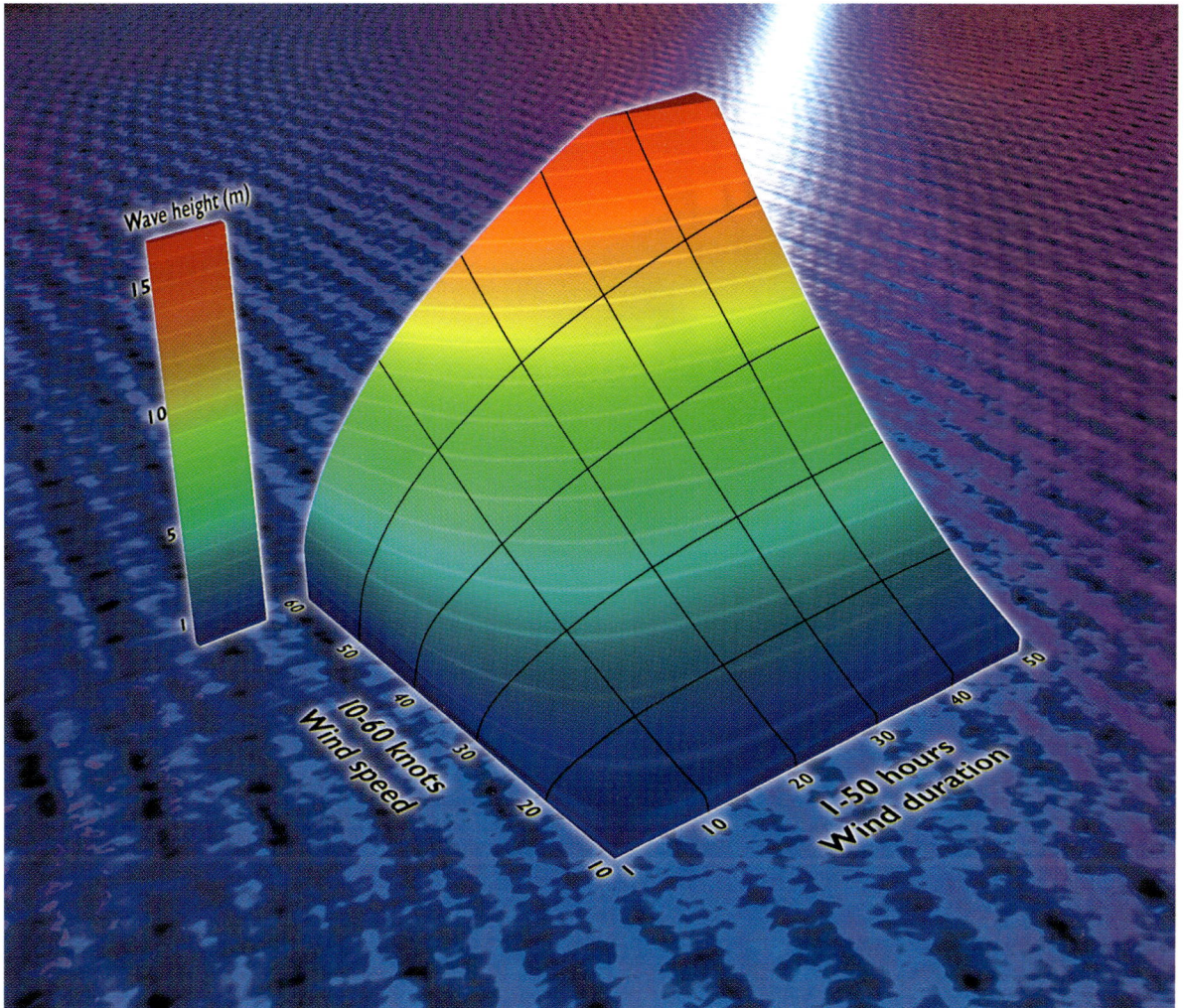

Figure 4.7
Potential wave heights in deep-water long-fetch environments for a range of wind speeds and durations. (Based on US Survey Corps of Engineers data.)

remains more or less constant in time (that is, for increasing duration).

A *gentle breeze* will produce small waves, which do not continue to grow with increasing duration; for example, 10 knots will produce a 0.5 metre wave in about 6 hours. This will grow to 0.8 metre in 24 hours. In 50 hours the wave height will still be about 0.8 m. Using data not shown in the figure, at 200 hours (the 'maximum condition') the wave height will still be less than 1 m.

A *fresh breeze* of 20 knots will produce a 1 metre wave in less than 4 hours, which will grow to 2 metres in about 16 hours. Unlike the 10 knot example, the wave will continue to build (although slowly); by 50 hours it will have reached nearly 3 metres, and, again using data

not shown in the figure, will eventually reach a maximum of a little over 3 m.

A *near gale* of 30 knots will produce a 2 metre high wave in less than 4 hours, which will build to 3 metres in about 8 hours. If you're unfortunate enough to be still at sea, expect 4 metre high waves in about 16 hours and 5 metre waves in 35 hours. The maximum condition (200 hours or more) will result in a wave of about 6.5 m.

A *gale* will produce very threatening waves indeed: 2 metre high waves can be produced in less than 2 hours and 3 metre waves in about 4 hours. They will continue to build to 5 metres in about 10 hours and 8 metres in 35 hours. The maximum condition might produce a towering 12 metre wave.

Wave heights in lakes and small water bodies

Waves produced in lakes, harbours and estuaries are generally less fearsome than those at sea. Shallow water waves are also far less predictable than those of the open ocean. There are no simple rules (or equations!) to predict how waves will behave in shallow water — a situation analogous to predicting the nature of wind near the ground. However, although shallow water waves may be smaller, a wave of almost any size can cause problems if you or your craft are ill-prepared.

A word of caution for people planning to use these and any other wave height figures. It must be stressed (again) that they are a guide only and not an absolute reference. Although they are based on data from a wide range of environments, your particular lake or harbour could have an unusual bottom topography or be unusually sheltered (or exposed). It could be open to the sea at one end or be subject to strong tidal currents.

> Shallow waters tend to produce steep-sided waves, often referred to as short seas. Apart from being uncomfortable to ride on, they can be dangerous; be aware of any rapid changes in water depth. In particular, be aware of areas where relatively deep water rapidly becomes shallow water and thus where a 2 metre 'ground swell' could become a 3 or 4 metre high steep-sided 'breaker'.

The following two figures (4.8 and 4.9) relate to shallow water. They show wave height as a function of fetch (x-axis) and average depth. The figures differ only in wind speed. Figure 4.8 shows wave heights for 15 knots, a moderate breeze, and Figure 4.9 shows 35 knots, a gale. (Please note that fetch is a logarithmic scale; each division (0.1, 1, 10) is ten times longer than the previous one, or 1/10 if reading from right to left. The range of fetches — from small lakes to oceans — is so large that it is difficult to show meaningfully on a normal axis. So the far left-hand limit of the fetch axis is 0.1 kilometre (100 metres) and the far right is 1000 kilometres, although we have data for up to 300 kilometres only.)

As you might expect, shallower water produces smaller (and frequently steeper) waves. If the lake is very shallow, say 2 metres deep,

then you are unlikely to see waves over 1 metre high. However, if the depth of the lake is more than 10 metres, then you can expect waves of over 2 metres if there is sufficient wind.

Notice also that waves produced in 15 metre deep water are still nowhere near as large as those in 'deep' water; we can conclude that friction effects between the waves and the bottom of the sea, or lake, are significant to a very considerable depth — say 100 metres.

15 knot wind speed

A 15 knot wind will produce a maximum wave height of only about 1 metre, regardless of fetch or depth (recall that the figures relate to water up to 15 metres deep). If the lake is very shallow — say, less than 3 metres — you don't have much to worry about (unless your craft can't handle 1 metre waves!). Metre-high waves can be achieved with a combination of 15 knot winds/ water depth greater than about 12 metres and a fetch greater than about 100 kilometres.

Figure 4.8 can be used to estimate how wave height increases with increasing *downwind* distance in a lake or a semi-enclosed water body like Port Phillip Bay in Victoria. At the upwind edge of the lake (which in this case might correspond to Point Lonsdale on Port Phillip Bay if the wind was coming from the south) there would be few or no waves at all, but as one moved in the downwind direction, ripples would be seen forming, then small wavelets, then waves — generally the largest waves would be found farthest downwind. In Figure 4.8 this can be seen by following one of the lines (representing a constant depth) from the left of the figure to the right; in Port Phillip Bay this could be estimated by the 15 metre depth line from where the fetch is zero until it is about 50 kilometres. At this point waves might be 0.5–0.6 metre high.

35 knot wind speed

Recalling from Chapter 3 that a 35 knot wind can generate more than five times the force of a 15 knot wind, it is not surprising that wave heights increase rapidly with increasing depth and fetch. Looked at one way, the wind has more opportunity to impart its energy on the water if it is able to 'operate' on a greater length of water (that is, greater fetch). Similarly, the deeper the water, the less the influence of bottom friction,

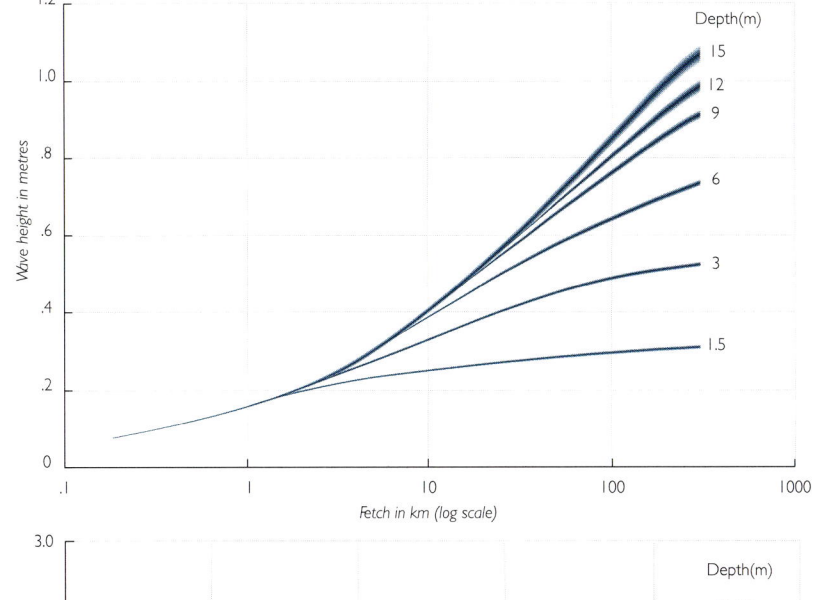

Figure 4.8
Potential wave heights for 15 knot wind speed for a range of water depths and wind durations.

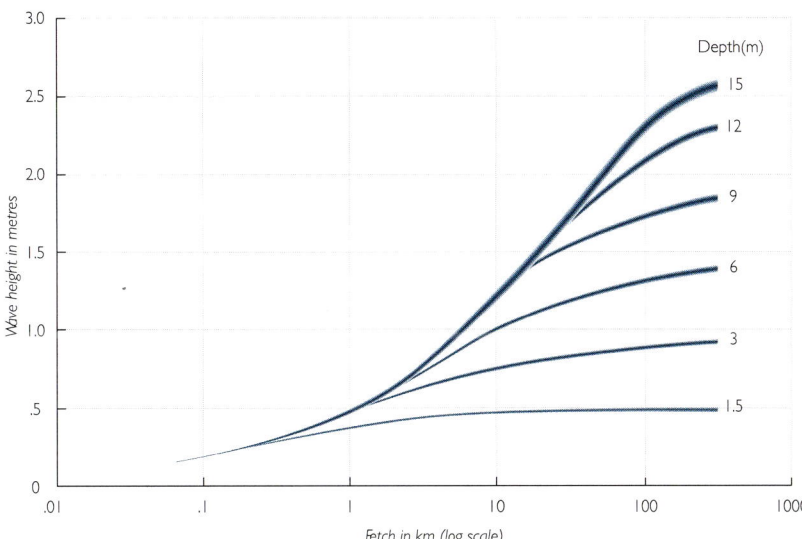

Figure 4.9
Potential wave heights for 35 knot wind speed for a range of water depths and wind durations.

and hence the greater the amount of wind energy that goes into building waves rather than stirring up the sand (or mud).

Gale force winds can produce waves above 1 metre, provided the fetch is greater than about 10 kilometres and depth exceeds 3 metres. Two-metre high waves are possible if the water is more than about 10 metres deep and the fetch exceeds about 70 kilometres. Returning to the Port Phillip Bay example, Figure 4.9 suggests that, in a 35 knot wind, waves might build across the bay to a height of 2 metres.

SHALLOW WATER WAVES

This section is intended as a guide to the sort of waves you might expect in shallow water; local conditions will vary greatly from place to place and, like wind patterns near the ground, are difficult if not impossible to predict without local knowledge.

Boaters will know that sometimes the most arduous part of their boating trip is leaving and returning to the shore, wharf, marina or estuary, especially where there are unusual underwater contours, abrupt changes in depth and groynes.

Swell waves will start to 'feel' the bottom when the depth of the water is less than half their wavelength — $d \leqslant L/2$ — and will begin to break (shoal) when the depth is about 80 per cent the wave height — d ≤ 0.8H. Boaters should consider this carefully: it is not unusual to see a 'friendly' 2 to 3 metre swell rolling in ahead of you when making for shore. If the water suddenly becomes shallow — and this may not be on the chart if it is the result of a moving sand bar — these friendly swell waves could become threatening breakers in a very short distance.

Shallow water waves are also influenced by reflection, diffraction and refraction.

Reflection

If a wave encounters a vertical or near-vertical barrier, such as a rocky cliff rising from deep water or a sea wall, it can reflect back upon itself with little loss of energy. If the approaching wave train is more or less regular, a pattern of standing waves may be set up in which the approaching and reflected waves interact to produce waves whose only motion is up and down. People who sail in lakes with vertical retaining walls will be all too familiar with these steep-sided bumpy waves that don't go anywhere.

> Although standing waves are not usually a serious problem for boating or sailing, their presence sometimes fools the unwary into thinking that there is much more wind than there really is. Sailboarders on short boards beware!

Figure 4.10 shows the concept of reflection and the generation of standing waves.

Diffraction

Imagine a train of regular swell waves of significant height moving across deep water and suddenly encountering an island that rises abruptly from the depths. Few would doubt that the waves on the seaward side of the island would be larger and more fearsome than those on the leeward side, but would you expect no waves at all in the lee of the island? In fact, most people would expect *some* waves. This is because, as shown in Figure 4.11, some of the energy of the approaching wave crests is propagated sideways — is diffracted — around the barrier, giving the appearance of the wave extending into the calm water.

Wave diffraction must be taken into account by engineers designing breakwaters and marinas, otherwise ships moored inside the safety zone might be damaged by diffracted waves during violent weather, especially if there are large waves.

Refraction

Wave refraction — or wave bending — is very common in shallow waters. It occurs when the depth starts to affect the wave significantly and, as we have seen earlier, this normally happens where the depth is less than half the wavelength.

As waves move into shoaling water (d ≤ 0.8H), friction slows them down; waves in the shallowest water move the slowest (recalling that, in shallow water, wave velocity is proportional to the square root of depth). Since different parts of the wave front will be travelling at different depths, the crests will tend to bend until the wave fronts become more or less parallel to the underwater contours.

○ reflected wave

● incoming wave

Figure 4.10

An illustration of wave reflection by walls and other hard structures, sometimes causing standing waves.

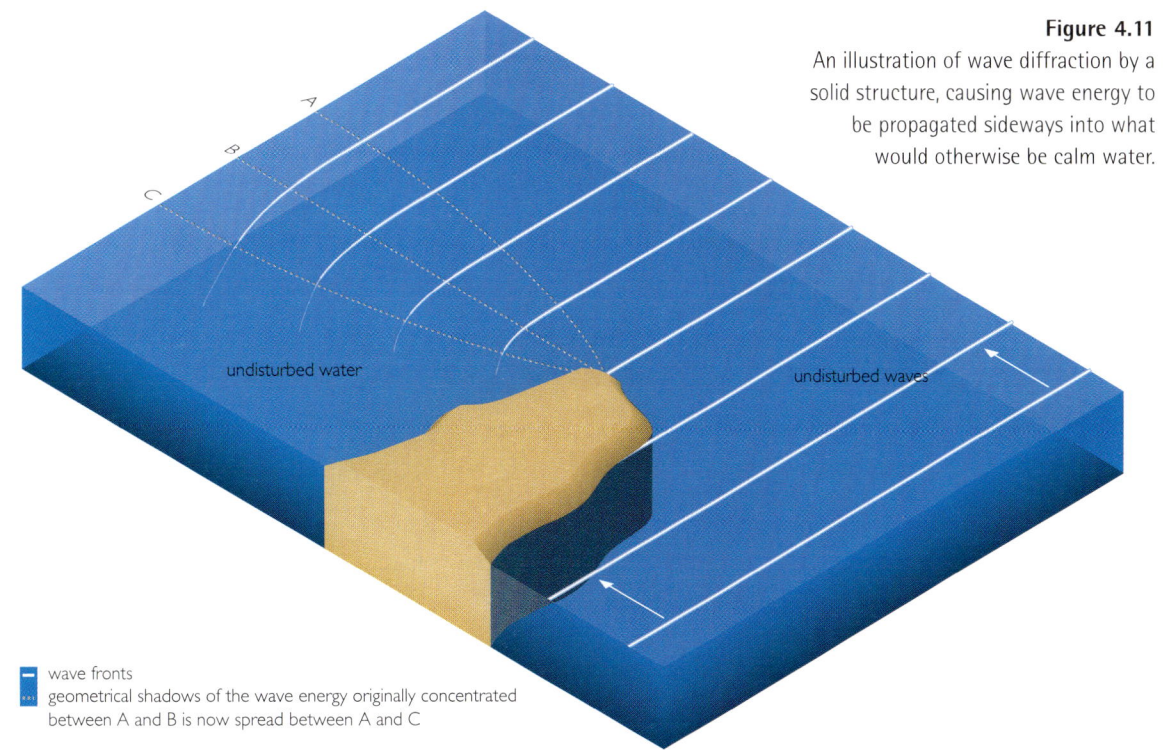

undisturbed water

undisturbed waves

■ wave fronts
⋯ geometrical shadows of the wave energy originally concentrated between A and B is now spread between A and C

The following three figures (4.12 to 4.14) show examples of wave refraction. Boaters are urged to study these, because — unlike wave reflection — refraction can have a very significant influence on wave *topography* and hence on boating safety. Each figure shows the approaching swell waves as white lines — representing the wave crests — and the undersea (or bathymetric) contours in brown. Each contour line joins areas of equal depth, just as contours on topographic maps depict mountains and valleys. Both the wave pattern and the underwater topography are marked, to show how the two interact.

Figure 4.12 is a simple example. It shows a set of regular waves approaching a straight shoreline at an angle; the inshore part of each wave is approaching shallow water and consequently is moving more slowly than the offshore part. The result is that the wave fronts tend to become parallel to the shore; depending on local geography, an observer on the shore might get the impression that the larger waves are approaching the shore directly (that is, parallel).

Figure 4.13 shows a set of regular waves approaching a headland. This diagram includes an impression of the bottom topography (depth contours) to illustrate how the waves are refracted according to depth and, importantly, how wave energy becomes concentrated on the headland as a result. So the old sailor's saying 'the points draw the waves' could well be interpreted as wave refraction.

Figure 4.14 shows a set of regular waves approaching an island with gently sloping underwater topography on all sides; the wave train from one direction will tend to warp around the island so that waves arrive nearly parallel to the beaches on all sides. The waves will, of course, be significantly larger on the side facing into the swell waves (the *seaward* side). The information in this figure may come in handy if you are trying to escape a large swell in the lee of an island!

WAVE MOTION AND CRAFT

Boat handling in inclement weather is covered in the next chapter, yet it is important, while considering seas and swell, to point out the implications for safe boat handling. These hold true for inclement and fine weather alike.

The main implication for boaters derives from

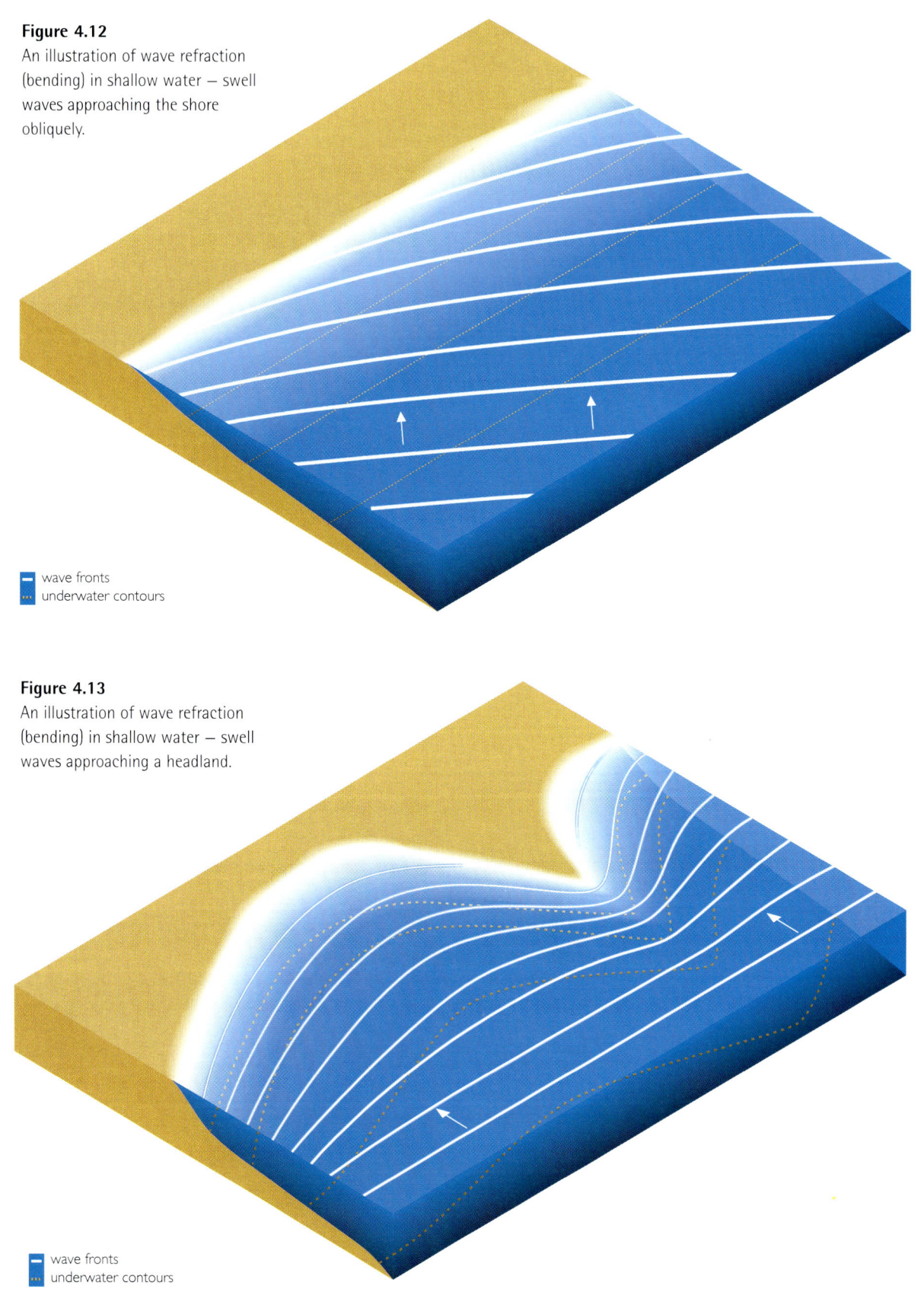

Figure 4.12
An illustration of wave refraction (bending) in shallow water – swell waves approaching the shore obliquely.

■ wave fronts
▦ underwater contours

Figure 4.13
An illustration of wave refraction (bending) in shallow water – swell waves approaching a headland.

■ wave fronts
▦ underwater contours

Figure 4.14
An illustration of wave refraction (bending) in shallow water — swell waves approaching a solitary island.

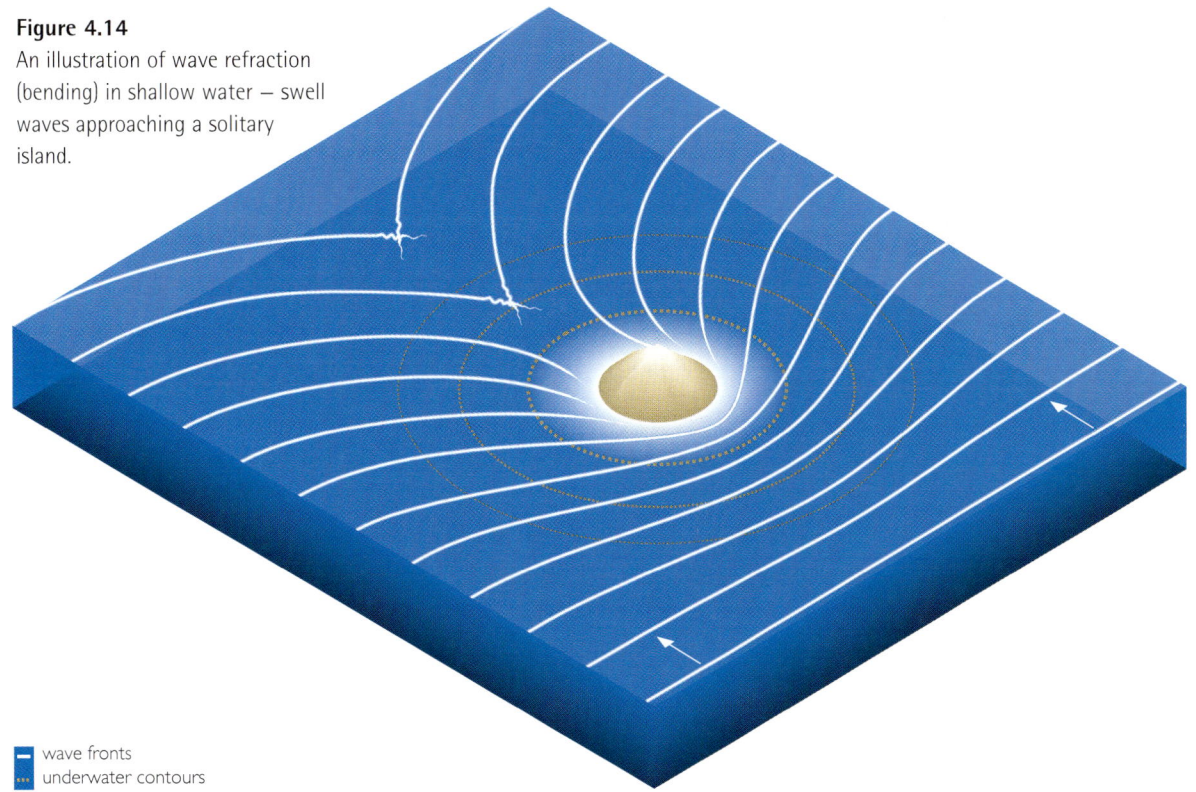

— wave fronts
··· underwater contours

a property of waves: that the water at the wave crest moves in the direction of the wave, but the water at the wave trough moves *against* the wave's direction. A boat of a length comparable to half the wavelength will therefore always experience an influence from the waves on its course.

Consider the case where the boat travels against the waves. When the bow is at the trough and the stern is on the crest, the bow accelerates while the stern is retarded. This makes it easier to keep course, since the waves tend to bring the craft back on its course. When the bow is on the crest and the stern in the trough, the bow is retarded while the stern tries to accelerate and overtake the bow. This tends to bring the boat off course. Since the boat is heading into the waves — which pass relatively quickly — the effect is of only short duration.

The situation with a following sea is quite different. Depending on boat speed and wave speed, it can take each wave a considerable time to overtake the boat, so the period during which the bow is retarded and the stern tries to accelerate can be quite long. This can cause powerboats, yachts and even trawlers to veer radically off course and end up sideways to the seas (that is, broaching). If this happens during the approach to a harbour with a shoaling sand bar, the wave may build up and begin to break, eventually crashing down on, or even overturning, the craft. Even in the open sea, yachts running exactly with the wind and therefore with a following sea can find themselves so far off course that the sails can catch the wind from the opposite side and bring the boom across the deck with tremendous force — often to the surprise of an unprepared crew, who may lose a member overboard. The effect, which depends on the ratio between boat length and wavelength and the ship's course relative to the direction of the waves, can be disastrous for boats and even larger ships.

The best way to be prepared for such situations is to have a constant lookout for the waves and anticipate the passage of each wave by counter-steering *before* the wave begins to act on the boat. Very little steering effort is required if this is done correctly. Once the correct moment for steering action is missed, tremendous effort is required to counteract the wave and avoid disaster.

CHAPTER FIVE

Ocean Currents, Sea Surface Temperatures and Tides

Currents in the open ocean are essentially surface phenomena operating in an active layer, typically 200 to 300 metres deep. Below this, water temperature drops rapidly with depth, and if we were to take the mean temperature of all the waters in the world ocean, throughout its depth, it would be little more than 3°C. The net effect of the strong gradient in temperature is to create a barrier to the transmission of energy downwards from the surface because of the presence of a 'pycnocline', an accompanying strong gradient in the density of the water column.

Let's start with a map that shows an apparent relationship between the direction of surface currents in the ocean and surface winds. It would be easy to conclude from it that the ocean currents are essentially wind-driven — and this is a fairly common view — but a closer look at monthly, or even daily, patterns of currents will show that this explanation is not generally adequate.

The wind patterns in Figure 5.1 are typical of mid-winter, whereas the surface currents are indicative of annual patterns (neither are particularly accurate for the Australian region). Since the major pressure systems of the world migrate north–south seasonally (see Figure 2.4) — and the major wind systems follow suit — both the surface winds and surface currents will also vary seasonally and so the figure cannot be very realistic. Local and regional currents will not appear in such a broad scale figure. In order to explain the patterns of currents better, we need to look briefly at the primary (wind stress and gravity) and secondary (Coriolis force and friction) current-producing forces.

Primary forces

'Wind stress' is the name given to the force of the wind on the sea surface (through friction). Wind stress is non-linear, which implies that the stress is not simply doubled, say, if the wind speed doubles. It is more likely that stress behaves as the square of the wind speed, so that to double the wind speed would result in increasing the wind stress by a factor of four, and to double it again in a sixteenfold increase of the stress over the original.

Gravity, in the context of currents, implies flow down an incline, where the incline is commonly set up by wind stress or by water density anomalies.

Secondary forces

The Coriolis force can be said to act on moving water in the same way as it does on moving air or any other body in motion on the rotating Earth (see Appendix V). As for wind, the Coriolis force deflects moving water (and air) to the left in the Southern Hemisphere. Its dependence on latitude gives it a zero value on the equator, increasing with higher latitudes. Again, the more rapid the flow, the greater the Coriolis force.

Friction generally occurs on contact with the sea bed or ocean bed, but in the case of surface currents one has to take into account the friction between different layers in the water body: inter-facial friction. Friction, like wind stress, is non-linear.

Surface winds and surface ocean currents

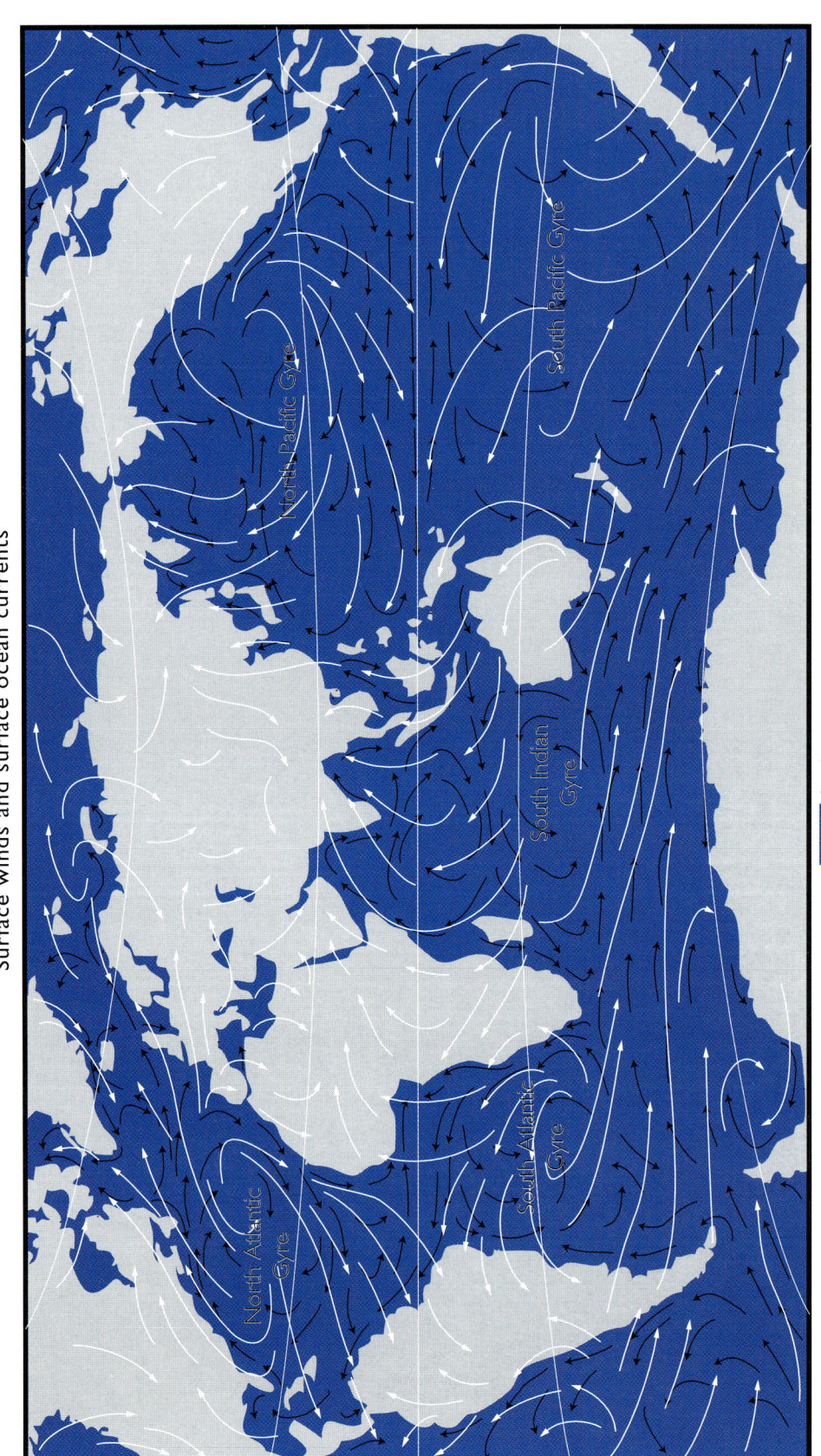

Surface Winds
Surface Currents

Figure 5.1
An example of the relationship between global scale wind patterns and surface ocean currents. (After the *Macmillan Atlas of the Oceans*, 1977.)

TYPES OF OCEAN CURRENTS

There are two basic types of ocean current: the first is wind-driven and results in accumulations of water — literally 'hills' of water — in the oceans; the second is a response to gravity acting down an incline in the ocean surface. Such gradients can be caused by wind stress but can also occur independently as a response to density differences in adjacent water columns. (Are you already seeing where satellite imagery might help in the explanation?)

Wind-driven currents (Ekman transport)

In 1893 a famous oceanographer named Fridtjof Nansen took his vessel, the *Fram*, on an expedition into the Arctic regions and became firmly frozen in the ice. During the three-year voyage he had observed that the motion of the ice was not along the path of the wind, but some 20 to 40° to its right (it would have been to the left had he and the *Fram* been in the Southern Hemisphere). In 1905 a physicist, Walfrid Ekman, explained this phenomenon, which is now known as the 'Ekman spiral' (the associated movement of the surface layer is known as 'Ekman transport'). Figure 5.2 illustrates the concept.

The principle here is that a steady wind blowing over a typical 'block' of surface water, perhaps 100 metres deep, sets the surface layer of the block in motion. Its direction will be deflected to the left of the direction of the wind (in the Southern Hemisphere) by the Coriolis force.

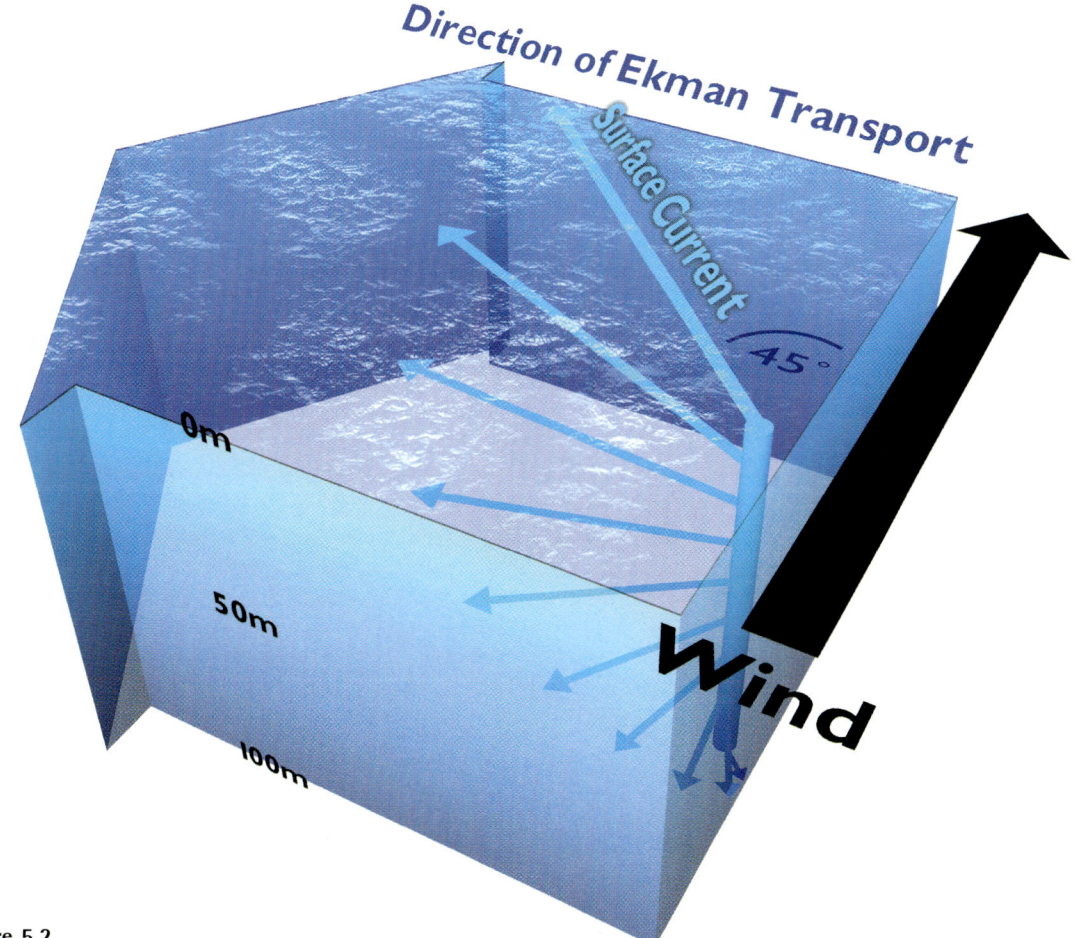

Figure 5.2

The concept of the Ekman spiral applied to the surface ocean in the Southern Hemisphere; overall movement of the surface layer (approximately 100 m thick) is about 90° to the left of the wind direction as seen from above. (Data courtesy G. Lennon.)

Observations suggest that this deflection may well be as much as 45° and that the speed of the surface current will be of the order of 2 per cent of the wind speed — that is, a steady wind of 10 m/s (19 knots) will give a surface current of 20 cm/s (0.4 knots).

Initially, only the top metre or so of the water will be involved, but in time the next layer down will begin moving. Starting from rest, and driven by a surface current flowing 45° to the left of the wind direction, the second layer will also be deflected by the Coriolis force and will adopt a direction even further to the left of the surface current. The strength of this flow will be slightly less than that of the surface layer since work has been done in overcoming *interfacial friction*.

Ekman's hypothesis suggests that, if a wind remains steady over five or more days, then this process will continue through the water column down to a depth of 100 metres or so, with each successive layer — driven from above by friction — continuing to be deflected further and further to the left, and flowing at a progressively slower rate. At 100 metres depth the flow will be negligible, being close to the depth of frictional resistance, which is the limit. Incredibly, when the flow has dropped to approximately 4 per cent of the surface flow, its direction will be opposite to the surface flow.

Overall then, by integrating the total effect of the wind operating on the upper 100 metres or so of the ocean, the *average* motion of the water is directed 90° to the left of the wind direction at the surface. In Figure 5.2 the total water body down to 100 metres is depicted as an arrow to reinforce this point.

This has a direct relevance to the mariner of a deep draft vessel in regions where steady winds are common, for obvious reasons. There are several circumstances where the navigator can be confused by taking bearings from the evidence of flow at the surface. This effect is less in lower latitudes and insignificant at the equator. In the Northern Hemisphere the same argument applies, but the deflection is to the right.

Ekman transport and upwelling events

Ekman assumptions require homogeneity and an absence of pressure or gravity gradients, so that 'pure' examples in the real world are rare. Ekman transport tends to be inferred indirectly from other evidence. For example, consider a steady wind over the continental shelf, in the Southern Hemisphere and with the coastline to the right. (Such circumstances sometimes occur along the south east of South Australia, near Kingston and Robe.) A fairly common situation in summer is for a high pressure system — that is, a blocking high — to remain near the Great Australian Bight for some days at a time, directing winds from the south to south-east along the continental shelf. Observations offshore and satellite infra-red images of sea surface temperatures confirm that, after 5 to 7 days of steady wind, an Ekman spiral has been established, creating an average current at right angles to the shoreline. The waters being transported offshore in this way will require replacement, and this can only be achieved by upwelling — drawing into the near coastal zone deeper offshore waters which bring cool and nutrient-rich water to the surface layers. Figure 5.3 shows a graphic example of this upwelling.

Such periodic upwellings are particularly important to the rock lobster fishery along the South Australian coast, since the predominant winds and currents are from the west and on-shore respectively and, following Ekman principles, would normally result in downwelling and the leaching of nutrients from the surface waters.

Ekman transport, in all its forms, has the distinction that the driving force is wind stress, and this is influenced and partially balanced by the Coriolis force due to the rotation of the Earth. Ekman transport is latitude dependent, and is zero at the equator, increasing with higher latitudes.

Ekman transport and the movement of pollutants

A knowledge of Ekman principles can be helpful in other ways. For example, in a pollution spill we need to track the potential hazard. First scientists ascertain the approximate density of the pollutant compared with that of sea water. It is then possible to suggest:

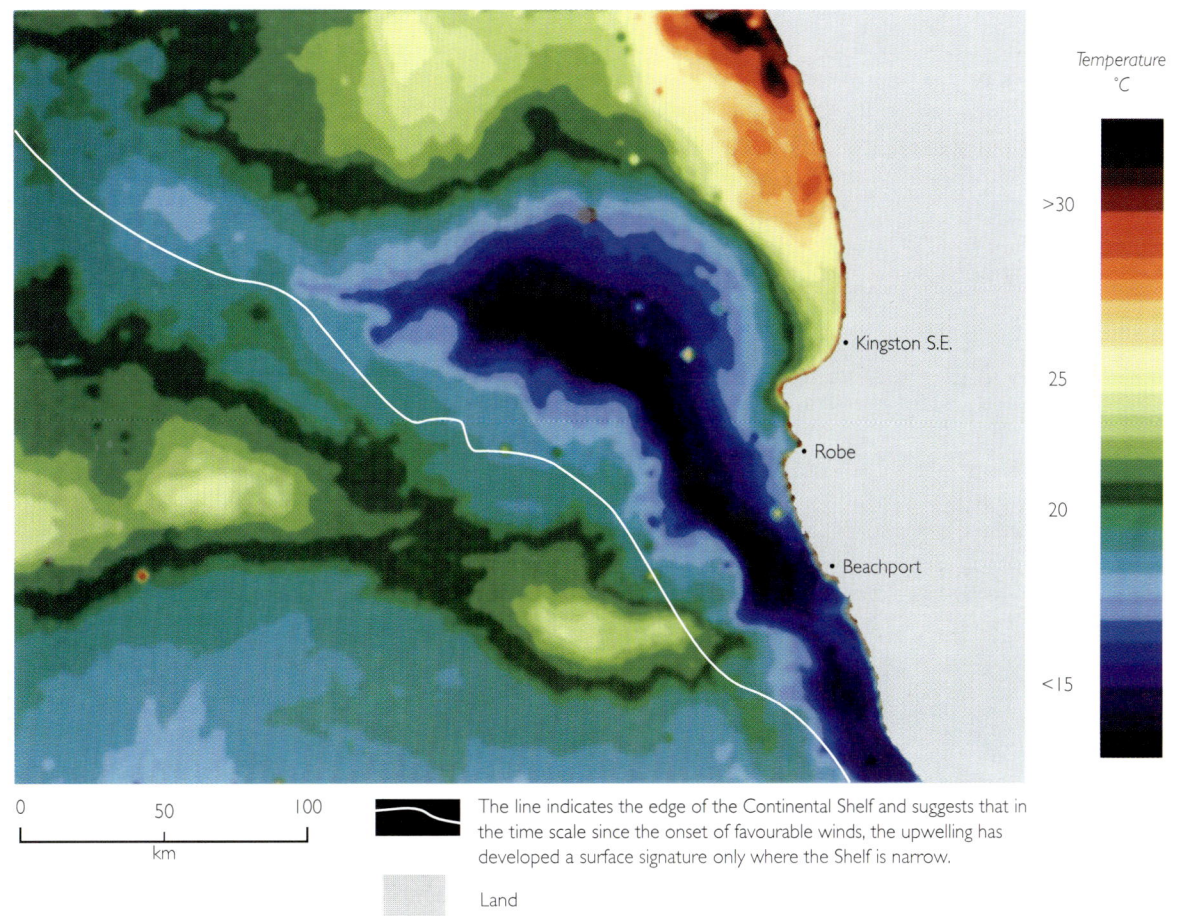

Temperature °C

>30

25

20

<15

0   50   100

km

The line indicates the edge of the Continental Shelf and suggests that in the time scale since the onset of favourable winds, the upwelling has developed a surface signature only where the Shelf is narrow.

Land

Figure 5.3
An upwelling event that occurred in the Bonney Coast Zone of South Australia in February 1989, captured by satellite from its sea surface temperature signature. (Data courtesy G. Lennon.)

- If the pollutant is *less dense* than sea water (for example, most oils), it will tend to remain on the surface and possibly travel at a speed of approximately 2 per cent of the wind speed and 45° to the left of the wind direction (Southern Hemisphere).
- If the pollutant is *denser* than sea water, it will tend to sink, then travel on the sea bed at a considerably slower rate and in a direction much further to the left (Southern Hemisphere), or even in the opposite direction to the wind, should the depth not exceed 100 metres or so. The pollutant will also be influenced by bottom topography (for example, by channelling).
- If the pollutant is *of similar density* to sea water, and therefore more easily mixed through it, then a more diluted concentration can be expected to travel at varying rates but principally at right angles to the wind direction.

The role of tides in this context is largely one of mixing. Drift/dispersion from tides is usually small, since tidal currents are generally oscillatory, as we shall see. It is, of course, necessary to take ocean currents into consideration (see Figure 5.9).

Gravity-driven currents (geostrophic flow)

The word geostrophic means 'Earth turned', and is a process in which the rotation of the Earth plays a key role. There is room for confusion, since Ekman transport is also dependent upon the rotation of the Earth and in a sense is also geostrophic. However, the convention is to reserve the title 'geostrophic' for processes involving the balance of Coriolis force against a pressure gradient, whether it be an incline in elevation or density, so that gravity will operate in some way. In the context of our subject here, the incline might well be created by Ekman transport.

Of course geostrophic flow is not restricted to the oceans. Perhaps its most common occurrence is in weather, and it is often featured on normal weather maps, where contours of atmospheric pressure (isobars, or lines of equal pressure) associated with high and low pressure cells operate, in principle, in a similar way to those of Figure 5.4.

Imagine a simple wedge-shaped incline (slope) of Australian coastal water in which a particle moves from an initial position at the top of the gradient. Under gravity, the particle of water begins its descent, directly parallel to the slope, and it begins to accelerate. Since the incline is on a rotating Earth, and in the Southern Hemisphere, as soon as motion begins a weak Coriolis force comes into action. It operates at right angles to the direction of motion, so that the particle's path is deflected slightly to the left. The particle is still accelerating so that the Coriolis force increases, again, at right angles to the new direction of motion and so on until an equilibrium is achieved.

At this stage the Coriolis force, acting in a direction up the incline, exactly balances the gravitational force acting down the incline. Then the motion of the particle would be steady and at right angles to the incline. Should the particle decelerate due to friction, then the Coriolis force would become slightly less and would no longer balance gravity. The particle would move slightly down the incline, and accelerate. The Coriolis force would increase and a balance would again be achieved. This stable equilibrium defines geostrophic flow (in very much the same way as the geostrophic wind depends on a balance between the force due to a gradient in barometric pressure and the Coriolis force).

Figure 5.4 represents this scenario diagrammatically. The incline is shown as H′, height above a datum (reference), divided by X, distance. The steeper this incline, the stronger the geostrophic flow.

Figure 5.5 shows a hypothetical analysis of currents in the Tasman Sea. In most circumstances an oceanic front exists across the Tasman Sea, usually appearing near the coast between Coffs Harbour and Sydney, and aligned in the direction of the North Island of New Zealand (as shown in Figure 5.9). As with atmospheric fronts, this also follows a wavy path, with perhaps four or five 'waves' in the breadth of the Tasman Sea. The front separates tropical ocean water — warm and of low salinity in the north — from cooler and perhaps more saline water in the south.

Figure 5.4

A schematic representation of the forces involved in geostrophic flow as applied to the generation of currents due to a slope in the water surface.

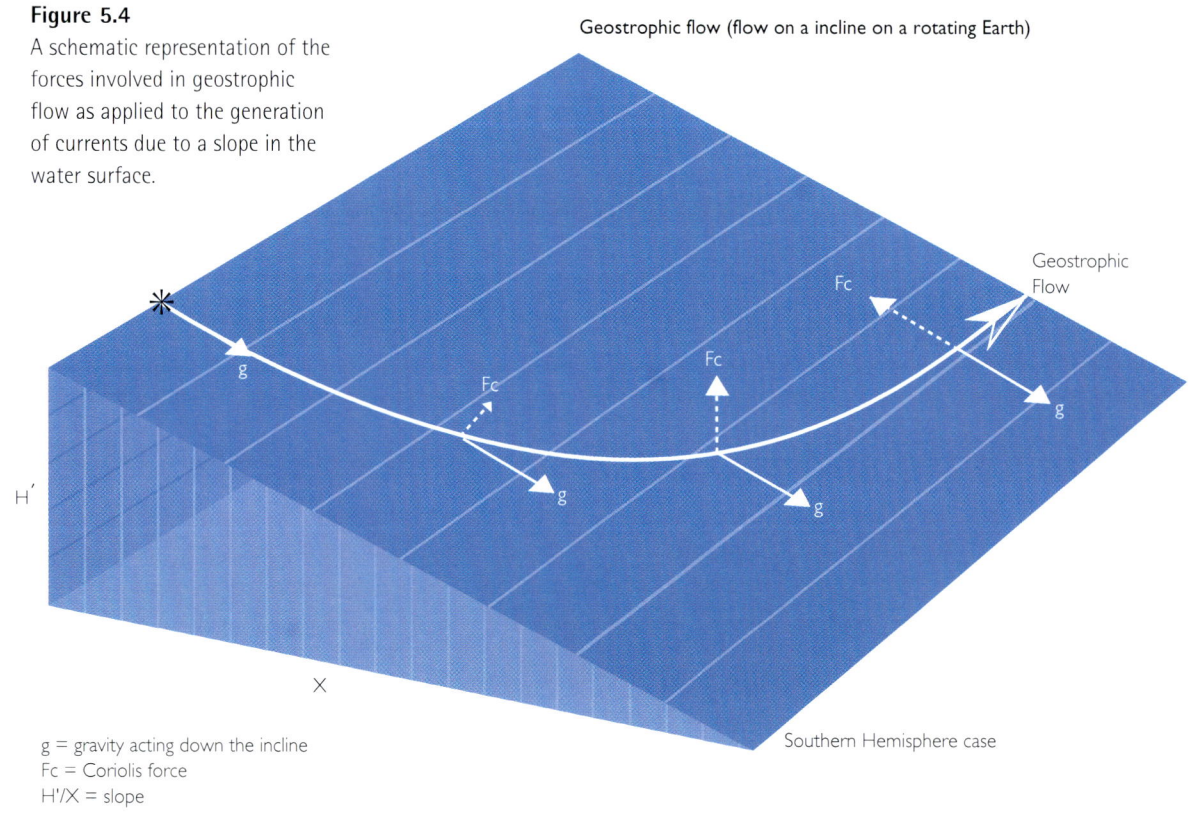

Geostrophic flow (flow on a incline on a rotating Earth)

g = gravity acting down the incline
Fc = Coriolis force
H′/X = slope

A hypothetical analysis of currents off NSW

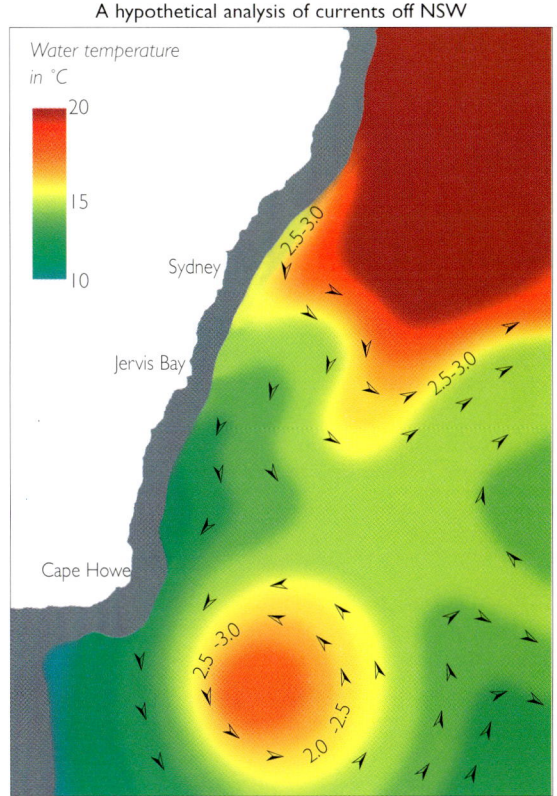

➤ Examples of Geostrophic currents in
2.0 knots. Arrows could be drawn around
all the contours, but this figure only
shows a selection. Current strength is
proportional to the slope of the sea
surface as explained in the text and is
shown as white numbers. Directions
of currents are shown by directions of
arrows.
—— Continental Shelf edge

Figure 5.5
A hypothetical analysis of surface currents from a
consideration of surface topography inferred from water
temperature and salinity data. If heights can be measured
directly from satellite or other source, the equations for
geostrophic flow can be used directly to infer currents.
(Data courtesy G. Lennon.)

In late summer, the wave nearest the Australian coastline commonly begins to develop and
elongate, eventually breaking off to form a near-
circular block of anomalously warm water south
of the front, and perhaps 5 to 8°C warmer than
the Tasman Sea proper. This block might be 300
metres deep and, being of low density compared
with the ambient Tasman Sea water, will stand
high (*literally*; if accurately measured from a satellite, the block would be seen to be higher than the
surrounding water).

In this example, the slopes (differences in
height) were calculated from temperature and
salinity measurements, rather than by direct
measurements of height from satellites, which are
becoming increasingly available. In Figure 5.5 the
eddy has a surface temperature 6 or 7°C warmer
than its surroundings. The warm water stands
higher than the surrounding water, resulting in a
geostrophic flow running at a significant 2 to 3
knots. Note that strong currents, which clearly
represent geostrophic flow, are also generated on
the incline of the front itself.

Flow around a cold block (a 'cold core eddy')
will be in the opposite direction to that around a
warm block (that is, it will be clockwise in the
Southern Hemisphere). A quick glimpse forward
to a 'real life' pattern of surface temperatures —
where both warm and cold patches are clearly evident — is provided by Figure 5.10, which shows a
satellite image off the coast of New South Wales.

The eddies have a long lifetime, perhaps up to
a year, during which they drift slowly southward,
so that at times there can be a series of three or
more slowly decaying eddies in existence in a line
down to the east coast of Tasmania.

> Competitive sailors should check the
> position of the eddies before and during a
> race. The pilots of commercial vessels should
> also be aware of the help or hindrance these
> features might give.

Estimating geostrophic flow (current strength)

It is possible to quantify the relationship between
strength of geostrophic flow and incline or slope
of a water surface. These days the slope can be
measured directly from satellite altimeters or it
can be inferred by calculating the densities of different blocks of water (as in the previous figure).
The density of water decreases as water temperature increases and as salinity decreases.

The traditional unit of mass — the gram —
was originally intended to be equivalent to the
mass of one cubic centimetre of water at 4°C. Sea
water can be considered as a diluted 'soup', containing dissolved salts and other chemicals, so
that the mass of a cubic centimetre of sea water is
likely to be greater than a gram, even though its
temperature near the surface is considerably higher than 4°C. The density of the water contained in

an eddy such as the one in Figure 5.10, when its temperature and salinity are considered, could realistically be 1.023 g/cm^3, while that of the cooler Tasman Sea water could be 1.026 g/cm^3. The eddy block may extend to a depth of 300 metres, so from this information we can calculate the level at which the block must stand above the adjacent Tasman Sea water.

Let's work to a reference point of 300 m depth. If we know this, and the densities of the waters, there is a simple equation to work out the height of the block above the ambient water. It relates the height, H^b, of the block, to the height, H, of the Tasman Sea water. Here we take H = 300m.

$$1.023\,gH^b = 1.026\,gH$$

Where g is the acceleration due to gravity.

But since g occurs on both sides of the equation, it can be cancelled so that:

$$1.023H^b = 1.026H$$

$$H^b = \frac{1.026 \times 300}{1.023}$$

$$= 300.88\,\text{m (i.e. 0.88 m above the general height of the Tasman Sea).}$$

What is implied here is that if we were to use a large manometer or U tube, more than 300 metres long, and if we were to fill one side of the U with water from the block and the other side with an equal mass of Tasman Sea water—at the same time preventing mixing—then we would find that the block water would stand high (in this example, by 0.88 m). In the ocean, mixing is not very efficient, so we do not need the walls of a U tube to prevent mixing. The block can then be associated with a gradient of 0.88 metres over the radius of the block.

In geostrophic flow, the Coriolis force ($F_c = fv$) is balanced by the acceleration due to gravity acting down the slope. Using the terms in the wedge figure (Figure 5.4), this means:

$$fv = g\frac{H'}{X}$$

where $\dfrac{H'}{X}$ is the slope

that is, $v = \dfrac{g}{f}\dfrac{H'}{X}$

where g is the acceleration due to gravity ($\approx 9.8\,\text{ms}^{-2}$), v is the speed of the flow.

The Coriolis force may be quantified as an acceleration as follows:

Coriolis force $(F_c) = fv$

where f is the Coriolis parameter $= 2\Omega\text{Sin}$ (latitude)

and v = the velocity of the object (water in this case)

and where Ω is the angular velocity of the Earth's rotation, that is,

$$\Omega = \frac{2\pi}{24 \times 3600}\,\text{radians s}^{-1}$$

Now we can apply the geostrophic flow formula to the case of the eddy in the Tasman Sea and calculate likely currents from the estimated surface slope. In the following example, the eddy has a radius of 50 kilometres (50,000 metres) and the latitude is 35° (a glance at Figure 5.10 suggests that this is a realistic example).

Geostrophic flow was defined as $v = \dfrac{g}{f}\dfrac{H'}{X}$

where $\dfrac{H'}{X}$ is the surface slope.

In the present case, the equation can be rewritten as:

$$v = \frac{g}{f}\frac{H^b - H}{X},\ \text{where } H^b \text{ is the local height of}$$

the block and H is the general height of the Tasman Sea.

If, as calculated earlier, $H^b - H$ is 0.88m, and $f = 2\Omega\text{Sin}$ (latitude),

where $\Omega = \dfrac{2\pi}{24 \times 3600}\,\text{radians s}^{-1}$

then $v = \dfrac{9.8 \times 24 \times 3600}{4\pi \times Sin35°} \times \dfrac{0.88}{50,000}$

≈ 2 m/s or about 4 knots of current.

Since the anomalous block is near-circular, and under geostrophic principles the current will reach an approximate steady state running parallel to the height contours, it follows that the result is a rotating eddy, with the flow in an anticlockwise direction in such a Southern Hemisphere case. Such a feature is often called a 'warm core eddy', for obvious reasons.

GLOBAL SCALE PATTERNS OF OCEAN CURRENTS

Now we come to the relationship between the physical principles just outlined and the large-scale circulation patterns in oceans at real locations. The East Australian Current (discussed later) is a good example, being (in part, at least) a geostrophic current—and one with a significant rate of flow.

Figures 5.6 and 5.7 describe in general terms the oceans between 60° north and south of the equator. The first diagram shows how surface winds produce surface currents that lead to the development of areas where water sits higher or lower than average. The second diagram—relating to the same areas—shows how this combines with geostrophic flow to contribute to the development of ocean gyres (large-scale rotating currents), clearly evident in Figure 5.1.

On either side of the equator, as in Figure 5.6, there are well-established wind belts: the north-east Trade Winds north of the equator and the south-east Trade Winds to the south.

The Trade Wind zones are very persistent wind systems, and, as such, have a full effect on the surface layers of the ocean. The earlier explanation of Ekman transport showed that, in such circumstances, one would expect to see the upper

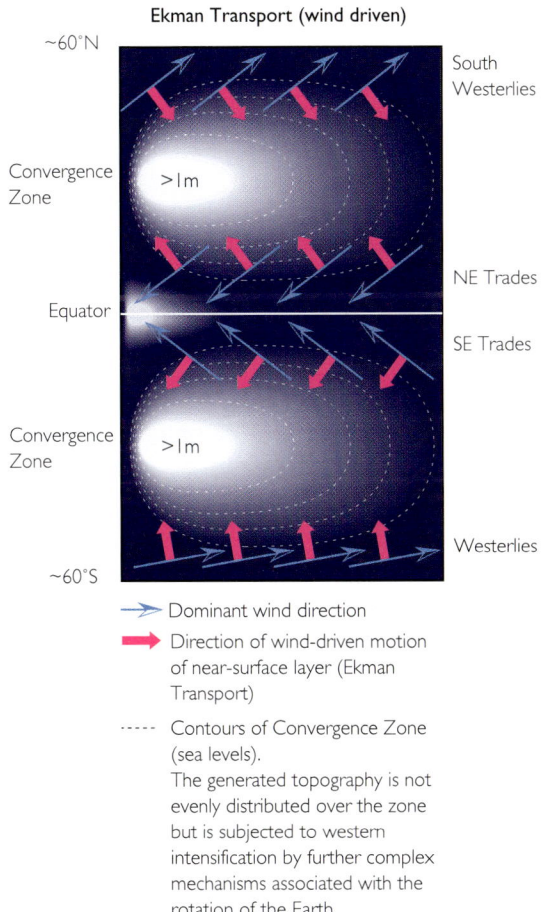

Ekman Transport (wind driven)

~60°N

South Westerlies

Convergence Zone

>1m

NE Trades

Equator

SE Trades

Convergence Zone

>1m

Westerlies

~60°S

→ Dominant wind direction

➡ Direction of wind-driven motion of near-surface layer (Ekman Transport)

---- Contours of Convergence Zone (sea levels).
The generated topography is not evenly distributed over the zone but is subjected to western intensification by further complex mechanisms associated with the rotation of the Earth.

Figure 5.6

The generation of surface currents on a hypothetical ocean between 60° N and 60° S of the equator Phase 1; the generation of surface topography in mid-latitude convergence zones by the action of surface winds and Ekman transport.
(Data courtesy G. Lennon.)

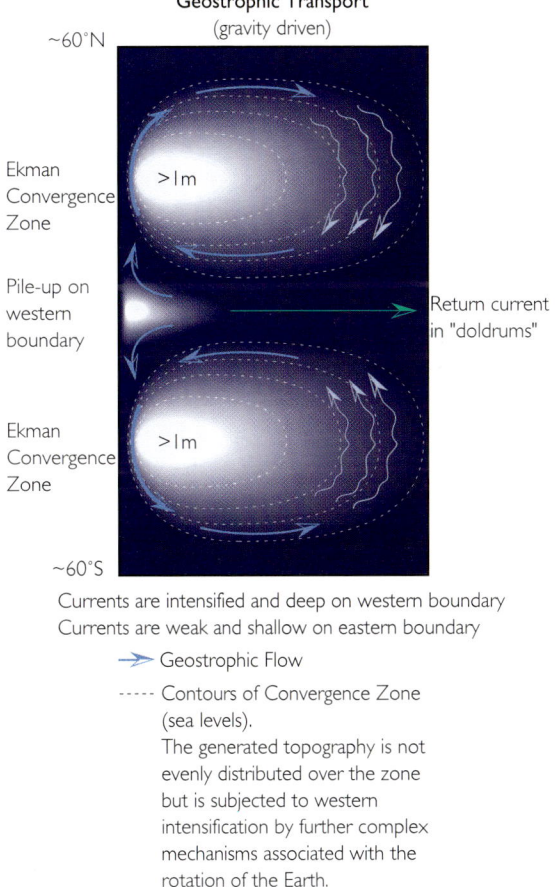

Geostrophic Transport
(gravity driven)

~60°N

Ekman Convergence Zone

>1m

Pile-up on western boundary

Return current in "doldrums"

Ekman Convergence Zone

>1m

~60°S

Currents are intensified and deep on western boundary
Currents are weak and shallow on eastern boundary

→ Geostrophic Flow

---- Contours of Convergence Zone (sea levels).
The generated topography is not evenly distributed over the zone but is subjected to western intensification by further complex mechanisms associated with the rotation of the Earth.

Figure 5.7

The generation of surface currents on a hypothetical ocean between 60° N and 60° S of the equator Phase 2; the generation of ocean gyres through geostrophic processes acting on surface topography in the form of mid-latitude convergence zones. (Data courtesy G. Lennon.)

100 metres of ocean moving at right angles to the direction of the wind: to the right in the Northern Hemisphere and to the left south of the equator. We believe that this actually does take place, as shown in Figure 5.6.

> In meteorology, winds (and wind direction generally) are named according to the direction from which they come, so that the north-east Trades blow from the north-east to the south-west. The opposite is true of oceanography, so that a current that flows in the direction of the north-east Trades would be called a south-westerly current.

It is also well known that, in each hemisphere, at a higher latitude, there is another well-developed and near-permanent wind zone: the Westerlies. These winds dominate the climate of northern Europe, for example, and bring much variability to its weather. Something similar occurs in the Southern Hemisphere in the latitudes of the 40s and 50s, though (in the absence of settlement there) this region belongs to the intrepid mariner.

As in the case of the Trade Winds, these zones are indicated in Figure 5.7, and all the requisite circumstances are present for the development of Ekman transport. The direction of this transport is indicated by the red arrows, from the north-west in the Northern Hemisphere and more or less in the opposite direction for the southern Westerlies. Consequently, the Ekman transport creates a 'convergence zone' in the 30s latitudes, both north and south, where there is an accumulation of water between the two wind zones, Trades and Westerlies, and where the ocean stands high as a result.

Surface gradients are small, typically less than 2 metres over 2000 kilometres. So how can we be sure that such a convergence does in fact take place? Confirmation is obtained mainly by two independent means.

The first, and traditional, method is by a calculation of 'dynamic heights'. This is a standard oceanographic procedure, followed when a research vessel is at work, to measure the temperature and salinity of water at as many locations and depths as possible. Researchers normally stop the ship frequently and pass a CTD (Conductivity, Temperature, Depth) Profiler through the surface layers, recording these values very rapidly, perhaps every second during the descent, or every metre or so of depth. Samples of water at different depths can be taken for more rigorous chemical analysis on board. At other times during the day, an XBT (Expendable Bathy-thermograph) may be used, falling through the surface layers while the vessel sails on. It usually records temperature only, although salinity can be assessed also by some systems. More complex computer-controlled towed systems are increasingly coming into use.

Then oceanographers calculate the mean density of the water column above a standard depth (or pressure), which may be deemed to be a depth of no movement and so considered to be a level datum. Although such an assumption is difficult to make, its aim is to ensure that no significant gradient in pressure or current or whatever is present. Based on the mean density, and with a similar method to that used for the east coast eddies, it is possible to calculate the height of the column based upon its average density, above the arbitrary level. When this is done for many points over our real oceans, a raised convergence zone does emerge.

More recently, evidence has been accumulated by a second means — satellites, the TOPEX/ Poseidon satellite being one example — which are capable of very accurate measurements (of the order of a centimetre or two) of the distance between the satellite in orbit and the Earth's surface. There can now be no doubt that the convergence belts do exist, as Figure 5.8 suggests.

Having established that surface topography does exist, let's now consider the consequent generation of geostrophic flow from this topography and its contribution to oceanic circulation or current gyres.

Figure 5.7 demonstrates that the topography generated by Ekman transport (Figure 5.6) is a perfect scenario for the establishment of geostrophic flow, in the way already demonstrated for warm core eddies of about 100 kilometres' diameter. Here we are dealing with much larger scales, perhaps twenty times larger. The end result is a current gyre of the type actually witnessed in the oceans. There is western intensification due to processes too complex to treat here, and of course there is some influence from the coasts in the real ocean, especially in the west, near the equator, where there is evidence of a pile-up in response to the wind flow, and where the Coriolis force is very small.

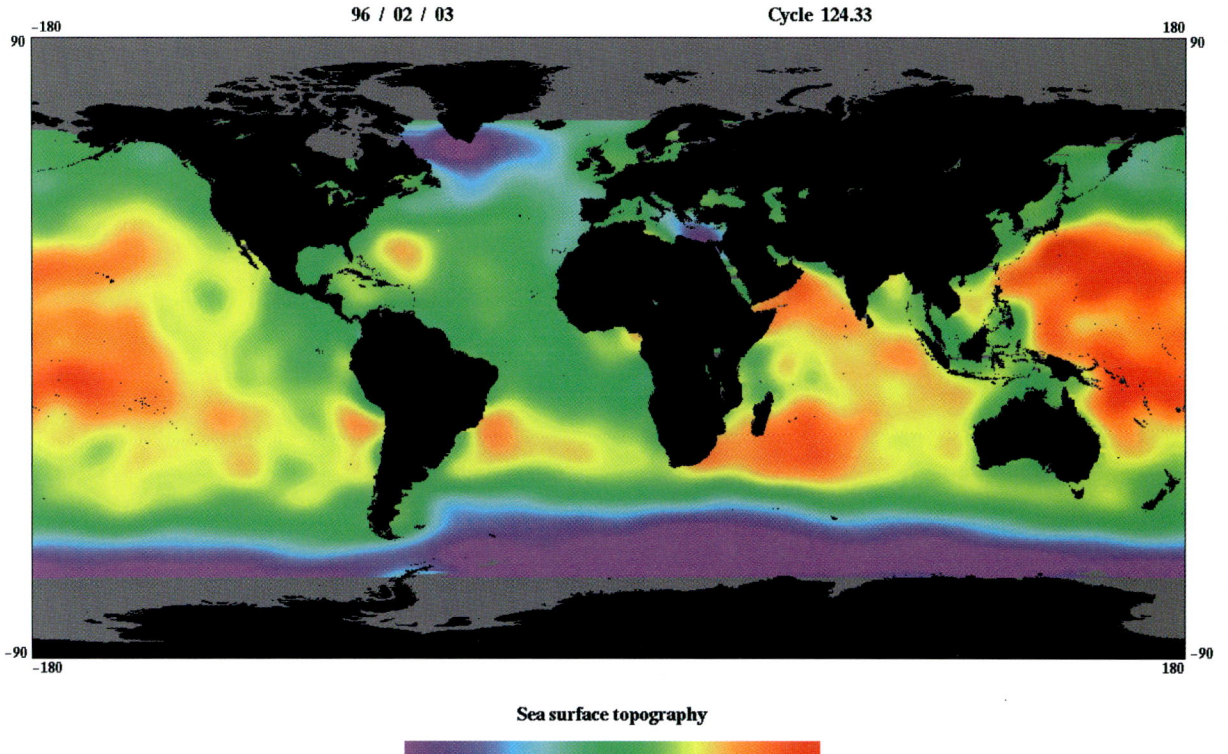

Sea surface topography

Centimeters

Figure 5.8

An example of a satellite derived topographic map of the oceans (all heights in cm). (Data courtesy Center for Space Research, University of Texas, Austin. Produced at JPL, with software developed by JPL and the University of Colorado, Boulder..)

One final comment on Figure 5.7: many features that respond to geostrophic and Ekman principles, especially those of small scale, have a limited life and disperse fairly quickly. The upwelling/downwelling events described earlier for South Australia depend upon transitory winds and 'struggle' to reach maturity within a week or so before the favourable conditions disappear. Geostrophic flow in many cases slowly erodes the incline that feeds it. But what is special about the ocean-scale circulation processes is that Ekman transport and geostrophic flow, when working in combination, are virtually mutually supportive, so lending permanence to the mechanism. The incline is continually refreshed by the convergent action of the two major wind systems, the Trades and the Westerlies, under Ekman conditions. Meanwhile, the resultant geostrophic flow does not inhibit the Ekman process.

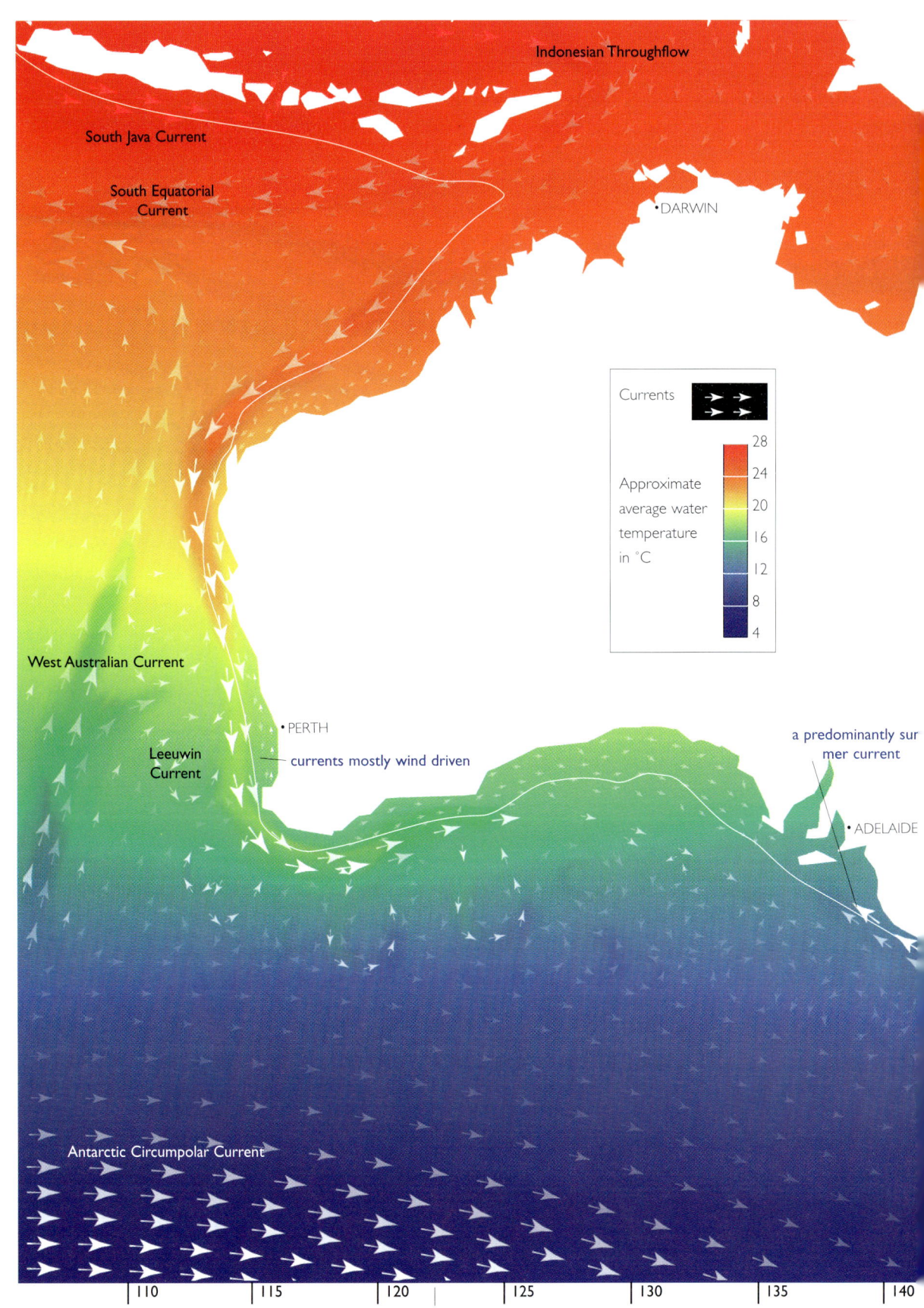

Figure 5.9

Typical patterns of the major surface currents in the Australian region.

(Based on data courtesy Dr George Cresswell, CSIRO Division of Oceanography, Hobart.)

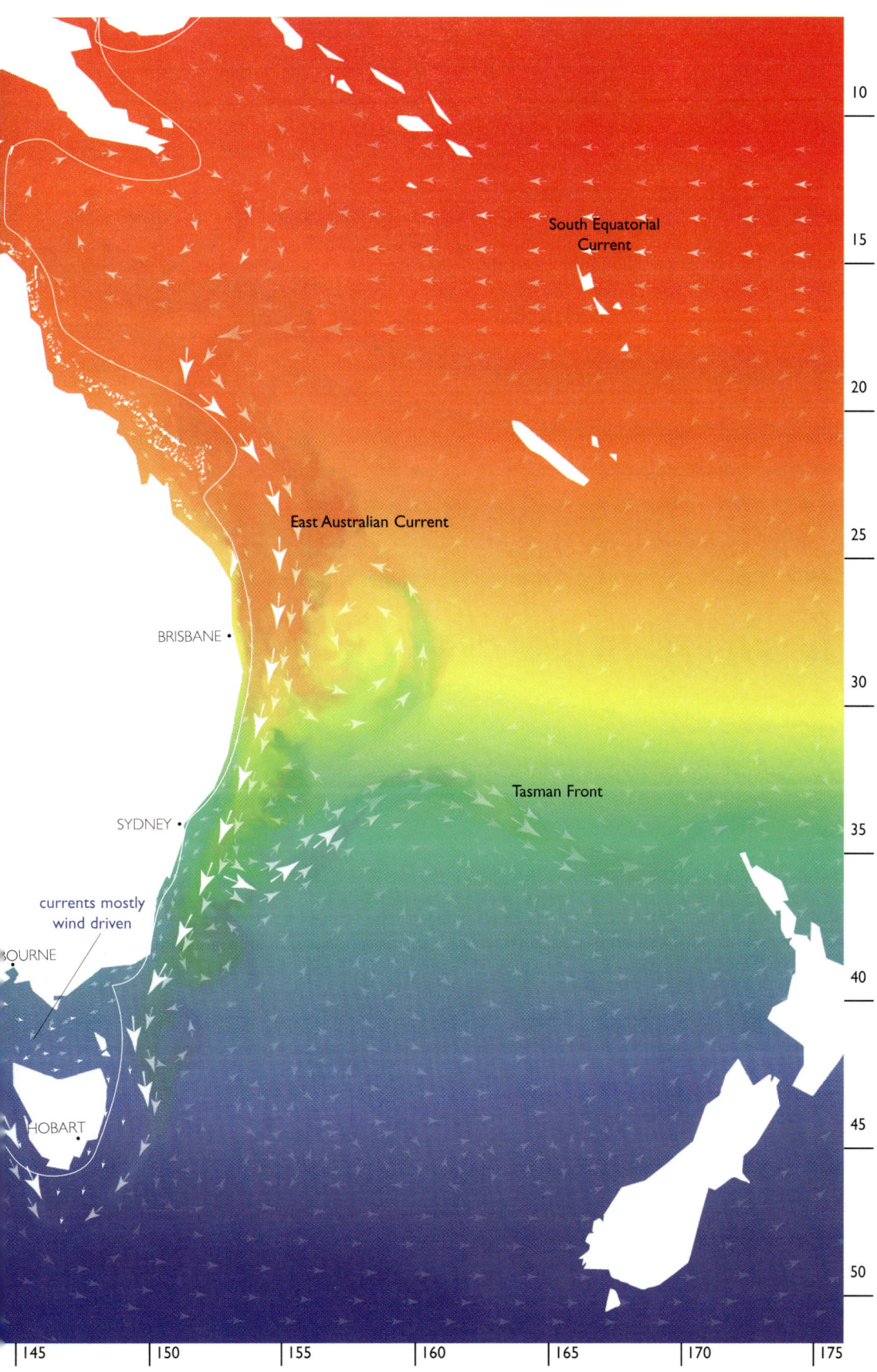

South Equatorial
Current

East Australian Current

BRISBANE •

Tasman Front

SYDNEY •

currents mostly
wind driven

OURNE

HOBART

10

15

20

25

30

35

40

45

50

145   150   155   160   165   170   175

MAJOR CURRENTS IN THE AUSTRALIAN REGION

At first glance, Australia may be considered to be influenced by four major currents: the South Equatorial Current, the East Australian Current (EAC), the Leeuwin Current and the West Australian Current. These are very large-scale, with properties that have been averaged over great distances and long periods of time. Hence, on any given day or at any particular place — especially if that place happens to be near the coast — the current may or may not be present locally, or may travel in a completely different direction, and may move faster or slower than the average, depending mainly on the wind.

Persistent patterns

Figure 5.9 illustrates temperatures and currents at the sea surface around Australia. Warm temperatures are red, ranging through yellow to deep blue for cool temperatures. Large arrows show where currents are both strong and reasonably persistent. Small, less prominent arrows show where the currents are weak, and variability is indicated by groups of randomly directed small arrows.

In the western South Pacific Ocean, the South Equatorial Current brings warm water toward north-eastern Australia and then splits: part goes north and then around Papua New Guinea; and part feeds in to the strong East Australian Current (EAC), which can flow at 4 knots or more. The EAC can reach south to Tasmania in summer and can spawn several anticlockwise eddies along the south-east Australian seaboard en route. These may also have current speeds of 3 knots. They tend to have diameters of up to 100 kilometres or even more and rotation times of 5 to 10 days at their circumference, with lesser times at their centres, and can keep rotating for more than eighteen months.

Part of the EAC also follows a meandering path east to pass to the north of New Zealand. The separation between the warm Coral Sea and the cooler Tasman Sea is known as the Tasman Front.

Water from the equatorial Pacific Ocean reaches the Indian Ocean via the Indonesian throughflow. This transfer of heat and water is believed to play a key role in climatic variability, in that it is linked with the El Niño events. It is also probably important in transporting the larvae of marine creatures. The throughflow certainly feeds the South Equatorial Current of the Indian Ocean and it plays some role in driving the Leeuwin Current. That current takes warm water south to Cape Leeuwin and then east toward the Great Australian Bight. Along the edge of the continental shelf, south of Western Australia, the Leeuwin Current can reach speeds of 3.5 knots. Varying amounts of water from the West Australian Current are entrained by the Leeuwin Current, and the summer Leeuwin Current carries more of the sub-tropical West Australian Current waters than the winter one.

The Leeuwin Current meanders and 'spins off' eddies all along its path. It is thought that the current off the west coast of Tasmania may be an extension of the Leeuwin Current.

In addition to these four major currents, Australia is also affected by the South Java Current, marked by its rapid seasonal variability, and the Antarctic Circumpolar Current, a broad flow with embedded eddies that is deflected southward by bottom topography south of the Tasman Sea. Unlike other current systems, which are near-surface features, the Antarctic Circumpolar Current reaches down to the ocean bed.

Satellite imagery of Australian currents

Images taken from satellites provide a much wider field of view than Earth-bound detection techniques; they can also provide infra-red 'heat' or thermal images which show surface temperatures over large areas.

Figure 5.11 shows one of these for the Leeuwin Current in June 1984. It clearly shows a tongue of warm water extending down to Geographe Bay and a fairly typical warm-water eddy to the north-west of Perth. (This should also remind you of realistic dimensions and patterns of a major current system. If you refer back to Figure 5.5 — showing inferred currents from patterns of sea surface temperatures — you will appreciate that, even in the relatively small area covered in Figure 5.10, there are likely to be quite complicated patterns of currents.) Figure 5.10 shows an equivalent image for the east coast of Australia.

Figures 5.12 to 5.15 show seasonal sea

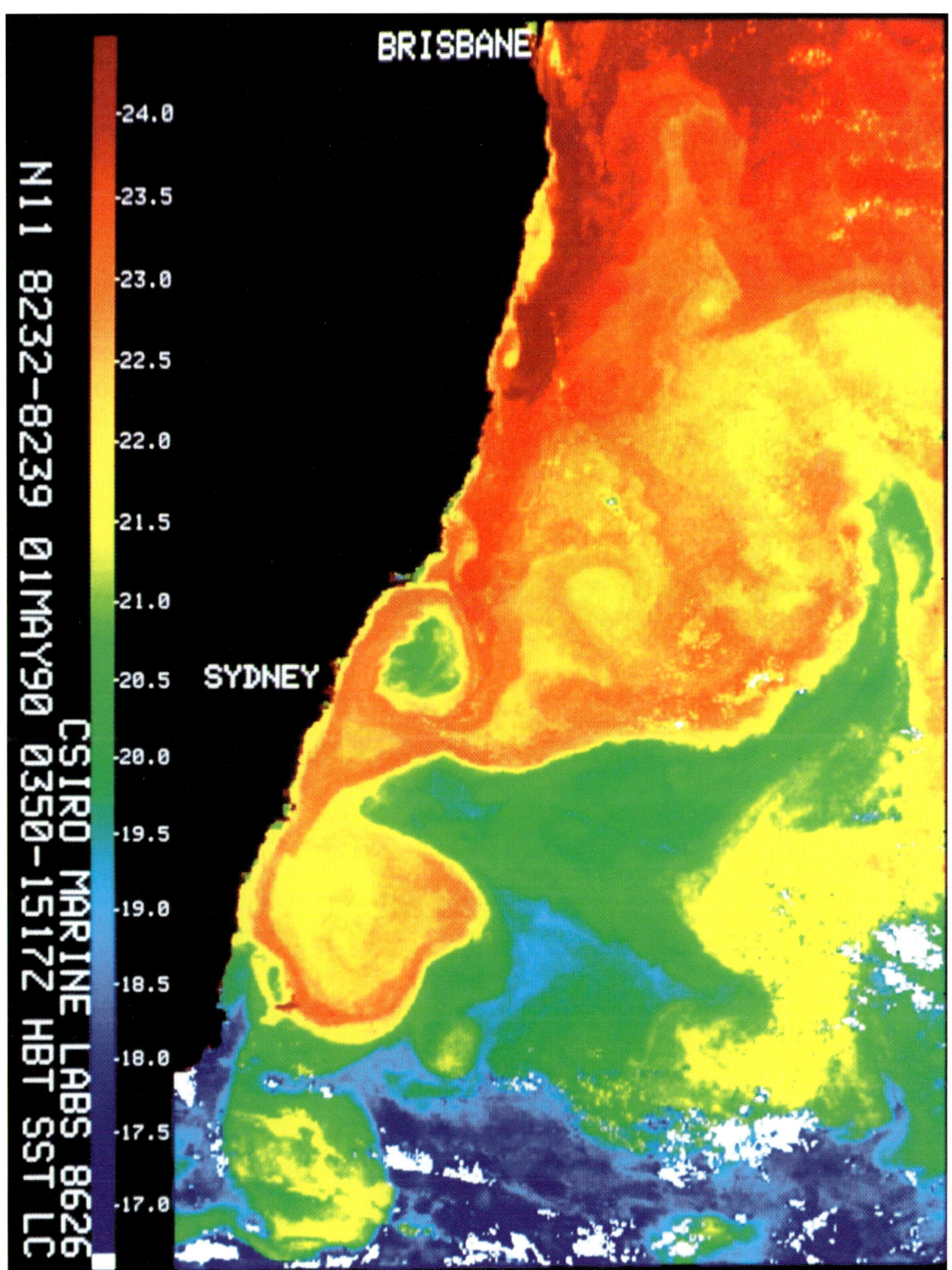

Figure 5.10
Thermal satellite image showing the East Australian Current in May 1990 (temperature in °C). The current is shown by warmer sea surface temperatures progressing south along the coast of Queensland and New South Wales. (Image courtesy CSIRO Marine Laboratories, Hobart, from data supplied by Curtin University of Technology.)

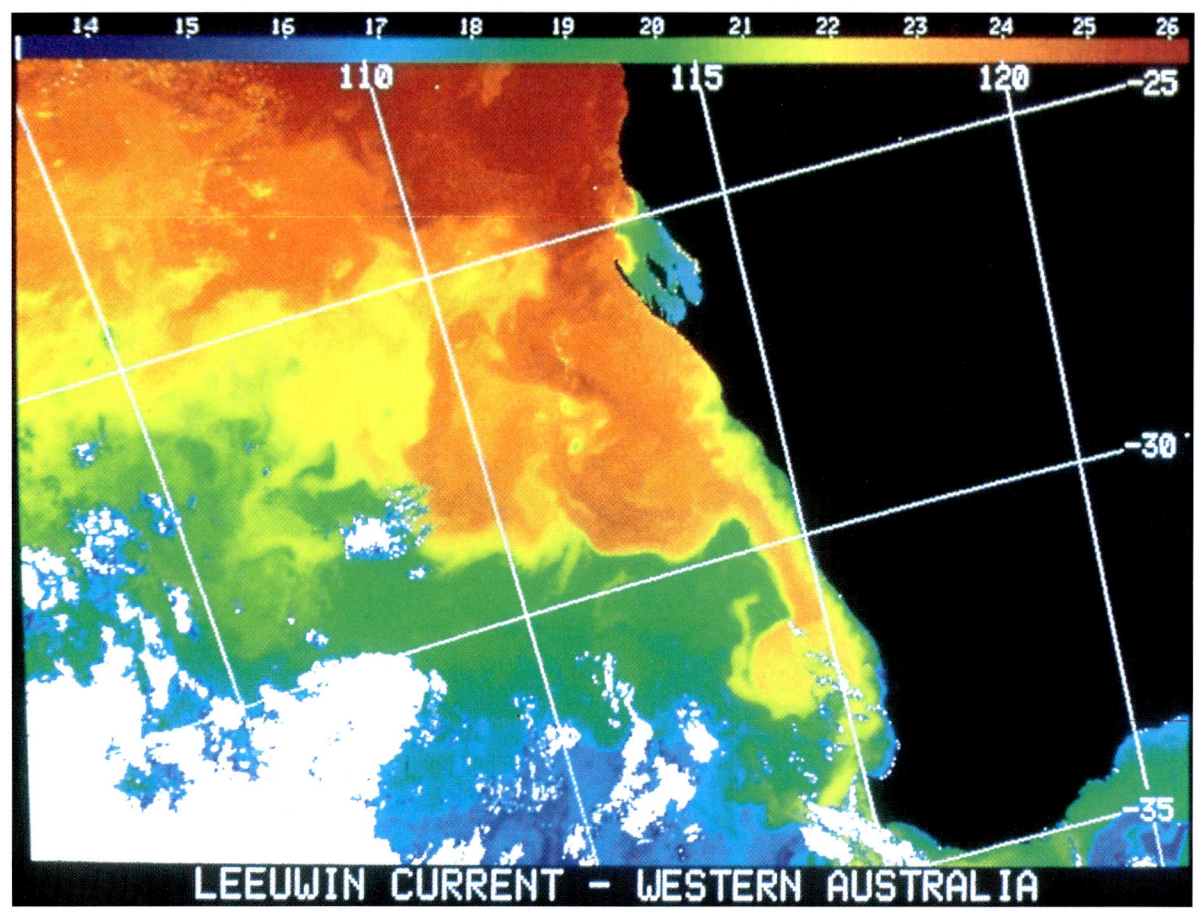

Figure 5.11

Thermal satellite image showing the Leeuwin Current in June 1984 (temperature in °C). The current is shown by warmer sea surface temperatures progressing south along the south-west coast of Western Australia. (Image courtesy CSIRO Marine Laboratories, Hobart, from data supplied by Curtin University of Technology.)

surface temperatures which have been estimated from thermal satellite imagery over a seven year period. Because these have been produced by averaging techniques, they include the net (broad scale) effect of currents and other longer term influences.

There is a good deal of information in these diagrams, which may be useful for a range of purposes. For example, swimmers seeking relatively warm waters will be interested to see that although Sydney and Perth are only 2° apart in latitude, the July water temperatures favour Perth by several degrees (presumably due to the Leeuwin Current); in summer, the average water temperatures are similar. At the cooler end of the scale, south-west Tasmanian waters range from a delightful 14°C in January down to a coolish 10°C in July — definitely wetsuit time. The winter waters off Sydney are not so fearsome; an unprotected swimmer would face danger after immersion of about 1.5 hours, assuming water temperatures around 15 to 16°C.

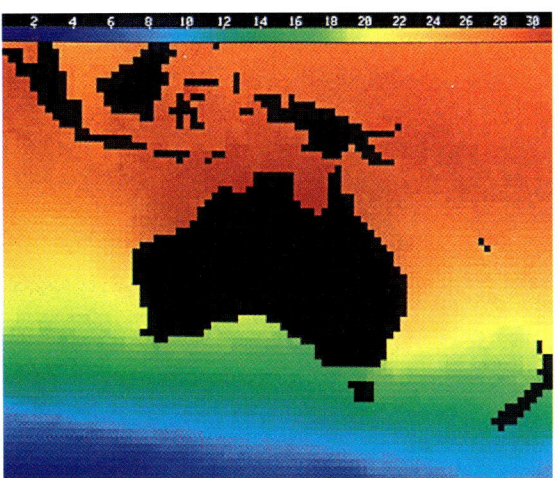

Figure 5.12
Average sea surface temperatures (°C) in the Australian region — January. (Image courtesy CSIRO Marine Laboratories, Hobart.)

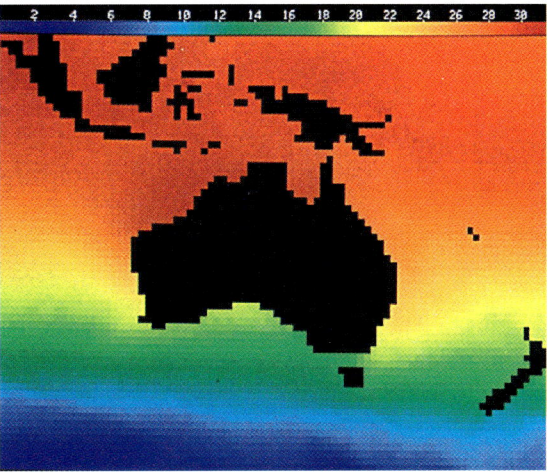

Figure 5.13
Average sea surface temperatures (°C) in the Australian region — April. (Image courtesy CSIRO Marine Laboratories, Hobart.)

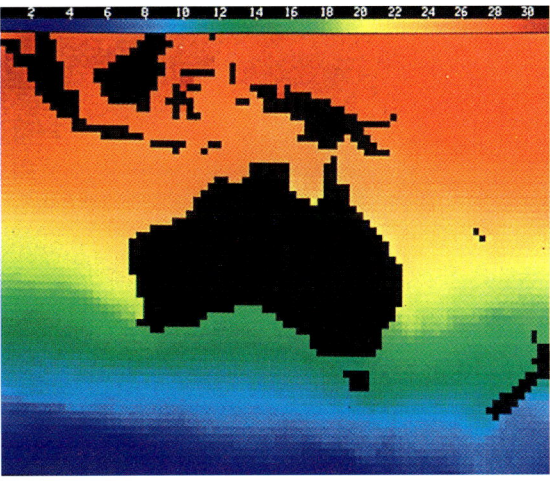

Figure 5.14
Average sea surface temperatures (°C) in the Australian region — July. (Image courtesy CSIRO Marine Laboratories, Hobart.)

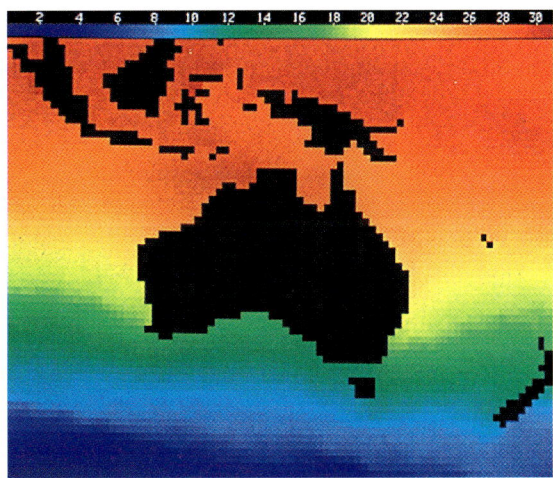

Figure 5.15
Average sea surface temperatures (°C) in the Australian region — October. (Image courtesy CSIRO Marine Laboratories, Hobart.)

TIDES IN THE AUSTRALIAN REGION

The tide is one of the most important examples of water movement and mixing, especially in coastal zones. It causes sea levels to follow complex oscillations, in some places through a range of several metres. These tidal movements can cause strong currents — ebbing and flowing in restricted channels — and circulating currents in open water.

In a practical sense, the tide can determine where, or whether, a bulk product can be exported. Take the export of ore through certain ports in West Australia, where the amount of the cargo loaded onto the bulk carriers is governed by the estimated under-keel clearance determined by the particular tide predicted for the day of departure. Vessels leaving ports with long-dredged channels are sometimes scheduled to leave an hour or two before high water so that off-shore, at some critical distance down the channel, under-keel clearance is tide-assisted, which decreases the dredging requirements.

In Europe, ships can be subject to traffic control, similar to that which controls air traffic. The movement of large and heavily laden vessels, such as super-tankers, may be individually scheduled to take advantage of the progression of the tide through shallow regions. Rather like a surfer on a wave, such a vessel may approach its berth by matching its route to the progression of the tide, riding the incoming tidal wave.

The tides can control many aspects of port operation and of navigation, and perhaps there is no more sensitive region in this respect than Torres Strait, an international navigational highway, where predictions of tidal currents (sometimes called 'tidal streams') often exceed seven knots. Tides also are important to the recreational sailor: they may determine whether a trailer-sailer can be launched or recovered; and tidal streams at sea can strongly influence the speed of all vessels and crafts.

Much of the energy for the mixing of coastal waters comes from the tides, and so the general health of many coastal inlets is dependent on them. The South Australian gulfs — Spencer Gulf and Gulf St Vincent — are highly dependent upon the unique spring–neap cycle there, which offers a still period at neaps (this is sometimes referred to as the 'Dodge' or 'Dodging Tide'). This temporary cessation of the tide, roughly at fortnightly intervals, allows stratification to occur in the water column. This in turn promotes bottom gravity currents, which transport an enormous amount of salt — the residual product of the high evaporation in that region — out to the Southern Ocean. Without such replenishment the gulfs would die.

Precise predictions of the tide are required for coastal zone management and infrastructure development in the coastal zone and for all users of the sea. In most countries these are prepared by an official agency, in association with the national hydrographic service. In Australia, tidal predictions are prepared, generally two years in advance, by the National Tidal Facility at Flinders University in Adelaide.

A simplified explanation of tide-producing forces

Our understanding of tidal physics owes much to Sir Isaac Newton. In 1687, he developed a theory of celestial mechanics in a treatise — *Philosophiae Naturalis Principia Mathematica* — occupying more than a thousand pages. A century after Newton, a French mathematician and astronomer, the Marquis de Laplace, developed Newton's work, but it was not until the nineteenth century was well underway that the early pioneers were confident enough to commit these theories to a practical service for mariners in the form of tide tables.

Equations are used sparingly in this book — they are included where they are straightforward and can be used for practical purposes. In the case of tide-generation forces, the most simplified mathematical explanation would involve complicated equations and, even so, would have almost no predictive ability. One reason for this is that the Earth, Moon and Sun (the relevant gravitational bodies) move in complex orbits which produce subtle and ever-changing resultant forces. These forces interact with each other, and with the water bodies they are influencing, and so change in space and in time, producing tidal patterns with periodicities ranging from hours to tens, even thousands, of years — and, importantly, patterns which will never be repeated exactly.

The concept of the equilibrium tide

'Equilibrium tide' is the name given to an idealised circumstance which would occur on an Earth entirely covered by a deep ocean, in the presence of the Moon and Sun as we know them. The response of the ocean to the changing gravitational fields would be complete and virtually immediate. Some books show this as a diagram of the Earth, under the influence of the Moon, displaying two bulging 'cheeks' — one on the side facing the Moon and the other on the opposite side. A point on the Earth's surface will tend to experience two high waters per day as the Earth rotates once per day through these deformations.

It is not hard to understand why there is a cheek on the side facing the Moon, but why is there one on the opposite side as well? The answer lies in the *difference* in the attraction from one place on the Earth's surface to another. It is convenient to associate the tide-generating force at a point with the difference between the attractive force at that point and the *average* force experienced by the Earth body as a whole, which may be assumed to be that at its centre. The latter has more relevance to the stability in motion of the Earth/Moon system, while it is the spatial variation in the attractive force that causes one particle of water to move relatively to another, and so causes the tides.

Let's explain this 'difference' using realistic numbers. Using Newton's principles, water at a point directly 'under' the Moon might experience an attraction toward the Moon of 0.000,003,493 grams (a very small force), whereas the one on the opposite side might experience a slightly smaller attraction of 0.000,003,268 grams, in the same direction towards the Moon. Attraction on a point at the centre of the Earth might have a value of 0.000,003,378 grams (also directed toward the Moon).

However, if the two surface forces are expressed relative to that experienced at the centre of the Earth, the results are quite different; the point nearest the Moon experiences a force directed towards the Moon of 0.000,000,115 grams, whereas the other point experiences a force directed *away* from the Earth's centre and the Moon of 0.000,000,110 grams.

Of course this argument is a gross simplification of the actual mechanism which generates the tides. In the first place, it considers only the effect of the Moon's attraction. In fact the Sun also is responsible for tide-generating forces although, due to its greater distance from the Earth, its effect is only half as strong as that of the Moon despite its greater mass. Then again the above argument considers only a simple rotating Earth in an otherwise static system, with the Moon notionally fixed above the equator. Quite major complications arise from the inclination of the Earth's axis of rotation to the orbit of the Moon, and also to the plane of the ecliptic which contains the Earth's orbit around the Sun. The interaction between these features causes the tide-producing bodies (Moon and Sun) apparently to move north and south of the equator. Then, since the relevant orbits are typically elliptic rather than circular, the distance of both Sun and Moon from the Earth changes periodically in convoluted sequences. Nevertheless the 'cheeks' argument is a suitable starting point for a study of the tides.

Geographic patterns according to Laplace

Laplace was able to improve scientists' understanding of tidal variations, and to identify three basic groups (in tidal literature often referred to as 'species') of tidal characteristics:
- a *semidiurnal* group tending to give two tides per day (as per the cheeks analogy)
- a *diurnal* group tending to give one tide per day (the result of declination of both the Sun or the Moon)
- a *long-period* group tending to give tidal oscillations with periods of months, years or even longer; for example, there is an important tidal oscillation that has a period of 18.61 years.

Figure 5.16 shows the geographical form of these three Laplace tidal groups. When considering this diagram remember the geographic distribution of the species applies to both solar and lunar tides.

An immediate impression is given of the geographical disposition of the three groups of equilibrium tide. One can imagine the Earth

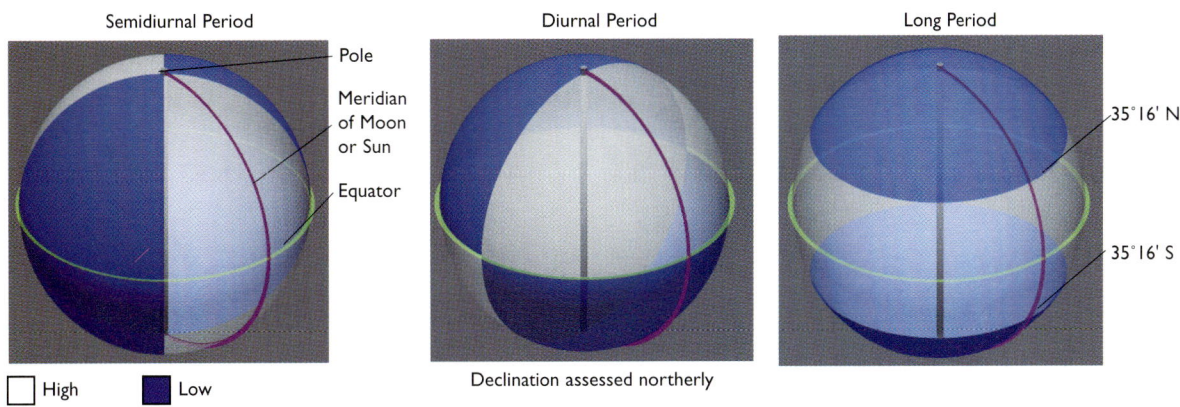

Figure 5.16
A geographical representation of the three tidal groups according to Laplace. (Based on original figure by G. Lennon.)

rotating through the patterns of each, which are locked in to the direction of the lines of centres, Moon/Earth and Sun/Earth. The number of tides that occur in a day in any particular place depends on the relative magnitude and the interaction of these types of tides — but, as hinted earlier, there is an almost endless array of possibilities.

Real world patterns

Actual tides experienced on the real Earth are very much more complex than might be suggested from Figure 5.16. Table 5.1 gives the tides, for a randomly chosen date, for six Australian locations as published in the official tide tables. The time differences of the morning high water, relative to the time of high water at Sydney, are compared with the differences in longitude. The table arranges the ports in order of longitude. If the tides were to arise from the rotation of the Earth through virtually static systems, as given in the 'cheeks' analogy, or even in the Laplace development, then one would expect, say, high tide to occur first at Brisbane followed by the other five ports in order of longitude.

The table shows that there is no such ordered sequence in the arrival times of the tide. In addition, the range in longitude is less than 15° (the rotation of the Earth in one hour), which suggests that the tide in all six places should occur within one hour. In fact the time span is nearly five hours! Clearly this evidence suggests that the equilibrium tide does not operate in the real world in so far as tidal levels are concerned. There are four main reasons for this:

- The continents impede the progression of an equilibrium tide through the oceans.
- Tides behave as shallow-water waves even in the deep ocean since their wavelength is much greater than twenty times the ocean depth (20 × 4 km). The equilibrium semidiurnal tide would have a wavelength equivalent to half the Earth's circumference, or approximately 20,000 km.

Note the velocity of a shallow-water wave is given by $v = \sqrt{gd}$. This gives a value for v of approximately 700 km/h. In order to keep pace with the rotation of the Earth, in low latitudes the semidiurnal tide would need to progress through 20,000 kilometres in 12 hours, approximately 2.4 times faster than it is able. The ocean would need to be at least 22 kilometres deep for this to be possible!

- The Coriolis force would also deflect the Equilibrium Tide from its preferred east–west path.
- Oceans have variable bottom topography and comprise many basins and partially enclosed gulfs, which possess individual natural periods of oscillation when disturbed.

All water bodies in the ocean are exposed to a sequence of tide-producing forces — semidiurnal, diurnal and long period. The forces pass through the water at the rate of progression of the equilibrium force field as the Earth rotates but the water fails to match this progression due to its speed limit — water near the equator will lag further behind the required rate than in polar regions. In the circumpolar canal of the Southern Ocean, the water moves virtually at the rate of the forces, but then there are other factors influencing its progress, notably the constriction of Drake Passage at the tip of South America. This, and the other factors mentioned above, conspire to complicate the issue, as shown in Figure 5.17!

Table 5.1
Time of morning high tide for 1 January 1994 for selected sites in eastern Australia.

Location	Longitude in degrees east	Difference in longitude from Sydney	Time of high tide (EST) for 1 Jan 1994	Difference in time from Sydney (minutes)
Brisbane	153.2	2.0	1144	+83
Sydney	151.2	0	1021	0
Hobart	147.3	-3.9	0944	-37
Cairns	145.8	-5.4	1123	+62
Melbourne	144.9	-6.3	0535	-286
Port Adelaide	138.5	-12.7	0652	-209

So why is the concept of the equilibrium tide relevant?

Although the world ocean cannot, for the reasons stated, respond to the vertical attraction of the equilibrium tide in an immediate sense, nevertheless a knowledge of the equilibrium tide is essential for all tidal calculations since:

- It still defines the forces, notably horizontal forces, that maintain the tidal waters in motion and which are used in tidal studies;
- A detailed analysis of the equilibrium tide identifies the hundred plus oscillations which describe the relative motions of the Earth, Moon and Sun and so the frequencies in which the tidal analyst must search for an ocean response; and
- It does provide a comprehensive time frame within which to relate the individual components of the tidal picture, based as they are partly in solar time as used in domestic life, and partly in lunar time based upon the Earth's rotation with respect to the Moon, whose time systems are mutually incompatible.

If one imagines that a particular location, at latitude 30° south, occupies the position in the central lower diagram from which the arrows radiate, then at the time when the Moon, at declination 15° south, transits (crosses the meridian of the location), the horizontal force applied to the surface waters is directed northwards and is represented by the blue arrow marked '0'. One lunar hour later, the direction of the force has been deflected slightly to the west of north as the blue arrow marked '1' suggests, and so on through the lunar day.

Note that a lunar hour is one twenty-fourth of the time taken for the Earth to rotate through one revolution with respect to the Moon. The latter takes approximately 24.8412025 solar or clock hours (24 hours 50 minutes and 28.33 seconds) so that a lunar hour is approximately 1 hour 2 minutes and 6.2 seconds of clock time.

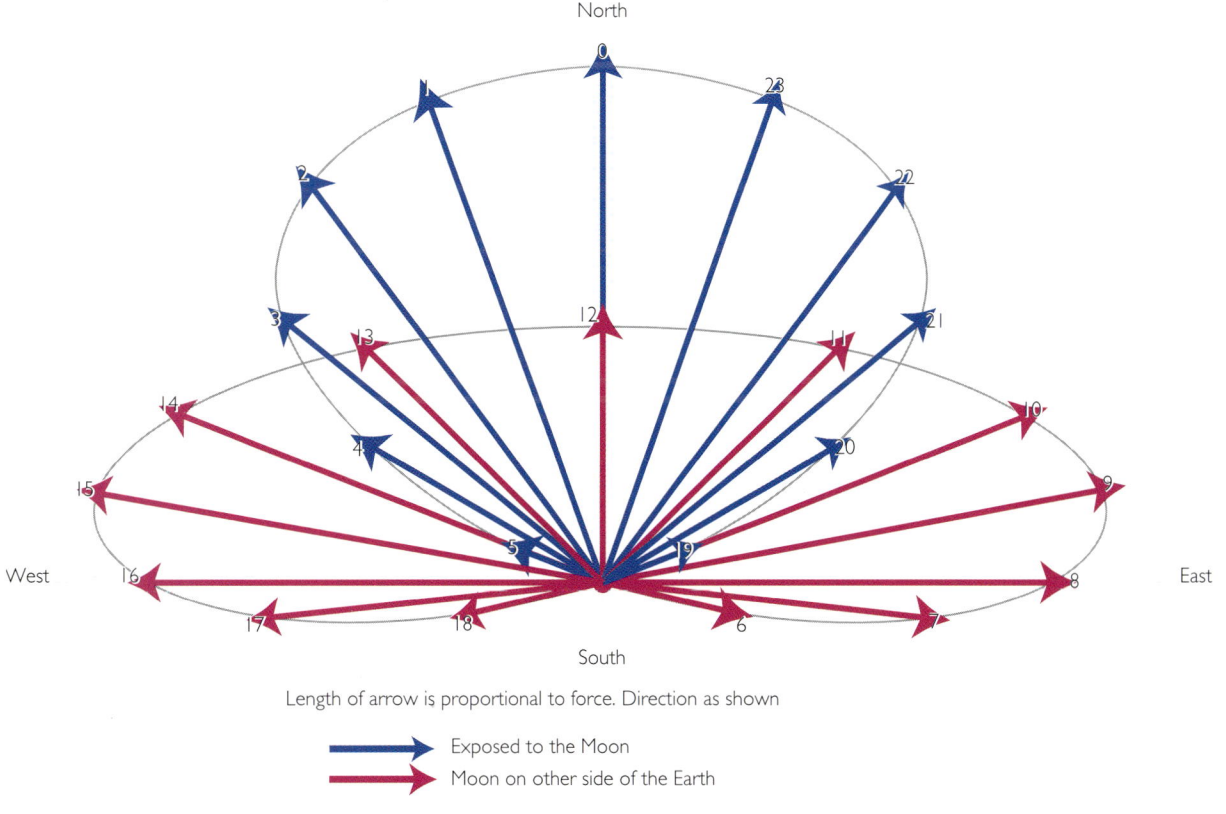

Length of arrow is proportional to force. Direction as shown

→ Exposed to the Moon
→ Moon on other side of the Earth

Figure 5.17
A typical cycle of lunar tractive forces which would be experienced over the course of one lunar day, in the open ocean, at latitude 30°S when the Moon has a south declination of 15°. (Based on original figure by G. Lennon.)

The blue arrows, 0 to 5 and later 19 to 23, represent the horizontal tide-generating forces applied to the ocean waters, shown at lunar hourly intervals, in that part of the lunar day when the Moon lies above the horizon as viewed from the location. The red arrows numbered 6 to 18 represent the forces that would occur when the Moon is below the horizon on the opposite side of the Earth from the location.

The length of the arrows is scaled to represent the relative magnitude of the forces concerned. The fact that there are two cycles (the blue and the red) which occur within the lunar day, indicates the presence of the semidiurnal tide. The difference between the blue cycle and the red cycle (e.g. that the number 12 arrow is smaller than the number 0 arrow) represents the influence of the diurnal tide.

The location will cross the line of centres of Earth and Moon at the time represented by the arrows numbered 0 and 12, but at '12' the Moon will be directly opposite, on the other side of the Earth. A high tide will be experienced at both times, representing the semidiurnal tidal influence, but the tides will be different in magnitude and character — shown by the difference in magnitude and direction of the arrows in the two sets. This difference indicates the influence of the diurnal tide; for example, the arrow for time 0 is longer than that for time 0 + 12 hours.

As one day gives way to another, the magnitude of the lunar force will change as the distance of the Moon changes due to the ellipticity of its orbit, and also, in the longer term, as its declination changes. Furthermore, a sequence of solar forces will also be in operation at the same location. As soon as the waters respond to these forces by movement, they will come under the influence of a slightly *different* set of forces, appropriate to their new position as the Earth rotates.

The influence of resonance

Resonance is a secondary feature of the physics of wave progression and applies to all water bodies. In an enclosed water body, such as a domestic bath or a swimming pool, there is a relationship between the length (l in metres) and depth (d in metres), acceleration due to gravity ($g \sim 9.8$ ms^{-2}) of the water body, and the period (T in seconds) of the resonant wave which, according to Merian's Formula, is defined as:

$$T = \frac{2l}{\sqrt{gd}}$$

A 2 metre long bath tub filled to a depth of 0.3 metre would have a period of resonance of about 2 seconds. This means that, if you were to disturb the bath water at one end, by repetitively plunging something into it, you would see that a very slow repetition would generate little response. Rapid repetition might create a lot of spray but still no great response. However, as the period of plunging approached the critical 2 seconds, a large standing wave would be generated, alternately high at one end and low at the other end, and then reversing. In the centre of the bath there would be little change of level. This standing wave is sometimes known as a 'half-wave oscillation'. An Olympic swimming pool 50 metres long and 2 metres deep might have a period of 22 seconds, but then one would need a very large plunger in order to conduct an experiment! The question of greater relevance here is 'What about the deep oceans?'

Suppose we take the semidiurnal lunar tide in the deep ocean and use it as the 'plunging factor' in the resonance equation, then examine what the characteristic length (l) might be. We fix the period at, say, 12 hours 25 minutes, or 44,700 seconds (24 hours 50 minutes is a lunar day), half the time for the Moon to reach the same spot on the horizon on successive days. The ocean's depth of, say, 4000 metres means that an enclosed basin in the deep ocean (the equivalent of the bath in the example above) would need to be approximately 4500 kilometres long to 'match' the 12 hour 25 minute period. For shallower seas, say 1000 metres deep, the basin or gulf would need to be half that, at about 2250 kilometres, and at continental shelf depths of 200 metres the length would be about 1000 kilometres.

Figure 5.18 reproduces the results of an early computer model of the major lunar constituent of the semidiurnal tide.

The diagram is provided not for its accuracy in reproducing the tides or its geography, but to demonstrate the *general* form of the tide in the deep oceans. It shows that the ocean responds to the tide-generating forces by subdivision into cells, called 'amphidromes', with an 'amphidromic point' in the centre. In each cell the tide rotates around this point, the tidal amplitude being zero at the point itself and growing with distance from it.

The lines join locations where the semidiurnal lunar tide occurs at the same time. These lines are marked by numbers which indicate the number of lunar hours which have elapsed since the Moon crossed the Greenwich Meridian. Note that one lunar hour represents one hour, two minutes and 6.2 seconds of clock time, this being 1/24 of the time taken for the Earth to rotate through one revolution with respect to the Moon. The lines also illustrate the manner in which the world ocean subdivides into 'resonant basins' matched to the astronomical tidal forcing function of semidiurnal period.

Figure 5.18
An early computer model of the progression of the semidiurnal lunar tide (M_2) in the World ocean. (Based on Y. Accad and C.L. Pekeris, 1978.)

The visual effect in the diagram is rather like a spider's web.

Take the cell in the lower central part of the diagram: in the Southern Ocean, south of the Atlantic. The line passing north from the amphidromic point is marked 0h — which indicates that this line joins points that experience high tide when the meridian of Greenwich (near London) on the rotating Earth passes below the Moon, that is, when the Moon transits at zero longitude.

The next radial line, moving clockwise, is marked 1h, which indicates that high tide occurs here one lunar hour after the Moon's transit at Greenwich, and so on, until, after passing the 11h line, we are back to the starting point, with an elapse of 12 lunar hours from initial transit (the equivalent 'solar' time would be 12 hours 25 minutes and 14 seconds). Greenwich is now experiencing lunar transit again, but, if the first transit was an upper transit (that is, if the Moon was directly overhead the Greenwich meridian at that time), the second transit will be a lower transit

(that is, the Moon will be directly underneath on the other side of the Earth, longitude 180°). Given that the tidal elevation is zero at the amphidromic point, increasing outwards, the tidal cells operate rather like the legendary storm in a teacup.

The tide tends to rotate clockwise around an amphidrome in the Southern Hemisphere, and anticlockwise in the Northern Hemisphere, so that the slope is on the left of the direction of progression in the Southern Hemisphere and vice versa. Consequently Coriolis principles are being observed in our geostrophic world. The scale of the cells, where unimpeded by the continental boundaries or bottom topography, is generally consistent with the scale lengths computed by Merian's Formula for depths of 4000 metres, namely 4470 kilometres.

It seems as though the ocean has 'protested' to the need to respond directly to the tide-generating forces, and consequently has divided itself up into cells that do approach the appropriate scale sizes. The cells then behave virtually as closed, but interacting, basins. It should be remembered that this diagram shows tidal effects in response to a single, but major, constituent of the lunar semidiurnal tide.

There are several constituents of the lunar semidiurnal tide and, for accuracy, these would also need to be considered, as would the solar influences. After this it is necessary to treat the diurnal tide in similar detail, followed by the long-period forces. The reproduction of the tide for the benefit of the mariner is truly a battle with complexity!

Tidal amplification in shallow gulfs

Before leaving resonance, it is useful to comment on features that affect the coastal zone rather than the deep ocean. Coastal bays and gulfs are small compared with the ocean. For example, the average depth of water may be as shallow as 40 metres. Merian's Formula indicates that, for the semidiurnal tide, the critical resonant length of gulf that will produce a resonant effect will be about 500 kilometres, while for the diurnal tide it will be twice this length. But bays and gulfs are not closed basins. They are open to the ocean at one end and often are smaller than the 500 to 1000 kilometres estimated above for closed basins. For a gulf that is open at one end, there is a modification to Merian's Formula, which defines its critical resonant length, l':

$$T = \frac{4\,l'}{\sqrt{gd}}$$

An open-ended gulf will generally be too small to respond directly to tide-generating forces, but instead it will co-oscillate with whatever the open ocean provides at its entrance. In general, the ocean oscillation is enhanced and amplified in the gulf, depending upon how closely the length of the gulf approaches the critical length for resonance. When the gulf is longer than the critical length (Figure 5.19), a node can exist in the gulf where there is no oscillation. Towards the head of the gulf, the tide is amplified and is much greater than the tidal range at its mouth, but it is in opposite phase.

When the gulf is shorter than the critical length, there can still be amplification, but somewhat less than in the other case, and without a change of phase. The formula for the open-ended gulf suggests that the critical length for a semidiurnal tide is approximately 250 kilometres and

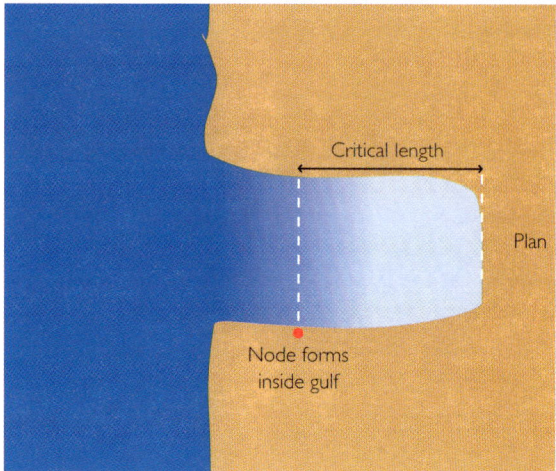

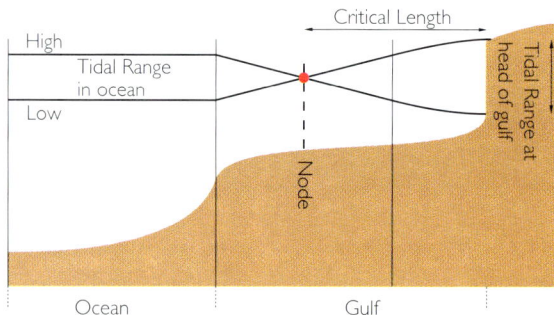

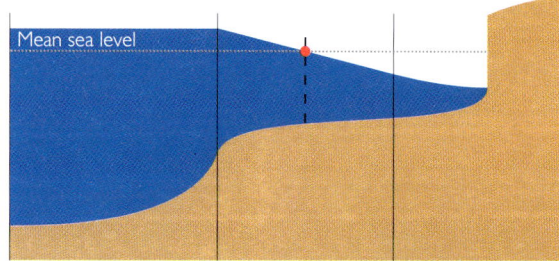

Vertical profile at time of high tide in ocean (low tide in upper gulf)

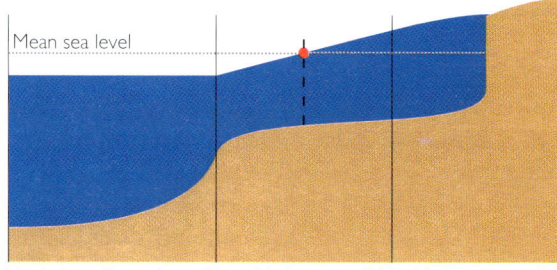

Vertical profile at time of low tide in ocean (high tide in upper gulf)

Figure 5.19
Tidal amplification in a shallow gulf open to the ocean, where the gulf is longer than Merian's critical length. (Based on original figure by G. Lennon.)

for a diurnal tide about 500 kilometres. It is not surprising, therefore, that the Gulf of Carpentaria, which has a length close to 500 kilometres, is very receptive to the diurnal tide which dominates over the semidiurnal component. A similar situation exists in Spencer Gulf in South Australia. Although still basically semidiurnal in character, Spencer Gulf is more receptive to the diurnal tide than is the adjacent and smaller Gulf St Vincent. Gulf St Vincent more closely fits the criteria for semidiurnal resonance and so enhances the semidiurnal tide.

Tide range and tide type

The range of tidal types in Australian waters is very great, arguably greater than that on the coastline of any other nation. At first glance this might seem surprising — largely because, as an island continent, Australia has relatively unimpeded exposure to deep water and it is only along its northern coastline that there is any significant area of shallow water — the condition normally associated with large tides. Again, Australia does not possess the large, wide estuaries common in the Northern Hemisphere and, apart from the Gulf of Carpentaria, the South Australian gulfs, and a few smaller features such as Shark Bay in Western Australia, there are few significant indentations of the coastline that might create unique and local modifications of the deep ocean tide.

Consistent with this geography, the tidal progression around the coast is generally slow, matching the large scale of the amphidromes of the deep ocean tide. Nevertheless, surprising anomalies occur, and some appear to be unique on a world scale.

Maximum astronomical tidal range

Figure 5.20 shows the tidal range, which is defined as the Highest Astronomical Tide (HAT) minus the Lowest Astronomical Tide (LAT). HAT minus LAT is the maximum range that can be produced by the astronomical factors described in

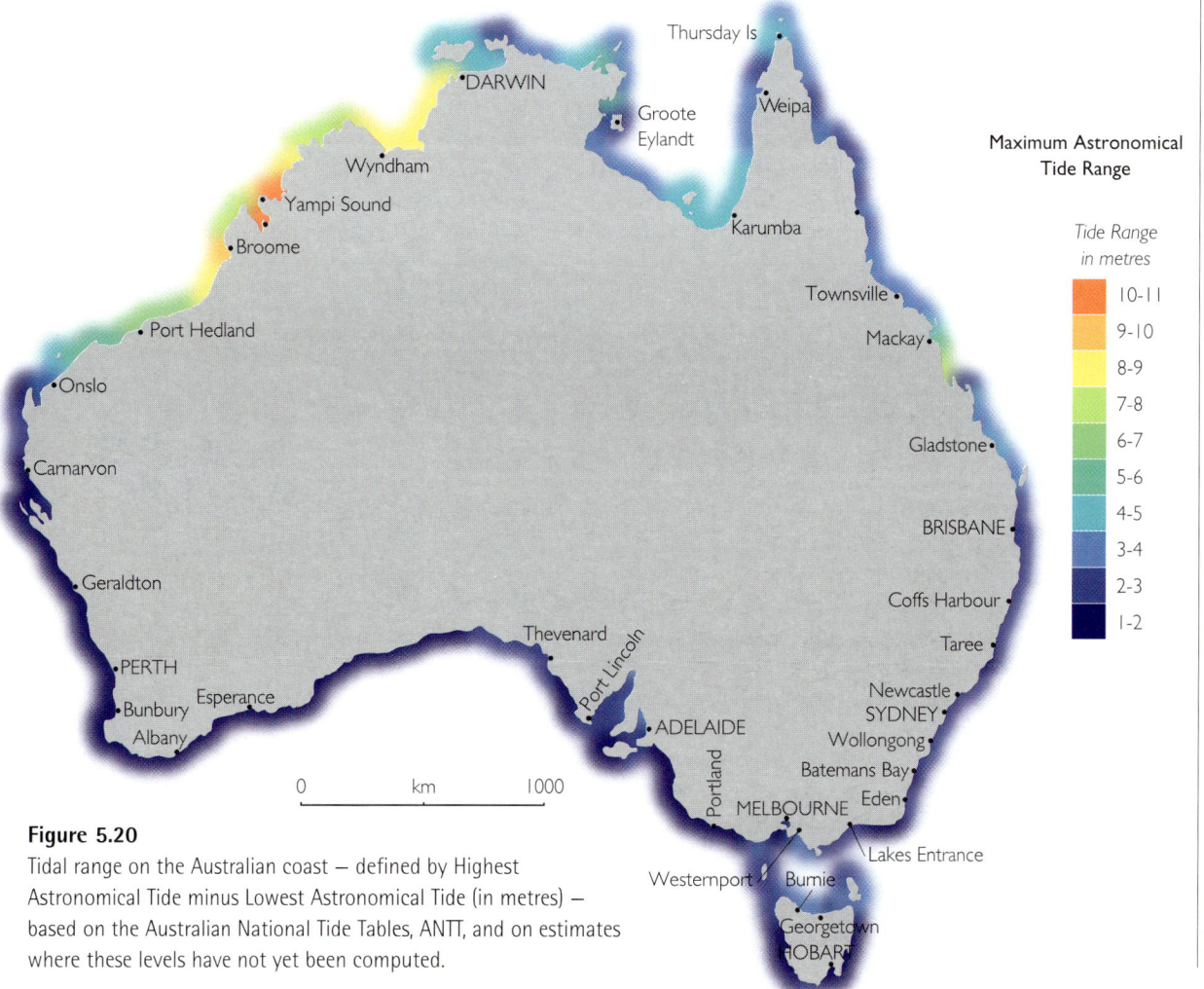

Maximum Astronomical
Tide Range

Tide Range
in metres

	10-11
	9-10
	8-9
	7-8
	6-7
	5-6
	4-5
	3-4
	2-3
	1-2

Figure 5.20
Tidal range on the Australian coast — defined by Highest Astronomical Tide minus Lowest Astronomical Tide (in metres) — based on the Australian National Tide Tables, ANTT, and on estimates where these levels have not yet been computed.

this chapter — that is, the maximum extent that the water level can rise and fall on any given day — but it *excludes* meteorological features such as storm surges. Tidal range is here based on about 80 observation sites around the coast.

Before looking at the patterns of tidal range, you might like to take a look forward to Figure 5.22, which contains a time series of observed sea level for one full year (1994). This figure has the HAT and LAT superimposed on the traces for individual stations and it is quite obvious that neither are reached all that often, even allowing for weather variables. What this shows is that, putting aside meteorological factors, the extreme highest and lowest tides are associated with relatively rare alignments of the Earth, Moon and Sun.

The datum (reference point) used for tidal measurements

If HAT for Broome, say, is given as 9.6 metres, or if the height of high water at Sydney on a particular day is quoted to be 1.7 metres, what does this mean? Above *what* datum? The convention has been to measure tidal heights above a low water datum. This in itself is apt to cause some confusion, since the datum point cannot be a level plane, but will need to change from location to location, as the tidal characteristics change. However, there is a strong practical reason for this procedure, since the hydrographer responsible for producing navigational charts uses the same datum, so that the soundings marked on the charts define the number of units (now usually metres) below this datum where the sea bed can be found. The common name for the datum is 'chart datum'. From the navigator's point of view, if tidal predictions give the expected height of sea level at a particular time to be x metres above chart datum, whereas the chart shows a sounding of y metres below chart datum, then the navigator may assume that the total depth of water available for navigation is $x + y$ metres.

However, the world community has recently been recommended to adopt LAT for chart datum, and Australia is already well advanced along this path. Consequently LAT is usually found to be listed as zero, or at least a very small quantity, and increasingly in the future will become zero.

Figure 5.22 will give coastal users a strong impression of how the tides vary in time for areas

of interest. More important, perhaps, it gives an impression of the usefulness (or otherwise) of the concept of HAT minus LAT (maximum tide range) for *characterising* a region. Some places commonly and consistently experience tide ranges that are reasonably close to the maximum (for example, Hobart), whereas others (Broome, for instance) show very strong, repeating patterns in time; a typical pattern is a number of days with small tide ranges, followed some days later by large ranges — close to the maximum. Interestingly, tide patterns like these also seem to show a strong relationship to the phases of the Moon.

Figure 5.20 shows that very large tide ranges are experienced across a broad area of coast in north-west Western Australia — for example, in Yampi Sound (10.9 m), Derby (10.5 m) and Broome (10.5 m). Large tide ranges are also found in a smaller area in central Queensland, particularly around Broad Sound (between Rockhampton and Mackay), plus a few scattered areas in the Northern Territory and in northern Queensland. Apart from these, much of the Australian coast experiences tides of less then 3 metres, with many areas only 1 to 2 metres.

Tide type (number of tides per 24 hours)

The classification of tide types is based on calculating the amplitudes (in metres) of the two major semidiurnal constituents (M_2 and S_2) and the two major diurnal constituents (K_1 and O_1), as shown in Table 5.2. The column on the far right of the table shows the semidiurnal/diurnal ratio

$$\frac{\left(M_2 + S_2\right)}{\left(K_1 + O_1\right)}$$

which is the basis of the classification of tide type; class limits are provided below.

But first we need to explain the terms: M_2, S_2, K_1 and O_1. These terms are in fact simple 'harmonic constituents' of the tide and are best explained each as a regular oscillation of sea level through a fixed amplitude specific to the location (in the form of a continuous cosine curve) and oscillating at a speed associated with a particular astronomical motion.

M_2, for example, provides the average lunar contribution to the semidiurnal tide. If we could accept that the Moon moves in a circular orbit around the Earth (at a constant distance) and always over the equator, then M_2 would describe the total lunar contribution to the tide. Other constituents are necessary to take into account the Moon's declination north and south of the equator and its changing distance from the Earth.

S_2 performs the same task for the solar tide and assumes that the Earth moves in a circular orbit around the Sun with its axis perpendicular to the orbital plane.

O_1 is the most important of the constituents which arise from the changing declination of the Moon.

K_1 is usually larger than O_1. It has a complicated origin with contributions from the changing declinations of both Moon and Sun.

In predicting the tide it is necessary to consider more than 100 constituents of this type, but these four are the most significant and together account for 70 to 80 per cent of the total tide.

The ratio	*Tide type*
> 4.0	**Semidiurnal** — predominantly semidiurnal, such as Broome;
1.5–4.0	**Semidiurnal with diurnal inequalities** — one tide usually larger than the other, such as Sydney;
0.5–1.5	**Mixed** — tidal oscillations drift between diurnal and semidiurnal characteristics in a repetitive sequence, such as Portland
< 0.5	**Diurnal** — single tide per day is the norm, such as Karumba.

Figures 5.24 and 5.25 will help you to understand the nature of harmonic terms and the manner in which they combine to reproduce the total tide.

In Table 5.2, the sixth column from the left gives the amplitude ratio of the solar to the lunar constituents: S_2/M_2. The equilibrium value of this ratio is 0.466. The column on the far right gives the ratio of the semidiurnal to the diurnal components of the tide. The equilibrium value of this ratio is 1.47. The greater the departure of these ratios from their equilibrium values, the greater is the departure of the tidal characteristics of the location from those of the equilibrium tide. In Australian waters these departures are greater than in most other regions.

The fact that in the waters of South Australia the tide due to the Sun is greater than that due to the Moon is a most unusual circumstance.

In the second case, the fact that the ratio listed in the column on the far right strays significantly from the equilibrium value of 1.47 illustrates the

power of resonance in favouring the diurnal frequencies in certain cases and the semidiurnal in others.

On an idealised Earth, that is, in the equilibrium tide, the above semidiurnal/diurnal ratio would have a uniform value of 1.466, which is significant. Consequently the tidal type categories have been selected to be symmetrical about this equilibrium value (1.5).

Figure 5.21 is a geographical representation of the information in Table 5.2. Colours indicate the predominant number of tides in a 24-hour period; areas with semidiurnal tides (two tides per 24 hours) are shades of blue and areas with diurnal tides (one tide per 24 hours) are shades of green. From the map it is clear that there are two transitional categories within semidiurnal and diurnal types. New South Wales seems to be the only state which does not experience pure semidiurnal or pure diurnal tides; virtually the whole state has *Semidiurnal with Diurnal Inequalities*.

A comparison of the tide type and tide range figures shows that while there is no simple relationship between the number of tides per day and their size, most areas that experience tides of 5 metres or more also experience semidiurnal characteristics. Similarly, most areas that experience diurnal types have tides of less than 4 metres.

Tide types in practice

Australian tidal types show a preference for the category 'semidiurnal with diurnal inequalities' (the S/D ratio in the range 1.5–4.0). This type extends through the east coast from the tip of Cape York right down to Hobart in Tasmania, with the exception of two small areas, south of Mackay and near Bundaberg. As confirmed in the time series figure below (5.22), tidal characteristics here are distinctly semidiurnal, but alternate high waters and alternate low waters are differentiated by the diurnal influences. This diurnal inequality is seen, for example, at Sydney where the S/D ratio is 2.57.

As this ratio increases above the equilibrium value (1.5), the diurnal influences become progressively less significant, and the inequalities of alternate tides become less marked. Contrast, for example, the difference between Cairns — S/D

Table 5.2
Amplitude, in metres, of the major semidiurnal and diurnal constituents of the tides for selected locations in Australia, leading to a classification of tide type.

Tide pattern	Semidiurnal (2 tides per day) occurs with or without declination		Diurnal (1 tide per day) occurs because of declination		Amplitude ratios	
Constituent amplitudes						
Cause	*Moon*	*Sun*	*Moon & Sun*	*Moon*	*Solar/ Lunar Ratio*	*Semidiurnal/ Diurnal Ratio (S/D)*
Constituent	M_2	S_2	K_1	O_1	$\dfrac{S_2}{M_2}$	$\dfrac{(M_2+S_2)}{(K_1+O_1)}$
Broome	2.368	1.475	0.255	0.156	0.623	9.35
Burnie	1.164	0.146	0.162	0.115	0.125	4.73
Georgetown	1.104	0.133	0.162	0.114	0.120	4.48
Mackay	1.661	0.624	0.390	0.196	0.376	3.90
Wyndham	2.304	0.988	0.652	0.338	0.429	3.33
Darwin	1.828	0.940	0.582	0.326	0.514	3.05
Westernport	0.918	0.232	0.227	0.153	0.253	3.03
Brisbane	0.713	0.196	0.211	0.118	0.275	2.76
Onslow	0.594	0.324	0.209	0.132	0.545	2.69
Sydney	0.501	0.126	0.148	0.096	0.251	2.57
Newcastle	0.504	0.127	0.163	0.091	0.252	2.48
Adelaide	0.506	0.510	0.253	0.169	1.008	2.41
Townsville	0.736	0.424	0.338	0.164	0.576	2.31
Thevenard	0.296	0.371	0.194	0.138	1.253	2.01
Cairns	0.568	0.336	0.309	0.152	0.592	1.96
Melbourne	0.235	0.055	0.097	0.068	0.234	1.76
Port Lincoln	0.240	0.257	0.242	0.168	1.071	1.21
Portland	0.126	0.137	0.180	0.130	1.087	0.85
Esperance	0.103	0.133	0.179	0.133	1.291	0.76
Hobart	0.252	0.012	0.222	0.149	0.048	0.71
Weipa	0.367	0.111	0.436	0.296	0.302	0.65
Albany	0.076	0.105	0.192	0.143	1.382	0.54
Rose River	0.263	0.066	0.509	0.284	0.251	0.41
Geraldton	0.072	0.048	0.174	0.123	0.667	0.40
Bunbury	0.057	0.053	0.173	0.120	0.930	0.38
Karumba	0.201	0.038	0.823	0.603	0.189	0.17
Equilibrium values of Ratios					0.466	1.47

Notes: The quantities listed here for the constituents M_2, S_2, K_1 and O_1 are *amplitudes*, which represent the maximum positive or negative contribution for the individual constituents. The *range* of the contribution of a constituent is twice the amplitude.

ratio 1.96, and Darwin — 3.05.

Truly semidiurnal tides (the S/D ratio > 4.0) are rare and seem restricted to the North West Shelf and along much of the coastline of Bass Strait. In places like Broome, Georgetown and Burnie, it is difficult to recognise the presence of any diurnal contribution, although at Broome K_1 contributes a diurnal oscillation half a metre in range.

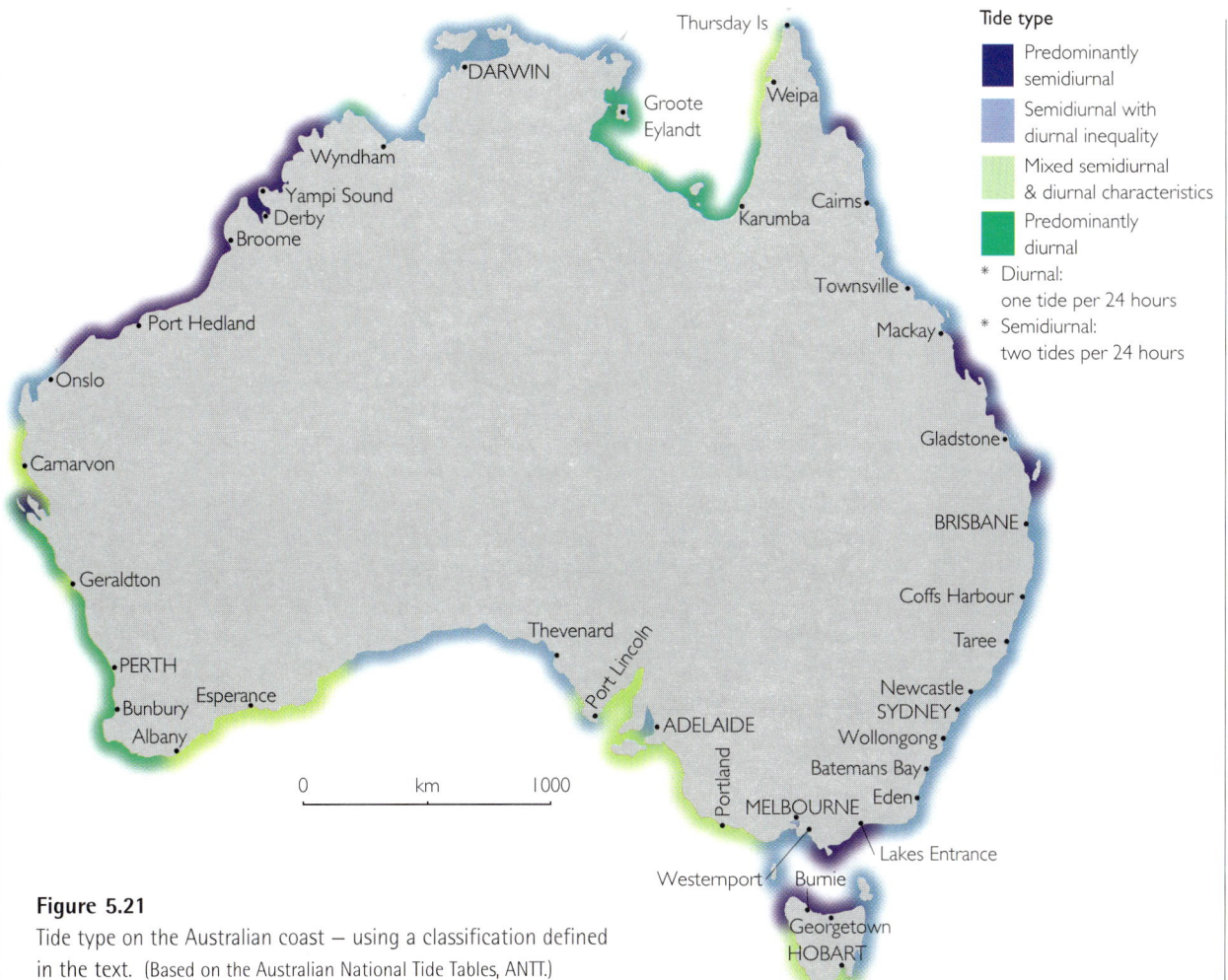

Tide type

- ☐ Predominantly semidiurnal
- ☐ Semidiurnal with diurnal inequality
- ☐ Mixed semidiurnal & diurnal characteristics
- ☐ Predominantly diurnal

* Diurnal: one tide per 24 hours
* Semidiurnal: two tides per 24 hours

Figure 5.21
Tide type on the Australian coast — using a classification defined in the text. (Based on the Australian National Tide Tables, ANTT.)

As the S/D ratio falls below the equilibrium value, the diurnal tide becomes more significant, as in the case of Weipa (which has a ratio of 0.65). Here, for several days at a time, the semidiurnal tide virtually disappears, only to return when, in the fortnightly cycle of K_1 and O_1, a diurnal beat phenomenon erodes the contribution from K_1 and O_1, thereby allowing the semidiurnal component to be exposed.

As the S/D ratio falls below 0.5, any semidiurnal influence is difficult to identify, and the tidal type becomes predominantly diurnal, as in the case of Rose River (0.41), which is located close to Groote Eylandt. At Karumba, with a ratio value of 0.17, the characteristics of the plot make it difficult to believe that there is any semidiurnal contribution at all, but, as seen from the list of constituent amplitudes, M_2 persists and continues to contribute an oscillating range almost half a metre in magnitude.

In other cases not covered by this survey, the likely tide type can be assessed if the amplitudes of the four major constituents are available. The Australian National Tide Tables (ANTT) are published each year by the Hydrographic Service, through the Australian Government Publishing Service. The ANTT includes a listing of up to 20 constituents for approximately 450 locations in Australasia. From this it is possible to select a set of harmonic constituents to represent virtually any location on the Australian coastline.

Patterns in time

Sea level traces for 1994

The variability of tides over a calendar year is illustrated in Figure 5.22, which selects twenty-five representative coastal locations.

The year plotted here is of little relevance for prediction since the patterns do not repeat. Whatever causes combine to create the tide occurring

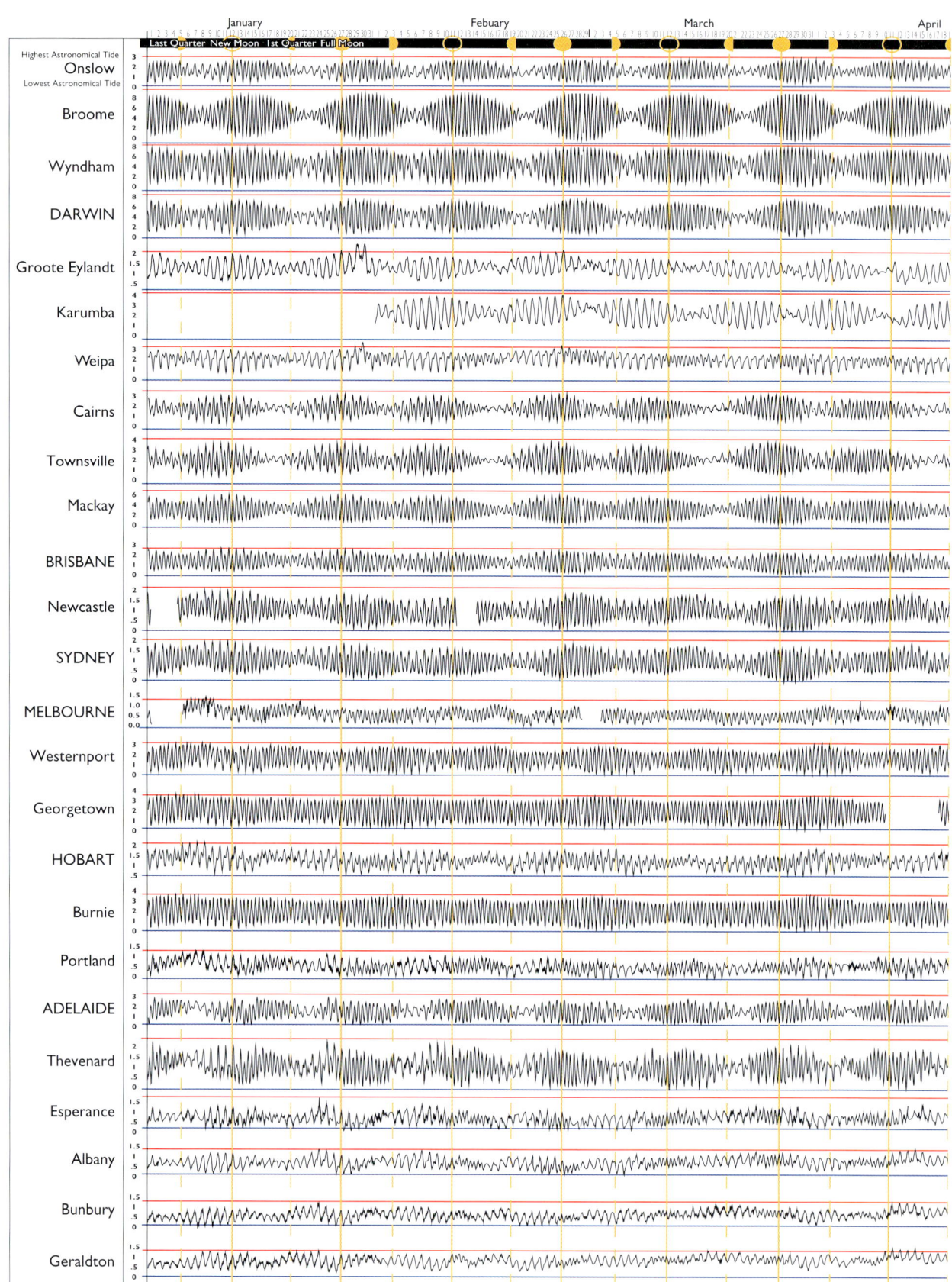

Figure 5.22

The variability of tidal characteristics on the Australian coastline illustrated by plots of observed sea level at selected locations over a calendar year (1994). (Tide data courtesy National Tidal Facility, Flinders University; Moon phases data courtesy Geodetic Operations Australian Survey and Land Information Group, Canberra.)

Observed sea levels (tides plus meteorological influences) for 1994 for 25 key sites*

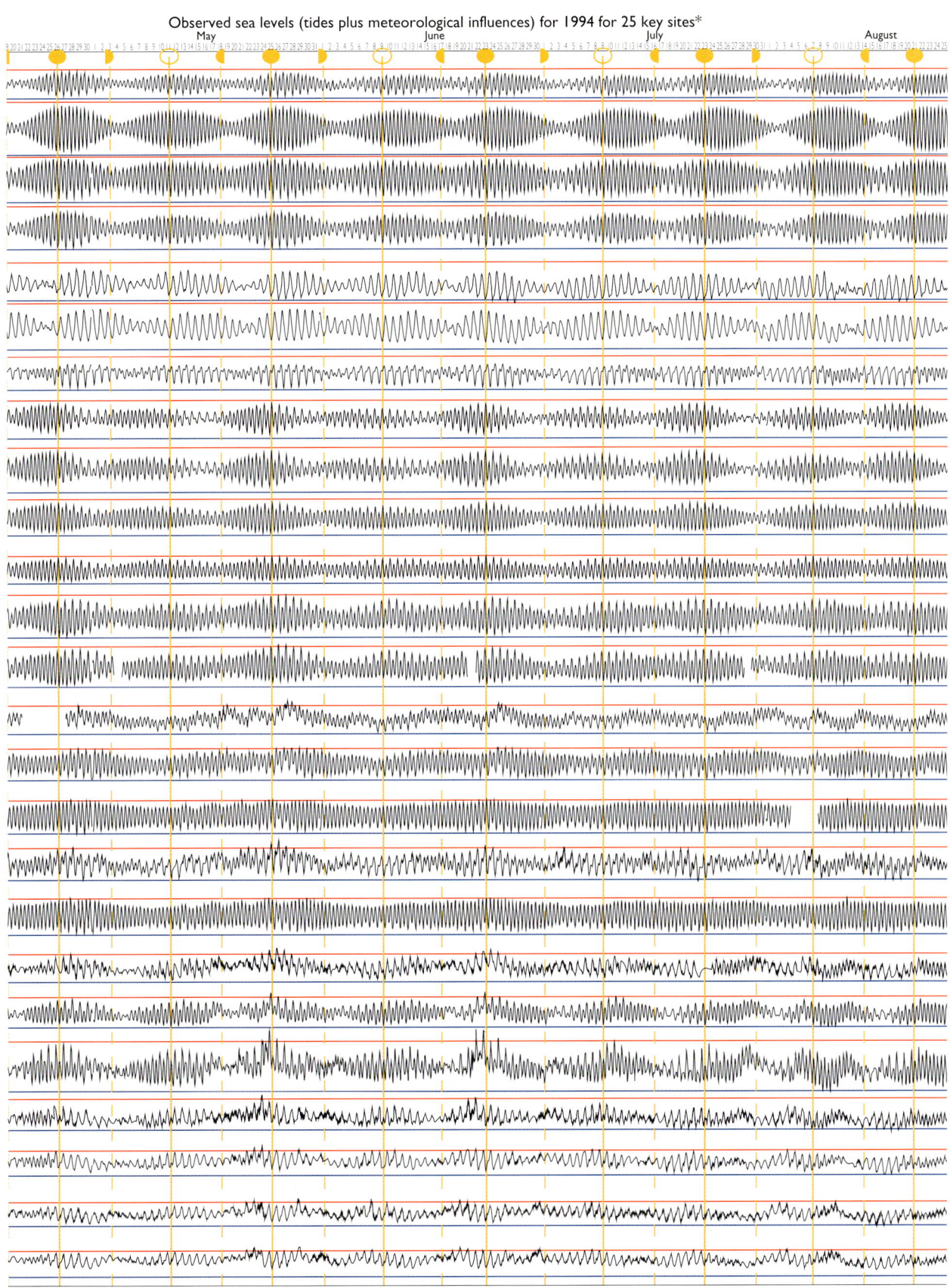

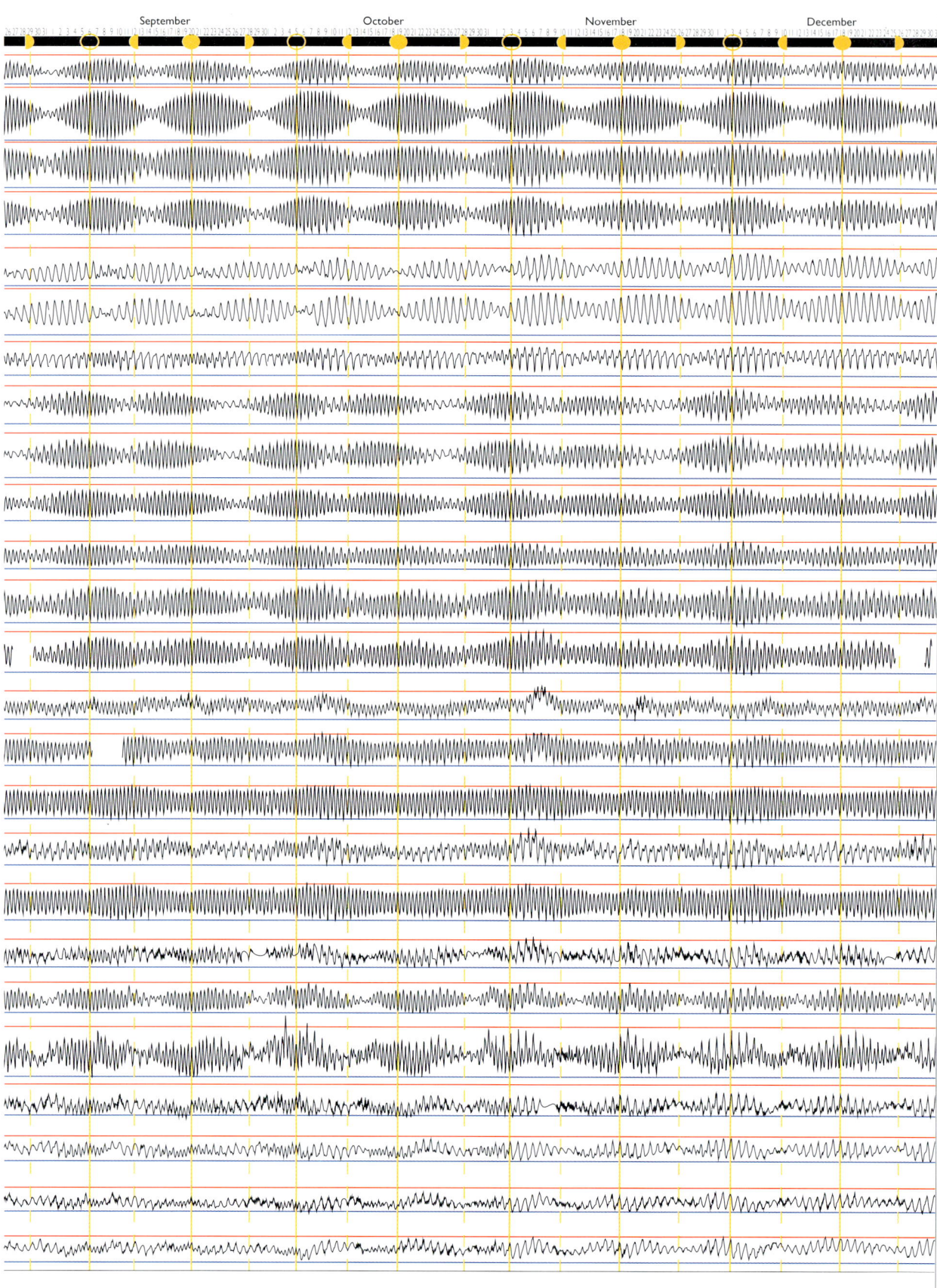

today at Sydney, or elsewhere, will never again be repeated. There might be a tide of the same height occurring at the same time as today's tide, but this will be the result of a different combination of the tidal constituents.

> The published tide tables cannot be applied to another time, either future or past. They are unique to the dates given, as are the particular combinations of the Earth, Moon and Sun.

Other cautions are appropriate here. Firstly, several of the plotted time series have gaps — Karumba for January, or Newcastle in mid-February. The records are incomplete because the tide gauges malfunctioned during these periods. Tide gauges have become much more reliable in recent years, so that some of the plots are continuous through the year. Secondly, the diagram is intended to demonstrate the range of tidal characteristics. Although there is a clear variation in the tidal

range from place to place, the vertical scales differ greatly (so that 25 sites could be placed on one page) — a point clarified in Figure 5.23, which shows tidal traces for Broome, Sydney and Bunbury on a common vertical scale.

The fortnightly spring–neap cycle

Possibly the most common feature of tidal variance worldwide is the fortnightly spring–neap cycle. Looking back to the time series figure (5.22), you can see that some places, such as Broome and other sites in northern Australia, seem to show a smooth and regular fortnightly cycle which is tied to the cycles of the phases of the Moon — spring tides occur soon after full and new moons and neap tides occur soon after the 'quarters'. Figure 5.22 also shows that, in many places, the tides are not so orderly with respect to either the spring–neap cycle or phase of the Moon.

The spring–neap cycle relates to the interaction of the two major semidiurnal constituents, M_2 and S_2, from the Moon and the Sun respectively, which differ in speed by 1.0158958° per mean

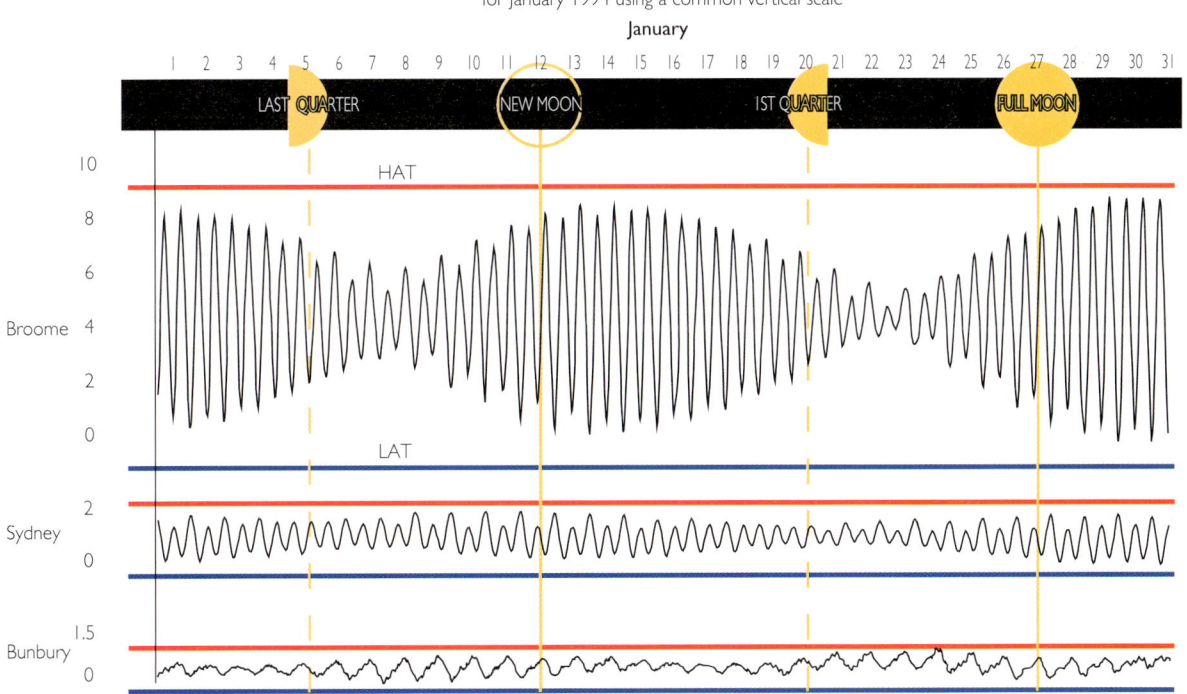

Figure 5.23
Observed sea levels (tides plus meteorological influences) for Broome, Sydney and Bunbury for January 1994 using a common vertical scale. (Tide data courtesy National Tidal Facility, Flinders University; Moon phases data courtesy Geodetic Operations Australian Survey and Land Information Group, Canberra.)

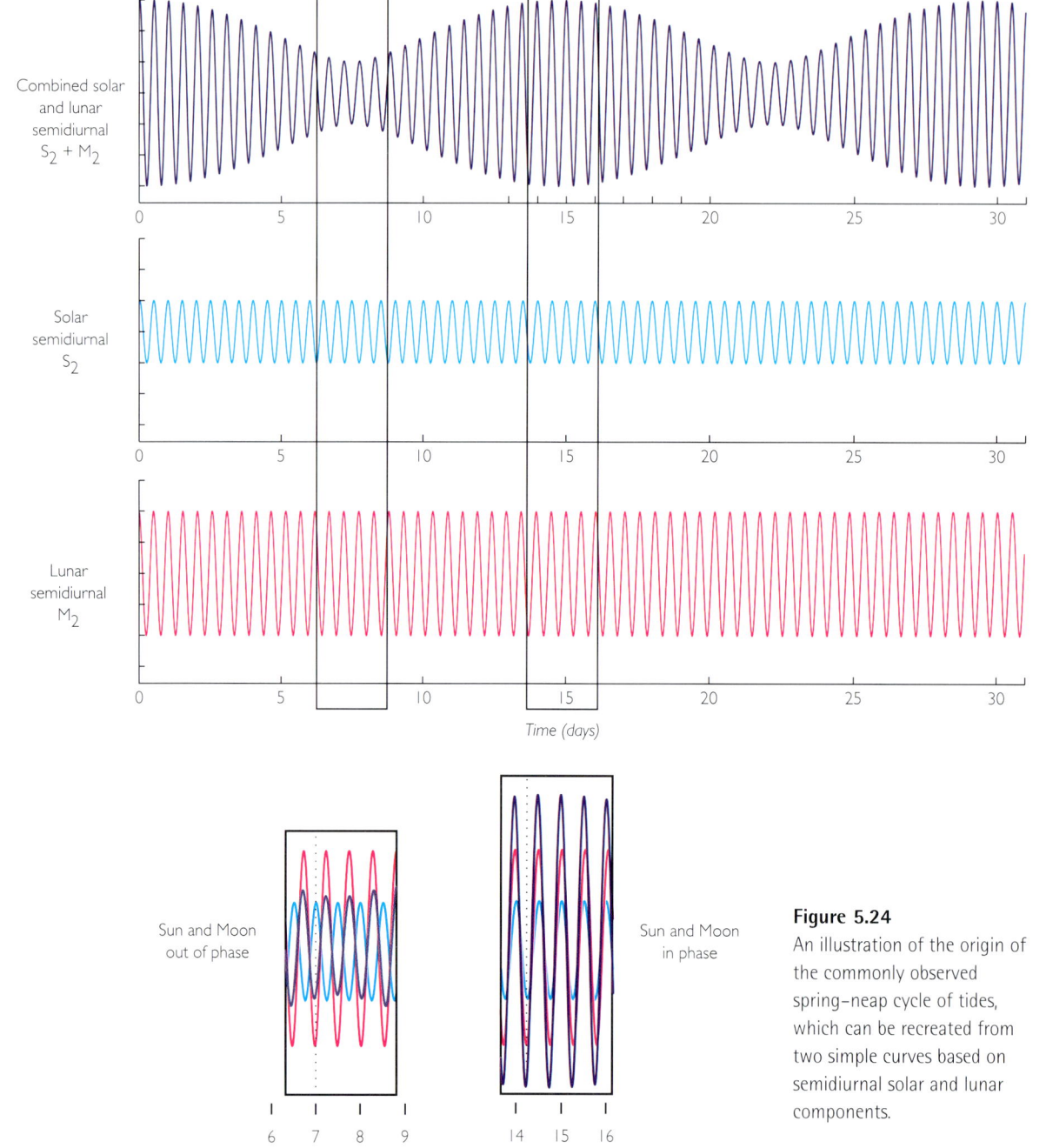

Figure 5.24
An illustration of the origin of the commonly observed spring–neap cycle of tides, which can be recreated from two simple curves based on semidiurnal solar and lunar components.

solar hour (28.9841042° and 30° respectively) — so that, with the passage of time, their contributions pass from opposition to conspiracy and back in the fortnightly time scale (that is, 360°/1.0158958° per mean solar hour = 14.77 days).

Figure 5.24 shows the contributions to the tide of the major lunar semidiurnal constituent, M_2, and of the major solar equivalent, S_2, over a month. The former is plotted in red and the latter in blue. The relative magnitudes of these contribu-

tions are the same as in the equilibrium tide. At the beginning of this hypothetical month, the two oscillations are in phase, giving the maximum positive contribution simultaneously. The combined effect is shown in the upper, purple, diagram. As time progresses the slightly different speeds of the two constituents cause them to become progressively out of phase, and after approximately 7½ days they are in opposition. At this stage M_2 will be giving a low tide signal when S_2 is attempting to give a high tide. Their combined effect at this time is seen in the upper diagram to be of considerably smaller range. This is the period of neap tide. Then the two constituents move back into phase and, after a further 7½ days, both simultaneously attempt to give a high tide: the period of spring tides.

The true spring–neap cycle, as shown in Figure 5.24, is a feature that arises from the interaction of the two major semidiurnal constituents. However, where Table 5.2 indicates the dominance of diurnal characteristics, it may be expected that the spring–neap cycle will be obscured. Yet by strange coincidence the two major diurnal constituents, K_1 and O_1, behave in a similar interactive manner. Their speeds differ by $1.09803300°$ per mean solar hour so that they beat in another fortnightly cycle of 13.66 days. This cycle obviously cannot keep pace with the phases of the Moon and so drifts in and out of the lunar phase cycle in periods of six months.

Again, returning to the interaction between M_2 and S_2, the true spring–neap cycle, this is really dependent upon a solar ratio near to the equilibrium case. Where S_2 becomes small (and so the solar ratio is also small), as in the case of Burnie and Georgetown, the interaction between M_2 and S_2 produces negligible results and the cycle disappears. Alternatively, where the solar ratio approaches unity, as in the case of Adelaide and Portland, the spring–neap cycle becomes very marked. When S_2 approaches the size of M_2, the semidiurnal tide disappears when the two are in opposition.

Meteorological influences

Not all changes in the level of the sea are due to tides, even when they appear to occur in a tidal time scale. The oceans and seas are in a constant state of flux as they adjust to many influences such as the weather. When observation suggests that the tide tables are inaccurate, these other influences are likely to be the cause.

First, the ocean acts like an inverted barometer. When barometric pressure increases, sea level is depressed, and vice versa. The effect is virtually instantaneous, so that it is useful to understand the simple relationship that links the two. The unit of measurement of barometric pressure is the hecto-Pascal (formerly millibar), and for every hecto-Pascal change of barometric pressure, the sea level will change by approximately one centimetre. Thus, low pressures are normally associated with higher sea levels than predicted, and vice versa.

Changes in barometric pressure are also associated with pressure gradients, and these, in turn, are associated with winds and wind stress on water. The latter can set water in motion, which may be impeded by the coastline or by friction in shallows, thereby creating other disturbances. *Storm surge* is the term used to describe sea level movements usually associated with tropical cyclones or other severe storms (wind stress is the most important factor — see page 111). Storm surges can be enhanced by dynamic resonance in cases where the progression of the cyclone over the sea approaches the speed of progression of a shallow water wave. A storm surge created by Typhoon Vera in 1959 caused 5000 deaths in Japan, and a similar event in Europe in February 1953 destroyed 46,000 houses and killed 1835 people in the Netherlands alone, with a further 300 deaths and 24,000 houses damaged in the United Kingdom.

Returning to Australia and the plotted time series (Figure 5.22), it can be seen that, in the northern half of the continent, there is little evidence of these disturbing events; smooth tidal oscillations tend to continue through the months, responding primarily to regular tidal forces. Where perturbations do occur — as seen in late January at Groote Eylandt and Weipa — they are rare and quickly over.

The same cannot be said for the south, from Melbourne to Geraldton. Visually the effect is magnified by the more expanded vertical scale on the one hand, and the much smaller range of tide on the other. The general effect suggests that the tidal oscillations are superimposed on a datum, which is constantly shifting through a range almost as large as the tide itself. At work here are the effects of weather systems, such as storm surges, which are driven from west to east across the Great Australian Bight. Coriolis principles apply, so that there is a continuous tendency for the disturbance to be deflected to the left, but the coastline does not allow this to occur. The term *coastally trapped waves* has been given to these features. In magnitude they can be greater than 1.5 metres, equivalent to the tide itself, and their period (the time taken for a single wave to pass a point on the coastline) can be 5 to 20 days. The response of larger bays, such as the South Australian gulfs, can be much greater since Merian principles then apply.

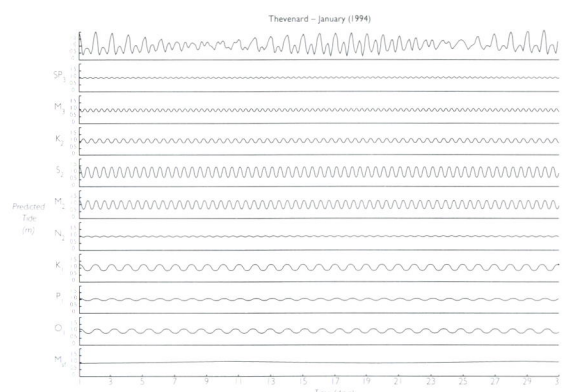

Figure 5.25

An example of the reproduction of a complex tidal record from a number of simple harmonic constituents.

As we have seen, the ocean contains fronts which separate very distinctive water properties, and these can spin off warm core or cold core eddies. Depending upon the depth of these, there may be quite significant differences in local sea level (see page 117). Where gulfs exist and restrict circulation, anomalous temperatures and/or salinities can be generated with similar consequences.

At a larger scale, seasonal climate may affect sea levels in ways that obscure the long-period tidal contribution. For example, tidal predictions make use of a constituent named S_a (solar annual) which has a period of one year. It is true that the equilibrium tide recognises a gravitational contribution of this period, but in practice the main contribution to S_a arises from density effects and seasonal wind stress, the latter being dominant in monsoonal regions.

Another, even larger-scale, feature exists — namely, the ocean–atmosphere interaction which, in a normal year, creates a warm patch of ocean water in the western Pacific near Papua New Guinea. This in turn creates a major low density feature which can be associated with significant differences in local sea level. In an El Niño year the patch will migrate eastwards, but when near PNG the patch may stand almost a metre high.

Tide tables

Tide tables base their predicted values on a complex calculation that aims to reproduce the complicated oscillations of the observed time series by the addition of a large number of simple, smooth and continuous curves like M_2 and S_2, as seen in Figure 5.25. Each of these smooth curves represents a single astronomical motion and together the combined model is arranged to represent the predicted tidal curve. Here the figure predicts tides for Thevenard, the port of Ceduna in the Great Australian Bight, for one month. The figure shows a handful of constituents, but accurate predictions of tidal amplitude *and* timing require 100 to 120 constituents!

CHAPTER SIX

Severe Weather

Forecasting weather is a complex process. It requires detailed knowledge, not only of surface conditions, but also of the temperature, moisture and winds in the upper atmosphere. A high pressure system is often taken to be a sign of fine weather; however, an upper air low or broad ascent of upper level moisture can produce rain even within a high. Onshore winds circulating about a high can produce coastal showers. Similarly, lows may be topped by stable subsiding air and fine weather may result.

RECOGNISING THE WARNING SIGNS

The best warning of impending severe weather can be obtained by listening to weather reports on the radio or on television. These are supplied by the Commonwealth Bureau of Meteorology, which has extensive data gathering networks. Even if your craft is well equipped with radar and atmospheric sensors (for example, temperature, wind, pressure), you cannot possibly hope to match the sophistication and coverage area of the Bureau of Meteorology. You can supplement this information by knowing your local area and by being aware of any unusual weather. *Keep a weather eye open.*

To the extent that you *can* avoid foreseeable disaster, here are some simple principles for improving your awareness of the weather:
- Observe cloud types and direction of movement; is the direction and speed of the low cloud different from that of the surface wind?
- Is there an advancing line of cloud or roll cloud (especially high cloud advancing from the west or south-west) or are there cumulonimbus clouds in the area?
- Is the wind speed increasing or has it suddenly and unexpectedly dropped?
- Is the wind direction unusual or has it changed unexpectedly?
- If you carry a barometer, is the pressure dropping or rising more quickly than expected?

Of course, clouds and surface wind changes are not the only signs of impending change, so all coastal users are advised to contact their local weather office before heading out to sea. It is also a good idea to contact them while you're out there (if you can — cellular phones and ship-to-shore radios are not common among sailboarders) especially if unexpected changes in wind and cloud occur. If you are not sure of what meteorological services are available, then contact: Bureau of Meteorology, GPO Box 1289K, Melbourne, Victoria, 3001 or your state/territory office of the Bureau of Meteorology.

This chapter contains guidelines for using clouds, surface features and other reasonably obvious signs to detect deteriorating weather conditions. However, be aware that the weather can deteriorate *without* such warnings, especially where the weather change is being caused by upper air disturbances; these may not be obvious or even observable from surface pressure maps.

CLOUD TYPES

Clouds are important in forecasting: they are frequently the most obvious signs of deteriorating — and improving — weather. The study of clouds is rather complex; coastal users who need more than the simple outline given here should refer to References, page 207.

Clouds are classified by their appearance (form) and by their height above ground (in the case of clouds with vertical development, the height of the base is used). Reference is commonly made to low cloud (0–2500 m), middle cloud (2500–6000 m), and high cloud (> 6000 m). The ten most commonly recognised cloud types are detailed in Table 6.1.

The word *cirrus* (or *cirriform*) refers to high clouds of a fibrous or streaky appearance that are composed of ice crystals. Clouds in sheets or layers are called 'stratus' (stratiform), and clouds with a heaped-up or puffy appearance are called 'cumulous' (cumuliform).

Clouds generally form when there is more water vapour in the air than it can hold at any given temperature. The ability of air to hold water vapour depends on its temperature — the higher the temperature, the more it can hold. Therefore, as the air cools, there is a temperature at which it is said to be saturated (with respect to water) and at which condensation takes place. Clouds are one form of condensation (fog, dew, and rain are others). Clouds may form as a result of mechanical turbulence (friction-induced mixing), convection (tendency for warm air to rise), orographic uplift (air forced to rise over mountains, and so on), and slow widespread ascent (for example, frontal uplift).

> Spend some time learning to recognise the different cloud types, then you can use them to help you assess changes in the weather.

Stratus cloud consists of 20–200-metre thick layers of very low cloud formed by small vertical movements in the atmosphere, some types of which are associated with cold stable air at low levels; so-called 'hill fogs' are a result of stratus cloud. It is generally uniform grey cloud, often with accompanying drizzle. Light turbulence is usually experienced beneath stratus.

Stratocumulus cloud (see Figure 6.1) may form as a result of capped convection, or more often by mixing below a subsistence inversion. It is grey to white patchy cloud in a layer, with noticeable dark areas and breaks in between. Light rain and drizzle may be associated. Light to moderate turbulence is expected beneath such cloud and smooth conditions above (if you happen to be flying).

Cumulus cloud is detached dense white cloud with vertical growth and a flat base, and if sunlit, has a bright white appearance. It can exist as small 'puffs' or massive columns. It is produced

Table 6.1

The ten commonly recognised cloud types and common cloud base heights (in metres)

Low cloud	Middle cloud	High cloud
stratus (150–600)	altostratus (2500–6000)	cirrus (6000–12,000)
strato-cumulus (600–1500)	alto-cumulus (2500–6000)	cirro-cumulus (6000–12,000)
cumulus (600–1500)*		cirrostratus (6000–12,000)
cumulonimbus (600–1500)*		
nimbostratus (150–2500)**		

Notes:

*refers to height of base — the tops can reach well into middle cloud levels

** the low figure for the base represents continuous drizzle and rain — altostratus can change into nimbostratus without any marked change in the appearance of the sky except a change from drizzle to rain; the cloud tops can extend upwards to 6000 metres (and some references classify nimbostratus as a middle cloud).

by vertical convection induced by instability (the tendency for air to rise). It is most common in summer over strongly heated land. Generally, where the cloud forms there is an updraft (thermal) and where the air is clear there is a downward movement. Since this is driven by solar energy, there is a tendency for cumulus to form into stratocumulus at night and over the ocean.

'Fair-weather' cumulus is formed at the level where the air becomes saturated (the condensation level — recall that temperature generally decreases as altitude increases), and appears as small isolated clouds with bases perhaps at 500 metres over sea, 1000 to 2000 metres over land and 4000 metres in arid areas. These clouds often form near-parallel lines oriented in the direction of the low level wind. Turbulence is generally light to moderate near the surface.

Towering cumulus cloud may form when there is considerable instability and thermals within the cloud. Where thunderstorms are associated, the cloud is called 'cumulonimbus' and is

Figure 6.1
Stratocumulus opacus – low level cloud. (Photograph from Mark Bedson's *Sky Chart.*)

Figure 6.2
Cumulus humilis – low level cloud also known as 'fair-weather cumulus' if significant vertical development does not take place. (Photograph from Mark Bedson's *Sky Chart.*)

Figure 6.3
Cumulonimbus calvus – low level cloud (the base is low level but the cloud is vertically developed and the top may reach high levels) often called the 'thunderstorm cloud'. (Photograph from Mark Bedson's *Sky Chart.*)

Figure 6.4
Cumulonimbus with anvil top – low level cloud (the base is low level but the cloud is vertically developed and the top may reach high levels) often called the 'thunderstorm cloud'. (Photograph from Mark Bedson's *Sky Chart.*)

characterised by great vertical development and an anvil shape at high altitudes; the anvil may form where there is a high level inversion and the upper winds spread the cloud. Occasionally, the top of the anvil can reach well above 15,000 metres (20,000 metres in the tropics). The turbulence associated with towering cumulus and particularly cumulonimbus can be moderate to severe, although not generally widespread. Lightning and thunder are often followed by moderate to heavy showers, snow (if cold enough) or hail.

Altostratus cloud is formed as a result of widespread slow ascent of air, and appears as a grey–bluish striated or fibrous veil blurring the sun. Widespread light to moderate rain and turbulence are associated with it. Altostratus can form into nimbostratus without any marked

change in the appearance of the sky; drizzle can turn to rain.

Nimbostratus cloud can extend upwards to 6000 metres and appear as grey sheets which often cover the whole sky and obscure the sun. Heavy continuous rain or snow is often associated with it. It can be thought of as stratus cloud that is thick enough and contains sufficient water and vertical movement to form rain. There is generally light turbulence beneath it (unless associated with tropical cyclones or strong fronts).

Altocumulus cloud consists of a layer or patches of flattened globules of cloud in groups, waves or lines, including the familiar mountain wave cloud, as well as 'billow' clouds, where air at different altitudes is moving at different speeds. Often these clouds indicate a moderate to strong

Figure 6.5
Altostratus translucidus – middle level cloud identified by 'frosted glass' appearance of the sun. (Photograph from Mark Bedson's *Sky Chart*.)

Figure 6.6
Altocumulus undulatus – middle level cloud. (Photograph from Mark Bedson's *Sky Chart*.)

Figure 6.7
Altocumulus castellanus – middle level 'turret' cloud which may develop more vertically and may serve as a precursor to thunderstorms and other forms of severe weather, including tropical cyclones. (Photograph from Mark Bedson's *Sky Chart*.)

Figure 6.8
Cirrus fibratus – high level cloud.
(Photograph from Mark Bedson's *Sky Chart*.)

Figure 6.9
Cirrocumulus – high level cloud. (Photograph from Mark Bedson's *Sky Chart*.)

wind aloft. One form of altocumulus, *altocumulus castellanus,* may serve as a precursor to the early development of severe thunderstorms and even tropical cyclones.

Cirrus cloud occurs at great heights where temperatures are often below –40°C. It appears as detached delicate fibrous lines or tufts. The familiar 'mares' tails' form as ice crystals fall from the source cloud and are drawn out by winds of different speeds below. When cirrus spreads it may form cirrostratus, which can be up to 3000 metres in depth; more commonly cirrostratus is formed by the gradual widespread ascent of an air mass. Cirrocumulus appears as a layer of thin white puffs or ripples.

CLOUD FORMATION

There are four basic cloud forming processes: mechanical turbulence, convection, orographic uplift and slow widespread ascent.

Mechanical turbulence

Mechanical turbulence arises from the interaction of the wind on the land surface. As described in Chapter 3, friction effects can be felt up to heights of more than 1000 metres, depending on the stability of the air and surface roughness. The atmosphere is said to be stable when the upward movement of an air parcel is suppressed; it is unstable when the upward movement of air is encouraged, or even accelerated. The stability depends on the rate at which temperature changes with altitude. If a dry air parcel rises, its temperature will decrease by approximately 0.98°C per 100 metres increase in altitude; at a particular height temperatures are low enough for the condensation level to be reached, and cloud begins to form. The condensation level also naturally depends on how much water vapour is in the air.

Under certain atmospheric conditions, mechanical turbulence transports water vapour to considerable heights and to the condensation level. Clouds formed by this process are initially stratus — sheet clouds without definite form. Often these develop wavelike upper and lower surfaces, where the upper vertical limit of such clouds often corresponds to the beginning of a stable layer.

Figure 6.10 is a representation of cloud formation by mechanical turbulence and Figure 6.11 shows a case where turbulence combines with a cool surface to generate stratus.

The latter kind of cloud formation will be enhanced by anything that causes the surface to be cooler downwind than upwind — this can happen when wind is blowing offshore or over a large inland lake. In both instances one might expect some drizzle and deterioration of the weather if stratus can be seen actively forming.

Convection

Convection occurs when the surface of the Earth is strongly heated by sunlight, causing the air to rise. It is often assisted by mechanical turbulence as well. If you were to sit out in the country on a hot summer's day you might see an eagle appearing to circle endlessly around a fixed point on the ground. The eagle has found a strong convective current or 'thermal' and is 'hitching' a ride up to several kilometres or more. At the top of this thermal there is a good chance that cumulus is actively forming. Cumulus can form with or without a wind blowing; when a strong breeze is blowing the shape of the thermal, if seen from the side, becomes distorted and the top of the thermal drifts downwind.

Like the earlier example of stratus, the vertical development of cumulus can be limited by a

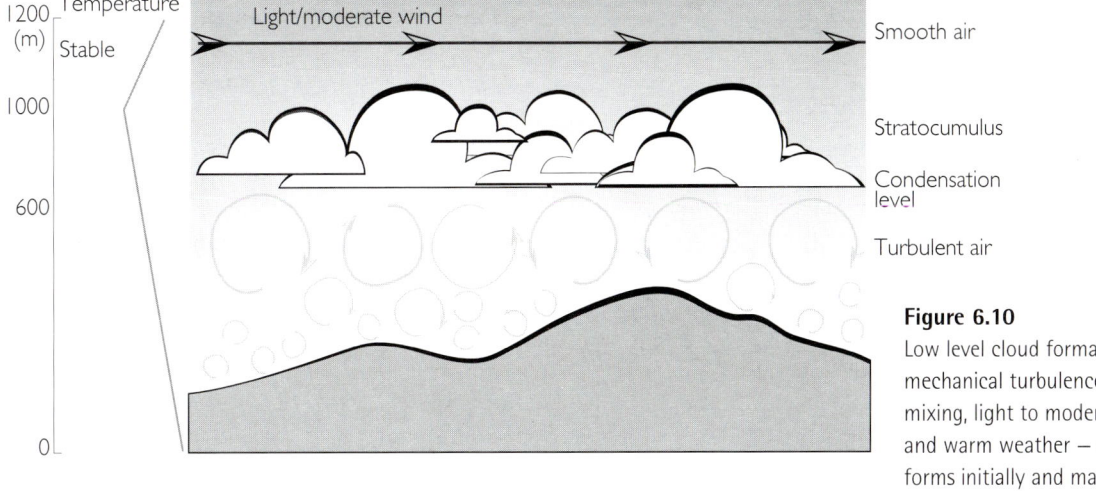

Figure 6.10

Low level cloud formation by mechanical turbulence and deep mixing, light to moderate winds and warm weather — *stratus* forms initially and may develop vertically to *stratocumulus.*

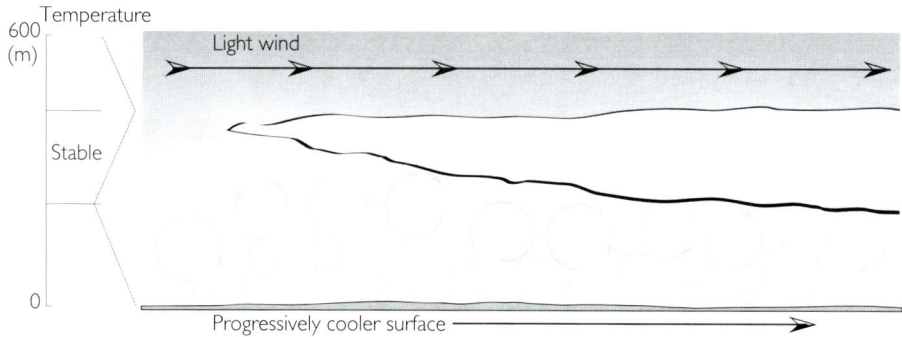

Figure 6.11
Low level cloud formation by mechanical turbulence and deep mixing, light to moderate winds enhanced by a progressively cooler surface — *stratus*.

stable air layer aloft — recall that, in a stable atmosphere, temperature increases with height — and where this happens stratocumulus may form rather than cumulus. Where a stable layer does not limit the vertical growth, the cloud can grow vertically to a height where ice crystals form, giving the top of the cloud a smooth fibrous appearance. This appearance is characteristic of cumulonimbus — the thunderstorm cloud.

Orographic uplift

When moist air is forced to rise over a mountain range, it may cool sufficiently for condensation to occur. Frequently you can observe stratus forming downwind of the range, as shown in Figure 6.12. Orographic uplift can also result in a more or less stationary lens shaped (lenticular) cloud at or above the mountain top in a stable airflow. In the case of a fairly steep sided mountain, and with strong wind aloft, you may see the mountain wave phenomenon, where the wind assumes a wave-like structure with a sequence of lenticular clouds with more or less even spacing, between 5 and 50 kilometres. Anyone sailing in a strong offshore wind ought to be aware of the presence of an upwind barrier (mountain, cliff, and so on) as in extreme

cases there might be a reverse-flow rotor which could obviously be potentially dangerous.

Slow widespread ascent

So far we have looked only at cloud formation associated with fairly small-scale vertical movements of air. Vertical movements can also take place over great distances, associated with lows, highs and fronts.

The boundary between adjacent air masses is called a 'frontal zone', and is characterised by fairly abrupt horizontal and vertical gradients of humidity, temperature, wind speed and wind direction. Where a cool air mass is advancing and wedging under a warmer mass, it is called a 'cold front'; in Australia these usually come from the south-west, moving north-easterly, although they can approach from the west or north-west off south-west Western Australia and south of Australia. Although warm fronts are not all that common in Australia — they usually occur at and above 50° latitudes — they arise from warm air advancing and rising over a retreating cooler mass.

Two typical sequences of approaching warm and cold fronts are shown in Figure 6.13. The slope from the side of a cold front is generally

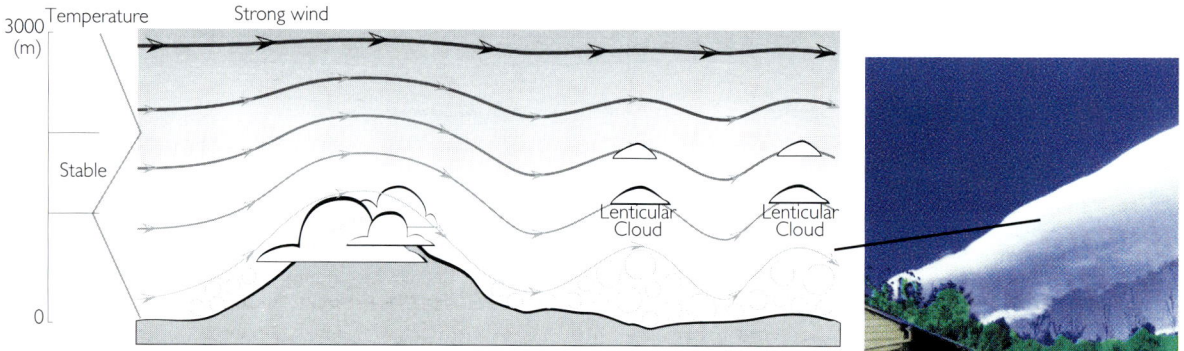

Figure 6.12
Middle level cloud formation by orographic uplift combined with strong upper winds — mountain wave (lenticular) cloud.
(Photograph courtesy R. Badham.)

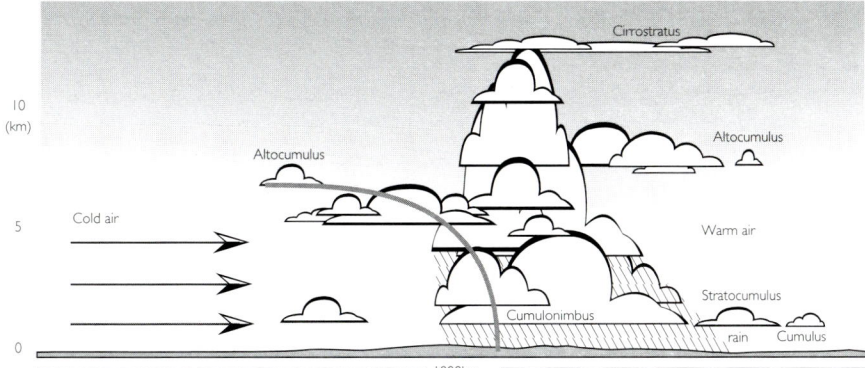

Figure 6.13

A typical sequence of clouds associated with an approaching warm front (upper) and cold front (lower).

much greater than for a warm front. This can result in violent storms, especially if the warmer air mass is unstable and the cold mass is moving quickly. In the case of the warm front, cirrus may form 600 kilometres ahead of the front; somewhat less for a cold front. Although various factors can influence the rate at which a front advances, commonly their speed is 15–30 km/h or so. Thus, using clouds as an indicator, one may have at least a day's warning of an approaching warm front, but frequently only a matter of some hours' warning of a cold front.

SEVERE THUNDERSTORMS, HAIL AND TORNADOES

Severe thunderstorms cause a great deal of damage. They are thunderstorms that produce wind gusts over 48 knots (89 km/h), hail larger than 20 mm, or tornadoes. They may be accompanied by heavy rain, causing flash floods. They may occur near intense cold fronts or in the area between a deep low pressure and a large high pressure system (see Chapter 2). Very localised storms often go undetected due to their small size and short life cycle and yet can produce devastating wind, rain and hail.

Thunderstorms develop in an unstable environment where there is adequate moisture to maintain deep convection and some lifting mechanism to start the process. Strong surface heating, an advancing sea breeze circulation or a front are examples of the sort of catalyst needed to start their development. During a hot afternoon the early warning sign may be rapidly growing cumulus cloud.

Thunderstorms contain strong vertical winds, both ascending and descending, and for this reason are usually avoided by pilots. At the surface, the descending currents or downdrafts are most important. Generated by negative buoyancy resulting from evaporative cooling by rain or hail, the downdraft accelerates toward earth to form a gust front, in which winds may exceed fifty knots. Because winds spread out from the downdraft, their direction will depend on the location of the

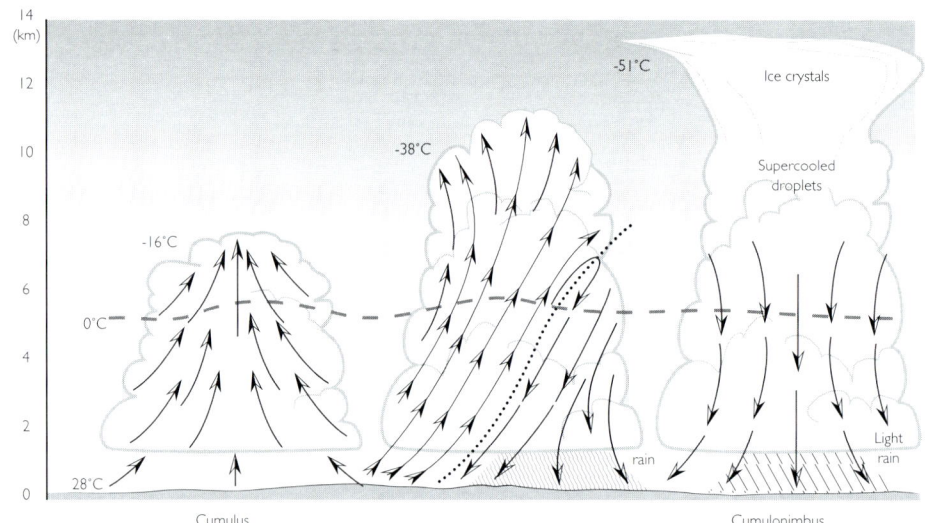

Figure 6.14
A schematic representation of the stages of development of a thunderstorm system.

storm relative to the observer — quite frequently, though, the dominant direction of the downdrafts is similar to the overall direction of movement of the thunderstorm.

Figure 6.14 shows the evolution of a thunderstorm; note, in the early stage, that surface winds tend to 'feed' the cumulus. At the mature stage, moderate to strong updraft and downdraft winds can occur. At the advanced stage, the upper part of the cloud consists entirely of ice (a cirrus anvil), giving it a wispy appearance. If the thunderstorm is moving, expect a major wind direction change as it passes over you. The cumulonimbus cloud, with its characteristic anvil top, is one of the easiest cloud types to recognise.

One reason thunderstorms can be so dangerous to mariners is that they may not be forecast by weather services, particularly if the forecasts have a long lead time. It is well worth noting their development and movement, and avoiding them. All thunderstorms are potentially dangerous, as they may change speed and direction quickly. The 'gust front' containing severe winds may extend ten or more kilometres ahead of the storm.

Thunderstorms can affect any part of the country at any time of the year, though the majority tend to occur during September to March. The most severe storms recorded occur in an area stretching from south-east Queensland, through coastal parts of New South Wales and Victoria, and into South Australia. In most thunderstorm events, the violent winds are short-lived; if sailing it is wise to reduce sails for the short period preceding and during the event.

Violent winds are not the only problem: on average, three people are killed each year by lightning. If lightning begins while you are in the open, seek shelter under jetties or bridges and (between strikes) make sure masts and stays are grounded. If strong or gale force winds occur, seek shelter in the lee of islands or headlands — using the principles set out in Chapter 3 — and be aware of the possibility of wind flow reversal behind the barrier.

Figure 6.15 shows data on severe thunderstorms in Australia. Unfortunately data are not available for the surrounding waters; don't assume that, because the colours stop at the coast, the thunderstorms will, too! This figure is based on information from thirty-nine sites over twenty years and so can only provide a coarse picture. The Sydney region, for example, experienced approximately thirteen severe thunderstorms per 100 km^2 grid cell over the twenty years. Given the paucity of data, it might be better to interpret the patterns in a relative way (that is, disregard the numbers altogether) and just note the areas that are more likely to experience severe thunderstorms.

Better than any knowledge of data, though, is a good working knowledge of the warning signs, and of what to do if you are threatened by a severe thunderstorm.

Warning signs of severe thunderstorms

Early signs

- Castellanus cloud — vertical 'turrets' on middle level cloud (see Figure 6.7), which is sometimes an early indicator of thunderstorm activity later in the day.

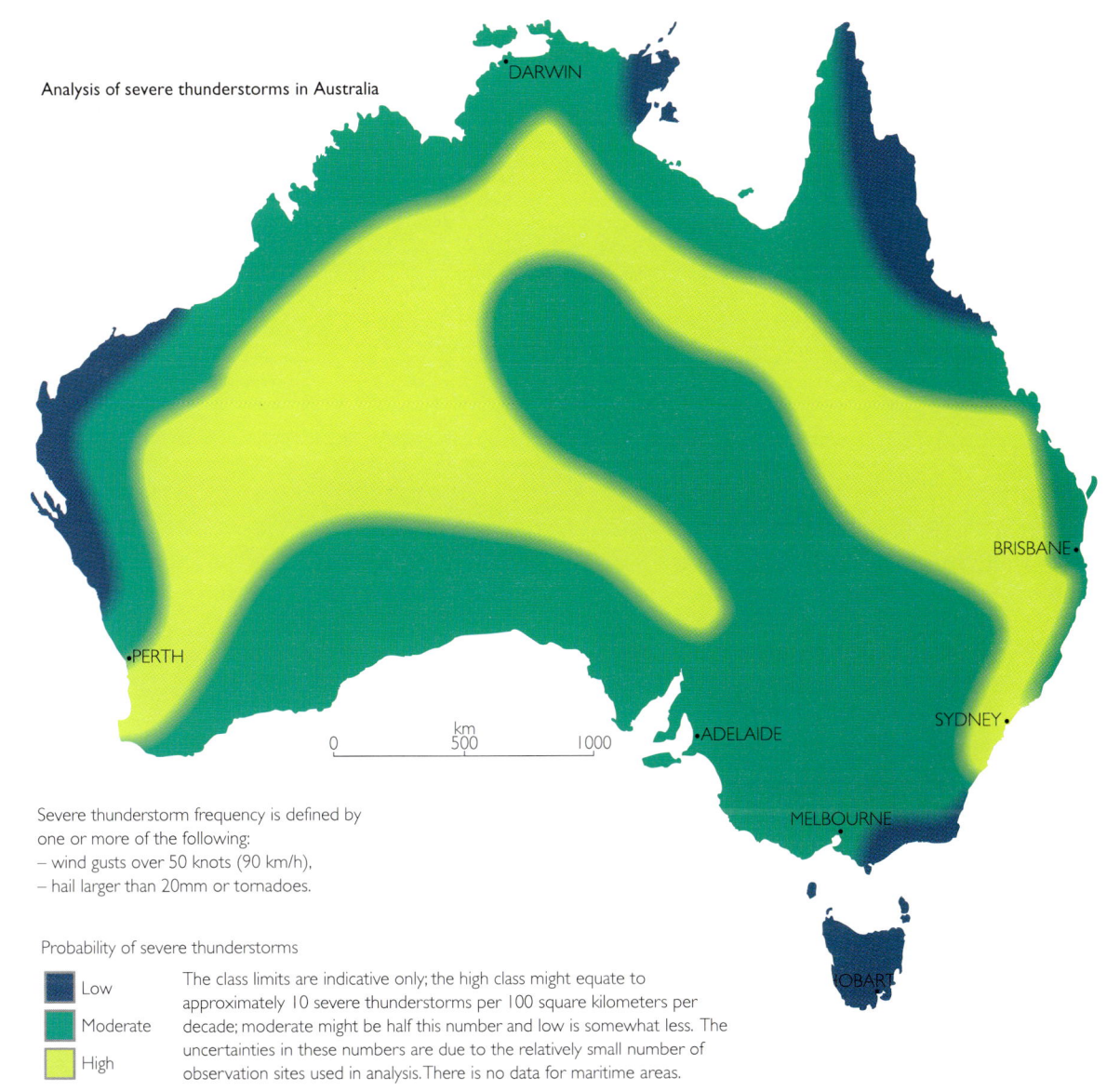

Analysis of severe thunderstorms in Australia

Severe thunderstorm frequency is defined by
one or more of the following:
– wind gusts over 50 knots (90 km/h),
– hail larger than 20mm or tornadoes.

Probability of severe thunderstorms

Low
Moderate
High

The class limits are indicative only; the high class might equate to
approximately 10 severe thunderstorms per 100 square kilometers per
decade; moderate might be half this number and low is somewhat less. The
uncertainties in these numbers are due to the relatively small number of
observation sites used in analysis. There is no data for maritime areas.

Figure 6.15
Severe thunderstorm frequency: defined by one or more of wind gusts over 50 knots (90 km/h), hail larger than 20 mm, or tornadoes. (Image based on data from 1972–91, supplied by Chris Ryan, Severe Weather Service, Bureau of Meteorology.)

- Rapidly growing cumulus in mid-morning, which can continue into full-blown cumulonimbus later in the afternoon (Figures 6.3–6.4).
- Characteristic cirrus 'anvils' which blow out from the top of fully developed cumulonimbus. These are easily recognised and may be observed, even when the storm itself is far away and possibly below the horizon (Figure 6.4).
- Lightning flashes that can be seen at a long distance at night, even when the storm itself is not visible.

- Thunderstorm static on radio receivers, which is relatively easy to identify; it will increase as the storm approaches.

Near the storm

The strong vertical movement that is typical of severe storms may be evident in the rapid growth of cloud and in any overturning within the cloud. Downward protuberances may be present. Funnel clouds extending down from the cumulonimbus indicate the potential for a tornado or waterspout to develop. Sometimes thunderstorms appear as well organised 'lines' of storms (see Tropical Squall Lines, page 175). In some situations there is a distinct 'gust front' and this may be accompanied by a roll-type cloud, as shown in Figure 6.16.

Judging storm proximity

Sound travels at about 330 m/s (640 knots), so it takes approximately 3 seconds to travel 1 kilometre. To work out the distance between you and an approaching thunderstorm, simply count the seconds between when you see the lightning and when you hear the thunder, and divide this number by 3: this gives you the approximate distance in kilometres.

Usually lightning travels along a tortuous path, with one end often much closer to the observer than the other. For example, if the initial clap is heard 3 seconds after lightning, and is followed by a rumble lasting another 3 seconds, this suggests that the closest part of the lightning stroke is about 1 kilometre away, and the most distant part 2 kilometres.

COLD FRONTS

As we have seen, fronts represent the boundaries of two air masses or other lines of significant weather. They can be a serious problem to users of coastal waters and oceans because the active cells may be extensive and organised, and at times rapidly moving. On weather maps, cold fronts are recognisable as a line of 'triangles'. (See also Figures 2.3, 2.16 and 6.13.)

Warning signs

The earliest warning of an approaching cold front is invading cirrus or cirrostratus cloud from the west; middle level cloud may increase with a lowering of its base. Rain may or may not develop. Wind may freshen, typically from the north-west but may be affected by local sea breezes or topographic influences. If the slope of the cold front is shallow and weak, the sequence of clouds may resemble that of a warm front (Figure 6.13) but in the reverse order — that is, low clouds first, followed by stratiform clouds.

Immediately ahead of the front one usually observes rapidly moving low cloud and the development of large cumulus or cumulonimbus. At the frontal transition zone there may be a single change or there may be several change lines, wind changes, thunderstorms or squall lines (Figure 6.16). The wind will normally back (swing anticlockwise) behind the change.

Figure 6.16
Approaching squall line — low level cloud often associated with thunderstorms and severe weather.

Squall lines and wind maxima may occur several hundred kilometres ahead of the main front as marked on a weather map, so beware of placing too much reliance on the position of single frontal lines in simplified newspaper and television maps. Figure 2.16 (lower panel) shows a dramatic example of a series of cold fronts embedded in the westerly flow. As an aside, the upper panel shows an example of a strong temperate cyclone which is frequently associated with wintertime cold fronts.

THE COOL CHANGE AND SOUTHERLY BUSTERS

Imagine a typical summer situation of a large anticyclone over Australia (its centre around 37°S), and another behind it, out in the Indian Ocean. The contrast between the converging airstreams of the two highs is rather pronounced — the northerly winds from the first high are usually hot and dry since the airstream blows over a hot land-mass, and the southerly air from the second is cool and moist. The change from the hot air of the first high to the cool air of the second is known as a cool change.

Cool changes are often accompanied by fresh and occasionally strong breezes, and by a rapid fall in temperature and rise in relative humidity. Remember that, depending on local conditions, such as the strength of sea and land breezes, the cool change may arrive up to twelve hours ahead of the normal cold front between air masses.

Southerly Busters are a special occurrence of the cool change; in Australia, they happen only on the coast of New South Wales and mark the sudden change from one anticyclone to the next. They are a type of front that occurs almost exclusively in spring and summer, when shallow fronts move up the coast rather than over the mountains. About thirty Southerly Busters are observed per year, most frequently in dry summers. They are characterised by 30–70 knot winds in a band 45–90 kilo-metres wide (centred on the coast) and moving northward at 30 km/h, sometimes much faster.

Warning signs

Southerly Busters are particularly dangerous because many do not have cloud directly associated with them and often there are few visible signs of their approach. Coastal users should be wary of any abnormal variations in the north-east sea breeze, particularly if you know that a front is expected to pass over south-eastern Australia. The north-easterly wind may weaken somewhat before a sudden change to strong or gale force southerly winds.

Watch for any line of cloud approaching rapidly from the south; do not wait until the last minute, as these systems move quickly and the southerly wind may arrive before the cloud line. They are occasionally preceded by a 'roll' cloud, which appears as a thin sheet of cloud being rolled up like a carpet, not unlike the 'squall line' cloud in Figure 6.16. They can sometimes be detected as a line of approaching 'haze' from the south.

LAND GALES

A land gale is simply a set of winds over the land in excess of 33 knots (63 km/h), usually as a result of strong surface pressure gradients (shown as very close spacing of the isobars on weather maps). Gales generally happen between a deep low pressure system and a large high, or near an intense cold front.

THE EASTERLY DIP AND EAST COAST LOWS

The easterly dip and the east coast lows are related. Both systems can produce heavy rainfall on the east coast. They are linked to the east coast ranges and to a strong temperature gradient between the ranges and the sea-surface temperature (SST) associated with the East Australian Current. Their incidence is highest in winter, when the land–sea temperature gradient is at its greatest and a trough of low pressure develops. The trough can be either onshore or offshore — the latter can be associated with very heavy rainfalls if the trough is parallel to the coast.

A small number of the offshore easterly dips develop into east coast lows. The east coast lows are defined as a closed cyclonic circulation, 50–1000 kilometres across, which develops at sea, in latitudes 20°–40°S, within 500 kilometres of the coast and with a pressure gradient of 4 hPa/100 kilometres. Gale to storm force winds are common, particularly south of the centre of the low.

It is important to note that east coast lows do not include tropical cyclones, which move into higher latitudes.

In 1967–92 east coast lows accounted for seven per cent of major Australian disasters during that time.

TROPICAL CYCLONES IN AUSTRALIA

The tropical cyclone season in Australia typically occurs during November to April. Although the nomenclature is not standardised internationally, a 'tropical cyclone' in the Southern Hemisphere is defined as a clockwise rotating low pressure system originating from the tropics, in which the ten-minute mean winds exceed gale force (34 knots; 63 km/h; Beaufort 8). Characteristically a large area of convective cloud and heavy rain is associated with the system; in the more intense tropical cyclones there is also a clear region — the 'eye' — near the centre. The strongest winds are in a band surrounding this eye, although, within the eye itself, winds are usually very light. The intensity of a cyclone is usually indicated by the barometric pressure in the centre of the system — lower central pressures will result in stronger winds.

'Severe tropical cyclones' — referred to internationally as hurricanes or typhoons — have surface winds in excess of hurricane force (64 knots; 118 km/h; Beaufort 12). Short wind gusts may be up to fifty per cent or more above the average speed.

Table 6.2 shows the tropical cyclone categories used in Australia. In the Australian context, a tropical cyclone would correspond approximately to Category 1 and 2 and a severe tropical cyclone to Category 3 and above.

Figure 6.17 shows the winds generated ten metres above the sea by Australia's strongest measured cyclone, severe tropical cyclone Orson, as it passed Woodside Petroleum's North Rankin platform on the Australian North West Shelf. Orson was classed as a severe tropical cyclone since surface winds exceeded hurricane force, with a category 5 rating as the central pressure was below 920 hPa. The maximum steady wind measured at the platform at a height of 36 metres was 224 km/h. Although measurements were not taken, short gusts would have exceeded 280 km/h.

As suggested by Figure 6.17, the zone of maximum winds occurs on the 'eastern' side of a cyclone centre and is shaped somewhat like a banana. This zone of maximum winds occurs as a result of the forward (that is, generally southerly in the Southern Hemisphere) motion of the cyclone.

Table 6.2
Classification of tropical cyclones in Australia

Category	Likely central pressure (hPa)	Likely max surface gust speed in knots (km/h)	Expected effects
1	> 985	< 67 (< 125)	Crop damage; craft drag their moorings; some house damage as roof tiles lift at about 50 knots (93 km/h)
2	985–970	67–92 (125–170)	House damage; significant damage to trees, caravans and signs; heavy damage to crops; small craft may break moorings
3	970–945	92–121 (170–225)	Roof and structural damage; major crop damage; some caravans destroyed and power failures — for example, tropical cyclone *Winifred*
4	945–920	121–151 (225–280)	As above but widespread — for example, tropical cyclone *Tracy* (Darwin, NT 1974)
5	< 920	> 151 (> 280)	Extremely dangerous and widespread destruction — for example, tropical cyclone *Orson* (NW WA 1989)

Note: Gale force = 34 knots (gusts ~48); storm force = 48 knots (gusts ~70); hurricane force = 64 knots (gusts ~91)

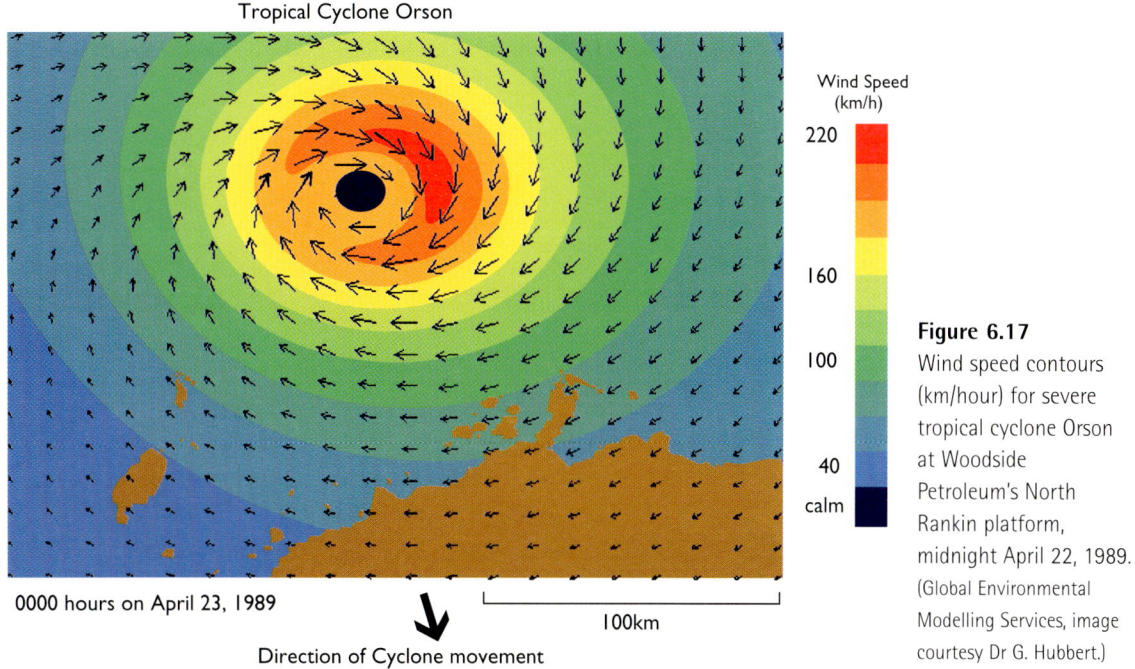

Tropical Cyclone Orson

0000 hours on April 23, 1989

Direction of Cyclone movement

100km

Figure 6.17
Wind speed contours (km/hour) for severe tropical cyclone Orson at Woodside Petroleum's North Rankin platform, midnight April 22, 1989. (Global Environmental Modelling Services, image courtesy Dr G. Hubbert.)

If the cyclone was stationary the wind field would be approximately symmetrical, rotating clockwise, with the winds increasing from very low speeds in the cyclone centre to a maximum at the 'radius of maximum winds'. Beyond this radius, which can range from 5 kilometres to more than 40, the wind speeds slowly decrease again and eventually blend with the large scale weather pattern at distances of up to 500 kilometres from the cyclone centre.

A few more words about this zone of enhanced wind: Figure 6.17 shows it concentrated on the east of the cyclone. In this example, the cyclone was heading more or less directly for the coast — in other words, south-easterly. But where would the zone have been if the cyclone had been heading south-west or due south? If you draw a line representing the trajectory of the cyclone and imagine yourself in its path, then the zone will be the right-hand semicircle. Therefore, if the cyclone had been heading south-westerly, the zone would be approximately from north-east to south-west. If it had been heading due south, then the zone would run from north to south (that is, the right-hand front side, as seen from above).

Figure 6.17 illustrates that, in this case, the area of enhanced wind is in the onshore direction. On the east coast, the area of enhanced wind is typically also in the onshore direction, although it can be in the offshore direction, depending on the cyclone's trajectory. The zone of enhanced wind can cause large destructive waves and significant coastal flooding from storm surge, which is discussed in more detail later; in some navigational books it is referred to as the 'dangerous semicircle'.

Heavy rainfall is also usually associated with tropical cyclones, and strong rain bands often extend outward from the centre for several hundred kilometres. This can cause dangerous flash flooding along the coastline when the rainfall exceeds natural drainage limits.

Conditions necessary for the formation of a tropical cyclone

Tropical cyclones form over water, not land. The pre-conditions for a tropical cyclone are:
- a large expanse of warm water (supposed minimum threshold is 26°C)
- an unstable atmospheric lapse rate for a considerable height (for example, twelve kilometres or more) above the water (specifically, a rising, expanding and/or condensing air mass remaining warmer than its surrounding air)
- a significant Coriolis force — that is, latitude greater than approximately 5° (north or south)
- weak vertical wind shear
- low level convergence/rotation in the wind field
- upper tropospheric outflow (divergence) above the surface disturbance.

Global tropical cyclones

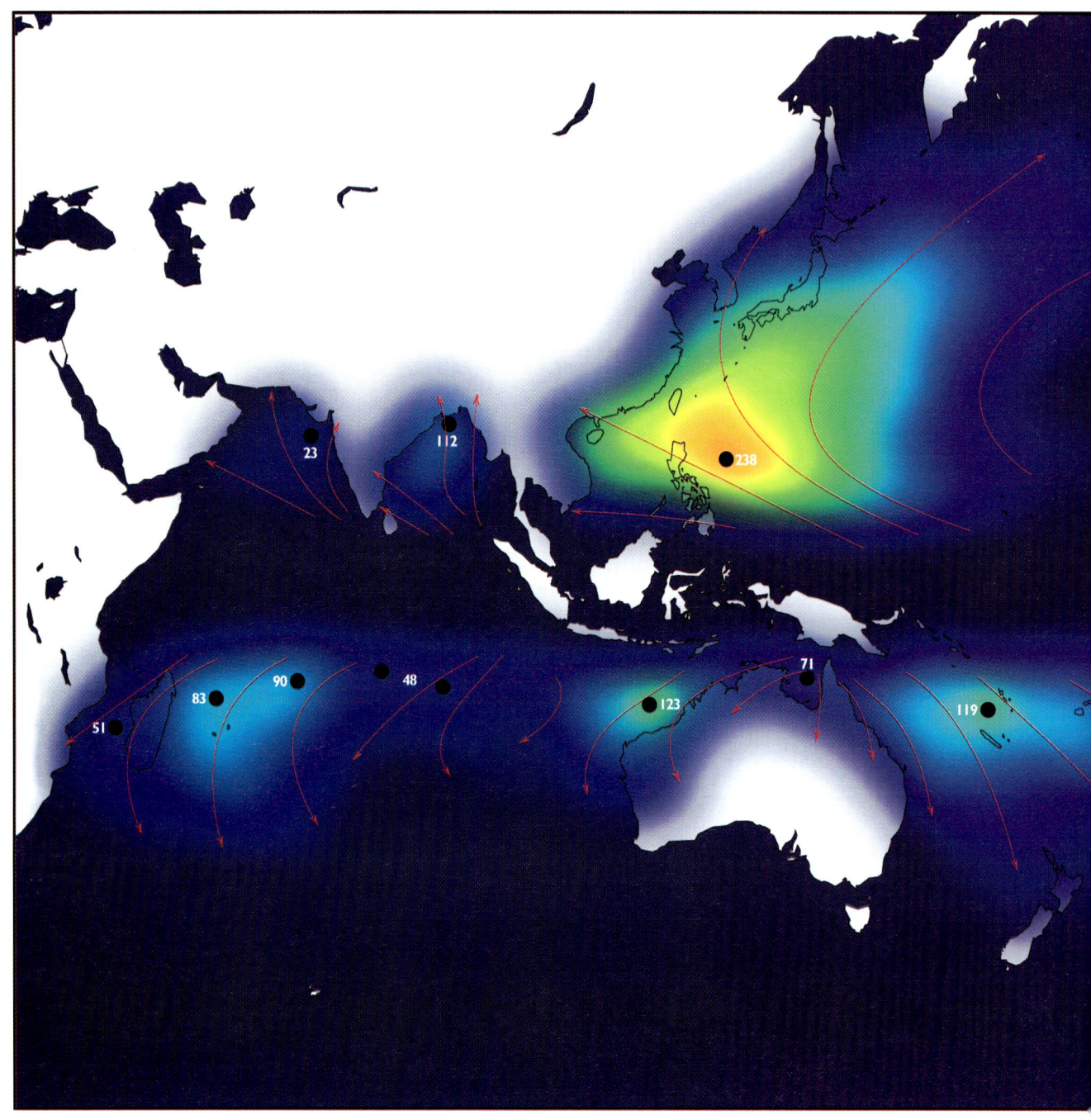

Pointers indicate the average trajectory of cyclones; these must not be taken as predictions of future cyclone events. Rather, they show indicative paths which can only be determined on the basis of many events and as shown elsewhere, individual cyclone events can vary widely from these 'averages'.

Figure 6.18
Global tropical cyclones: frequency of occurrence and indicative trajectories based on at least 40 years of data. (Original images and analysis by Frank Woodcock, Severe Weather Service, Bureau of Meteorology, Melbourne.)

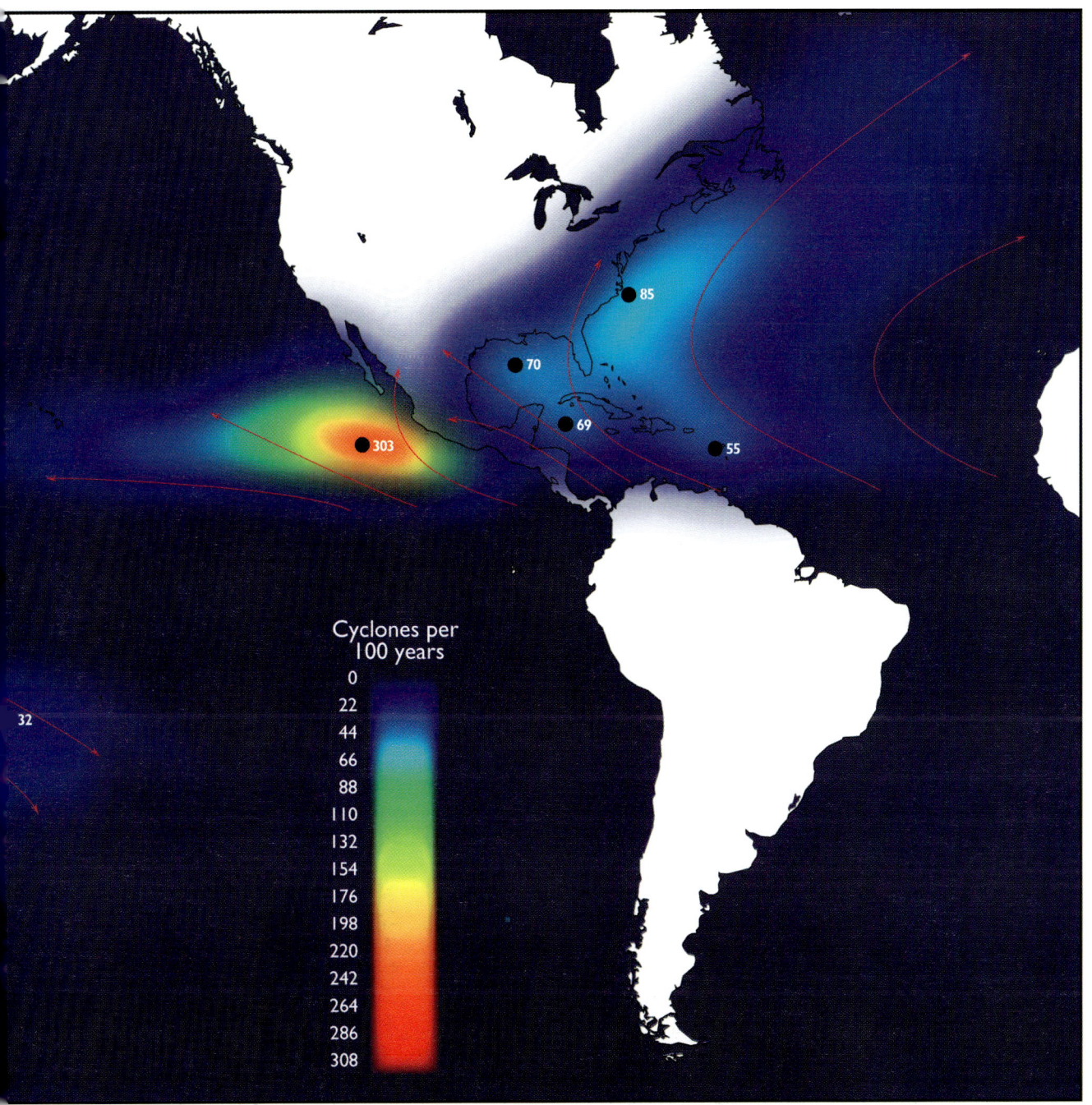

Cyclones per
100 years

0
22
44
66
88
110
132
154
176
198
220
242
264
286
308

Global cyclone analysis based on at least 40 years of record. Analysis shows frequency and indicative trajectories of tropical cyclones. White numbers show regional maxima. All frequencies expressed in terms of cyclones per 100 years expected to pass within 75 nautical miles (140 km) of any point. Some other maps give frequencies of 5° latitude/longitude squares (roughly 500 by 500 km) which may give higher values; eg, the maxima off WA in this figure is 123 cyclones per 100 years. A comparable map gives this as 250 cyclones per 100 years for the larger 500 km square.

Global patterns of cyclones

Figure 6.18 shows a global analysis of tropical cyclones based on approximately forty years of record. It depicts the number of cyclones per hundred years (in different colours) and the average trajectories of the cyclones (red curves). This figure should not be used as a way of predicting cyclones — individual, as opposed to average, cyclone events can, and do, vary widely with the data given in this diagram.

One obvious conclusion we can draw from the data is that cyclones rarely occur between 10°N and 10°S. The highest occurrence of cyclones is in the north-east Pacific, off the coast of Mexico — with a regional maximum of 303 cyclones per 100 years — and in the north-west Pacific, off the coast of the Philippines — 238 cyclones per 100 years. Although less frequent, the cyclone 'belt' off the east coast of the USA extends far further into the north Atlantic than the west coast cyclones do into the north-east Pacific; it is not uncommon for north Atlantic cyclones to continue from their point of origin to 55°N. The 'Philippine' belt also extends well to the north. Cyclones do not occur in the south Atlantic, or in the south-east Pacific, primarily because the water temperatures are usually below the threshold of 26°C.

Cyclones in the Australian region

In the period 1860–1960, tropical cyclones have caused more than 750 deaths and have wrecked 250 pearling luggers, a coastal steamer and two small freighters. The Commonwealth Bureau of Meteorology, in its publication *Understanding Cyclones*, puts the death toll higher than this — at 1500 for the period 1830 until the 'present' (presumably the early 1990s; publication undated). Even if you are not in tropical waters, you should be aware of the potentially large, or even huge, swell which, although generated in the north, can still affect southern waters.

In our part of the world there are two regional hot spots for tropical cyclones: off the coast of Queensland at about 20°S latitude — 119 cyclones per 100 years — and off the north-west coast of Western Australia — 123 cyclones per 100 years. The area of Australia's coast most affected by cyclones stretches from the New South Wales/Queensland border in the east to about Geraldton, or a little further south, in the west,

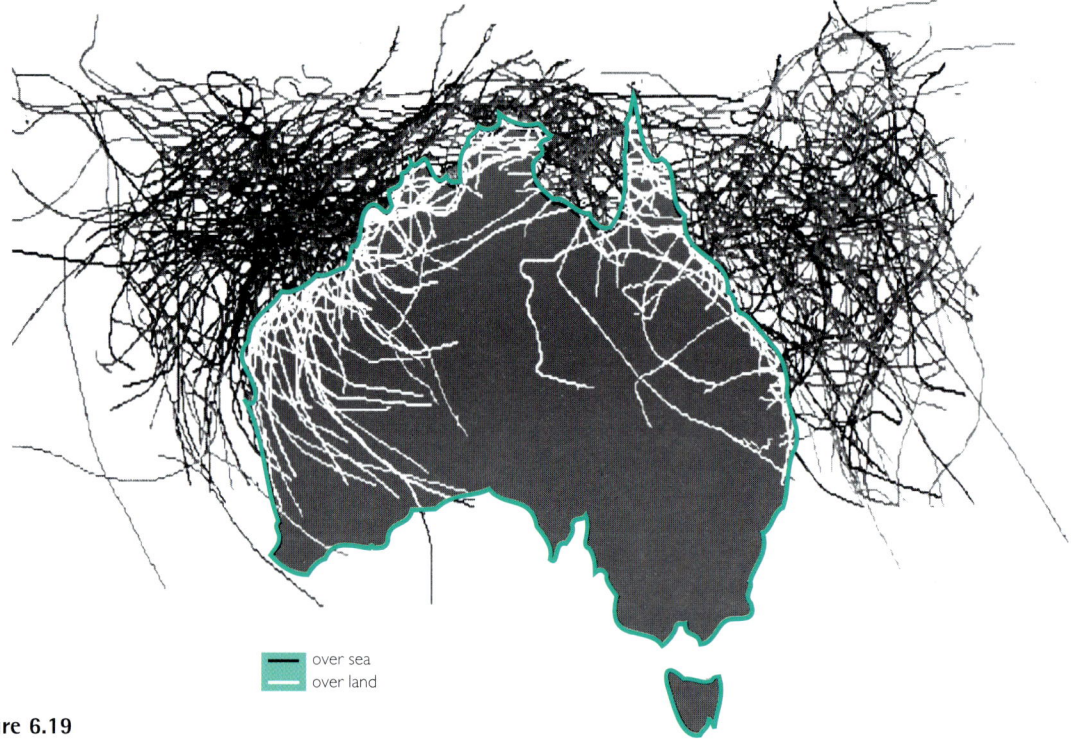

Figure 6.19

Tropical cyclones: record of trajectories in the Australian region based on 30 years of data. (Original image and analysis by Frank Woodcock, Severe Weather Service, Bureau of Meteorology, Melbourne.)

but as shown in Figure 6.19, individual cyclones do occasionally move well outside these 'limits'.

The following generalisations can be made about tropical cyclones in the Australian region:

- Cyclones form in the tropics at latitudes more southerly than 5°S during November to May inclusive, peaking in January and February. The latitude spread is considerable and depends on the location of the monsoon depression (see Chapter 2), which can occur anywhere from 5°S to 20°S, usually making its most southerly excursion in the middle of the season.
- Cyclones travel with a *mean* speed of about ten knots (individual cyclones can move much faster than this and they may become stationary as well).
- At some latitudes cyclones *may* change direction (recurve); west coast cyclones may move south-west until about 20°S, after which they may turn southerly or south-easterly; east coast cyclones, although less definite in terms of recurving, may move south-east until about 20–25°S, after which they may turn southerly or maintain their original direction (see the box).
- Cyclones usually dissipate after travelling over land for a significant distance.
- The average duration of a cyclone is 9 days, but it can range from 1 day to more than 20 days.

Figure 6.19 shows *actual* cyclone tracks in the Australian region for nearly the same period as the Figure 6.18. This 'spaghetti' map shows that individual cyclones do not necessarily behave predictably. For example, although cyclones do not travel far below 30°S, individual ones have continued south until almost the same latitude as Melbourne (38°S). Similarly, recurving – while more obvious on the west coast – is by no means certain for any cyclone.

Warning signs

When a tropical cyclone is imminent, local knowledge is of little use; it is best to rely on advice from professionals who will be tracking the system — initially by satellite, and by radar as it comes closer to the coast. Do not be deceived into believing that you can make reliable assessments about the location or movement of the centre by watching your barometer or by extrapolating movement from old satellite pictures. Be wary of any local radio reports that deviate from the official Bureau forecast — local 'experts' may be well-meaning, but their advice can be dangerous. Official warnings are updated as new information comes to hand.

A service known as Tropical Cyclone Watch Advice operates when a tropical cyclone or potential tropical cyclone threatens to cause gales on the coast within 24 to 48 hours. A Tropical Cyclone warning is issued when gales are expected to affect the coast within 24 hours.

Cyclones at sea

Cyclones at sea are a serious threat for sail and powerboaters. If you are ever caught in a tropical cyclone, the following information may save your life. Boaters are also advised to read the section on boat handling (pages 175–82).

If the full fury of a tropical cyclone is to be avoided, it is essential to have early information about its position and direction of travel relative to your vessel. Bulletins and forecasts are usually very dependable and reasonably up to date. At sea, an early indication of the approach of a cyclone is a long 'greasy' swell. Swell may be observed several days before the cyclone's arrival. In deep water, it approaches from the direction of origin (the position of the storm centre when the swell was generated). In shoaling water, this is a less reliable indicator, as the crests tend to run closer to the direction of the bottom contours.

When the storm centre is 500 to 1000 nautical miles away, the pressure may rise a little and skies will be relatively clear. Cumulus cloud, if present at all, is minimal and its vertical development appears suppressed. The barometer usually appears restless, pumping up and down, with movement of about one hecto-Pascal (1 mb).

Mares' tails (cirrus) appear when the storm is about 300 to 600 nautical miles away. Usually these seem to converge more or less in the direction from which the storm is approaching. If the storm is to pass well to one side of the watcher, the point of convergence shifts slowly in the direction of storm movement. If the storm centre will pass nearby, this point remains steady. Cirrus gives way to a continuous veil of cirrostratus. Below this veil altostratus forms, and then stratocumulus. These clouds become more dense and weather becomes unsettled. Fine mist-like rain begins to fall, interrupted by showers. By now, the barometer will have fallen perhaps three hecto-Pascals. (This cloud sequence is represented by Figures 6.8, 6.5 and 6.1, in that order. In a strong tropical cyclone, the cloud bank would be very much denser than shown in Figure 6.1.)

As the pressure fall becomes more rapid, wind increases in gustiness, and speed increases (to 22–40 knots, Beaufort 6–8). A dark wall of heavy cumulonimbus appears on the horizon — the 'bar' of the storm. When the bar becomes visible it appears to rest on the horizon for several hours. If the storm is to pass to one side the bar appears to drift slowly along the horizon. If the storm is heading directly towards you the position of the bar remains fixed. Parts of the bar drift across the sky, accompanied by rain squalls and wind of increasing speed. As the bar approaches, the pressure drops faster and wind speed increases. The seas, which have been gradually mounting, become tempestuous. Squall lines sweep past — one after the other — in ever-increasing number and intensity. With the arrival of the bar the sky becomes very dark, squalls become virtually continuous and pressure drops precipitously, with a rapid increase in wind speed. The centre may still be 100 to 200 nautical miles away in a fully developed tropical cyclone.

As the centre approaches, rain falls in torrents and the wind fury increases. The seas become mountainous. Visibility is minimal. If the eye of the storm passes overhead, the winds suddenly drop to a breeze, the rain stops, skies clear enough to let the sun through the relatively thin cloud cover, and visibility improves. The seas approach from all sides and are large and confused. The pressure drops to its lowest point: between 30 and 70 hPa below normal (that is, it could be anywhere between about 980 and 940 hPa or lower). As the wall on the opposite side of the eye arrives, the full fury of the wind strikes as suddenly as it ceased, but from the opposite direction. The sequence of conditions that occurred during the approach of the storm is reversed, and passes more quickly (as the various parts of the storm are narrower in the rear of a storm than on its forward side).

Where to set course in a tropical cyclone

We have seen that tropical cyclones have a zone of enhanced wind speeds on their 'eastern' (i.e. right-hand side, seen from above) sides, and a preferred origin and trajectory. Structurally, they comprise two semi-circles, each of which has distinctly different features.

The dangerous semi-circle and quadrant

The part of the cyclone where the wind speed is at its greatest, because the already furious wind is augmented by the forward motion of the storm, is called the 'dangerous semi-circle'. The most destructive part of the dangerous semi-circle lies ahead of the storm and is termed the 'dangerous quadrant'.

The dangerous semi-circle and quadrant tend to lie to the east of a southward-moving tropical cyclone (that is, at 3 o'clock looking from above), to the north-east of a south-easterly moving tropical cyclone (that is, at 2.30), and to the south-east of a south-westerly moving tropical cyclone (that is, at 4.30), as shown in Figure 6.20. In other words, the centre of the dangerous semi-circle lies approximately at 90° to the right of the trajectory as seen from above.

The higher wind speeds in the dangerous semi-circle cause higher seas as well. This is particularly treacherous for navigators because the directions of the wind and sea will carry a vessel directly into the path of the storm.

The navigable semi-circle

The part of the cyclone where the wind is decreased by forward motion of the storm, although far from friendly, is called the 'navigable semi-circle'. Any vessel caught in this part of the cyclone will tend to be blown away from the storm track because of the direction of the wind, and will find itself pushed further from the centre by any subsequent change in the storm's direction.

Figure 6.20 identifies the important characteristics of a tropical cyclone in the Southern Hemisphere from a navigational point of view.

Manoeuvring to avoid the storm centre

Certainly, the safest procedure for navigators is to avoid the storm centre. If action is taken

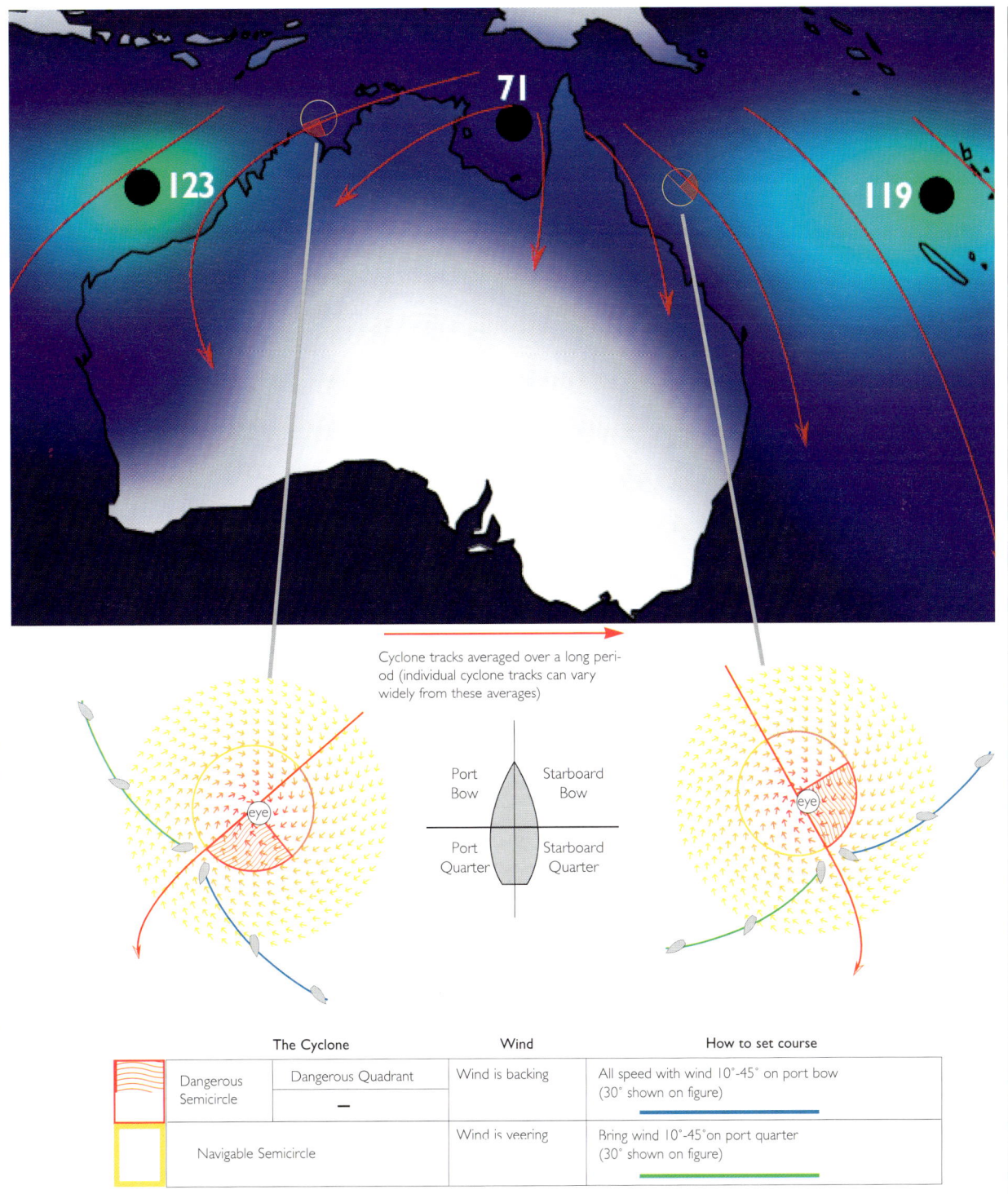

Cyclone tracks averaged over a long period (individual cyclone tracks can vary widely from these averages)

	The Cyclone		Wind	How to set course
Dangerous Semicircle	Dangerous Quadrant		Wind is backing	All speed with wind 10°-45° on port bow (30° shown on figure)
	—			
	Navigable Semicircle		Wind is veering	Bring wind 10°-45° on port quarter (30° shown on figure)

Figure 6.20

Tropical cyclones: a schematic representation of how to set course in a tropical cyclone in the Southern Hemisphere on the assumption that the cyclone follows an average trajectory.

sufficiently early, this is a matter of setting a course that will take the vessel well to one side of the probable track. You then continue to track the storm centre and revise course as needed.

However, this is not always possible and the action to take depends on your position relative to the storm centre and its direction of travel. It is *vital* to avoid passing within fifty nautical miles or so of the centre of the storm. It is preferable, but not always possible, to keep at a distance of at least 200 nautical miles.

If a storm is suspected in your vicinity it is best to continue on course until:

- the track of the tropical cyclone is known (by contact with meteorological sources), or
- the barometer has fallen 5 hPa below normal (corrected for normal daily variation), or
- the wind has increased to force 6 and the barometer has fallen at least 3 hPa.

If the wind is backing — direction changing anticlockwise — the ship must be in the dangerous semi-circle. Proceed with all available speed with the wind 10–45°, depending on speed, on the port bow. As the wind backs, the ship should turn to port, thereby tracing a course relative to the cyclone's course as shown. Keep a very close watch on wind direction and if the cyclone recurves, you will have to adjust your course continually (same rules apply). If obliged to heave to, do so with the head to the sea.

If the wind is veering it is likely that the boat is in the navigable semi-circle. Bring the wind onto the port quarter and hold course, making as much way as possible. If obliged to heave to, do so with the stern to the sea. As the wind veers, the ship should turn to starboard as shown in the diagram.

If the wind remains steady in direction, so that the ship seems to be nearly in the path, the ship should proceed with wind well on the port quarter with all available speed. When well within the navigable semi-circle proceed as indicated above.

If the boat is on the storm track but behind the storm, avoid the centre by the best practicable course, keeping in mind the tendency for tropical cyclones to curve southward then south-eastward in the Southern Hemisphere. If there is insufficient room to run, when in the navigable semi-circle, and it is not practicable to seek shelter, the ship should heave to with the wind on her port bow (Southern Hemisphere).

If you are in harbour when a tropical storm approaches, it may be best to go ashore and leave your boat to 'fend for itself' (boats are expendable, lives are not). If time permits, you may take your craft as far up a creek as you can, and leave it there. Riding out a tropical storm in a harbour or anchorage, particularly if the centre passes within fifty nautical miles of you, is unpleasant and hazardous. This is especially so if there are other ships nearby. Even if berthed alongside, or if special moorings are used, a ship cannot be entirely secure.

TEMPERATE CYCLONES

Temperate cyclones, lows or depressions in mid to high latitudes do not occur as isolated entities (see Figure 2.16). They are frequently seen in a well-ordered 'family' group, the parent or primary low pressure system and associated frontal band spawning new secondary (even tertiary and other) low pressure systems. The first stage is a wave or secondary low development that forms on a cold front trailing out from the parent depression. Given favourable conditions, the wave low will deepen into a mature depression with its own warm and cold sectors. Eventually, as occlusion takes place, the depression will weaken and fill. The time span for this cycle varies a great deal: new wave lows may be very transient and disappear within hours, while others may live for more than ten days as they move from west to east over the Southern Ocean. In their latter life they often track to the south-east, disappearing close to the Antarctic ice shelf.

As discussed in Chapter 2, many surface pressure systems — including depressions, secondary lows and frontal systems — are moved or steered by the prevailing upper westerly winds, and their speed can vary quite considerably. An average speed is approximately 20 to 25 knots. The frontal bands that move in the circulation of the low vary even more, and occasionally cold fronts can burst north and north-east towards southern Australia at up to 50–60 knots.

Depressions are associated with low level convergence and slow widespread ascent, with typically low and middle level stratified cloud (altostratus, nimbostratus). If warm moist air is drawn into the circulation (most often from the north and north-east) then considerable cloud and rain can be expected. Primary low pressure systems are mostly located well to the south of Australia while secondary lows form over waters around the southern half of the continent.

Primary low pressure systems are most frequently seen at around 50°–55°S, moving slightly closer to the equator during winter and to Antarctica in summer. The cyclonic circulation typically extends 1000 nautical miles from the centre. The central pressure of these lows is generally 980 to 960 hPa, though intense systems are occasionally seen with readings down to 930 hPa.

The wind distribution is not uniform or symmetrical. Gale force winds are common, especially close to active fronts and in the north-westerly wind segment (similar in principle to the tropical cyclone but with the enhanced wind zone displaced because of their different trajectory). In rapidly deepening secondary low development, storm force winds (50 to 70 knots) can be experienced, sometimes close to or across Tasmania, over the Great Australian Bight or near the southwest corner of Western Australia. These explosive developments, sometimes referred to as 'bombs', can see the surface air pressure fall by 12 to 16 hPa over a twelve-hour period. Cold air advection is common, with satellite images showing tightly packed large cumulus/cumulonimbus clouds producing heavy showers.

The huge extent of gale force winds makes it impossible to hide from or avoid the nasty conditions (wind and seas) that often develop across the Southern Ocean. Lost clipper ships last century and more recent losses during around-the-world yacht races bear testament to the conditions that can occur.

STORM SURGE

Storm surges are temporary rises or falls in sea surface height driven by surface winds and changes in atmospheric pressure during the passage of a tropical cyclone near, or over, a coastline. The severity of the storm surge depends on the strength and duration of the tropical cyclone and the coastal terrain. Further contributions to the height of the sea level are the effects of breaking waves at the coastline ('wave setup') and the tides. Severe storm surges can cause inundation of low-lying coastal plains and flooding of river systems and, combined with wind-generated waves, significant coastal erosion and destruction of property.

The storm surge and wave setup are directly dependent on the characteristics of the cyclone, including its intensity (as measured by its central pressure and radius of maximum winds), path and forward speed. For a given cyclone, the storm surge and wave setup will also be highly dependent on the local ocean bottom topography and coastline shape.

At sea

Cyclones can induce a temporary increase in sea level in open waters due to the 'inverse barometer effect' — a doming up of water across an area 50 to 100 kilometres wide corresponding to the area of lowest pressure in the cyclone.

In open waters, however, the most significant effect of a tropical cyclone is the large seas generated by the very strong winds. Figure 4.6 shows that very large seas can be generated by cyclonic winds. For example, if the cyclone is moving at 20 km/h and has a zone of gale force and above winds 500 kilometres wide, it would take 25 hours to pass over a fixed point on the surface (15–20 hours is often quoted); a combination of 60 knots (the maximum wind speed in the figure) blowing for 20 hours would produce 13 metre seas. Wind speeds in cyclones can far exceed 60 knots and so larger seas are possible.

Approaching land

In shallow coastal regions the storm surge generated by the lowering of surface pressure is usually small compared with the storm surge produced by the wind on the ocean surface. Surface winds exert a stress on the ocean, which results in movement of water in the direction of the wind. The Earth's rotation then deflects the current to the left in the Southern Hemisphere. When a coastal barrier blocks the deflected flow, elevated sea levels are produced at the coast. This is known as 'surface wind drift' or 'coastal current setup'.

A second effect produced by the wind, 'wind setup', is produced when a steady wind blowing over the sea surface along a coastline maintains a slope on the water surface at right angles to the coast.

The tropical cyclone winds therefore contribute to storm surges on the coast. As the cyclone approaches land, these wind-driven effects begin to dominate and can add significantly to the inverse barometer effect of open waters, to produce large storm surges and coastal flooding. The largest storm surge ever actually measured in Australia was seven metres, associated with tropical cyclone *Douglas Mawson* in 1923, although in 1899 a surge in Bathurst Bay associated with tropical cyclone *Mahina* was estimated to be fourteen metres.

As a cyclone approaches the shallower waters of the continental shelf, the water mass begins to 'feel' the bottom and waves begin to break, giving rise to wave setup, which further increases the water level. This effect is increased in areas with wide and shallow shelves like the North West Shelf and the Gulf of Carpentaria. Locally important factors include the shape of the coastline and near-coast bathymetry, as well as the presence of reefs, which usually act to reduce the effective fetch and hence reduce the size of the seas. The direction and speed of advance of the cyclone are also important; as mentioned earlier, the speed of the cyclone will alter the time available to develop a sea.

Cyclone landfall versus time of tide

The maximum storm surge usually coincides with the time of landfall of the cyclone. So, in this simplified explanation, the major effects of the cyclone will be felt over a period of 15 to 20 hours if it is moving at the average speed of about 20 km/h. Some areas of Australia's coast experience one tide per day (diurnal tides) but most experience two (semi-diurnal tides). Figure 5.20 shows that tide range varies from almost no tide up to about eleven metres at Derby in Western Australia.

The conditions for large tidal ranges and large storm surges are similar, in that both usually occur where there is a wide and relatively shallow continental shelf or a narrow coastal feature. Generally, therefore, places that experience large tidal ranges and are low lying are the most vulnerable to storm surge. If populated areas are near

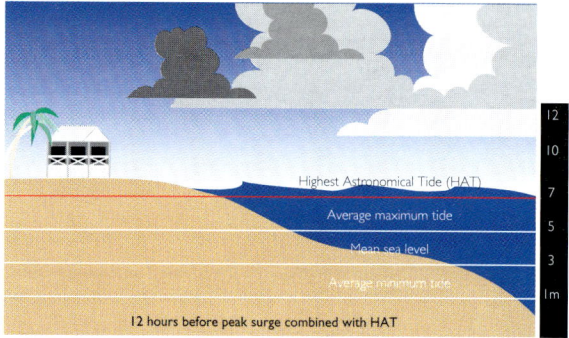

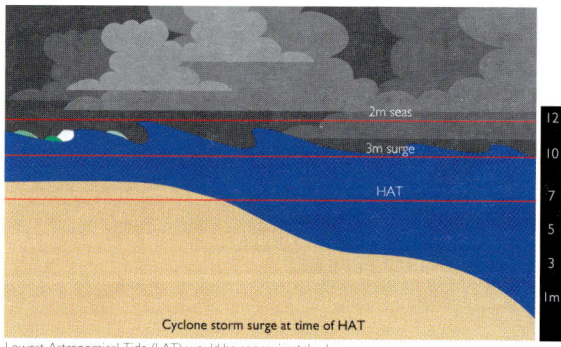

Figure 6.21

Storm surge associated with a tropical cyclone; the timing of the surge in relation to the timing of high and low tide is critical in determining the risk to low-lying areas of the coast.

deltas, estuaries and other areas not far above normal sea level, the potential for damage is very high indeed.

Let's assume that there is a single tide per day and that the storm surge begins building as the tide is dropping. If the two are in phase the low tide could well occur at the same time as the surge and the hazard will be reduced; if tide and surge are of equal size the hazard will be minimal. If, however, the surge — which may build for 6 to 7 hours in a 'typical' cyclone crossing — occurs at the same time as the high tide, its height may be added to the tide and the results can be disastrous. Figure 5.22 showed tide traces for a year for 25 key sites on the coast. Now consider the timing of the storm surge in relation to this figure. Depending on where you are, and in relation to the spring–neap cycle, the surge could be completely neutralised or alarmingly increased by the timing of the tide.

A very serious situation could occur if the surge approached a low-lying coastal area with a large tidal range at a spring tide. Figure 6.21 shows a worst-case scenario. It is a site with an 'average' high tide of 5 metres and average low tide of 1 metre above the local datum — that is, an average tidal range of about 4 metres. It has a Highest Astronomical Tide (HAT) of 7 metres and Lowest Astronomical Tide (LAT) of about minus 1 metre — that is, a maximum astronomical tide range of 8 metres.

The top panel of the figure shows a cottage 'safely' above the normal high tide, at about 5 metres. The middle panel shows a deteriorating situation. Middle-level turret clouds (*Altocumulus castellanus*) are building and the Earth, Moon and Sun are aligned to produce a very large tide — in this case 2 metres above normal high tide. The bottom panel shows a 3 metre surge topped with 2 metre seas to produce a 5 metre surge — large but certainly not extreme. The surge combined with the HAT has produced a 12 metre wall of water threatening all low-lying areas.

Although not shown on the figure, had the same surge occurred at the time of LAT, the result would be only 4 metres, well below normal high tide and out of danger. Even an average low tide of 1 metre combined with the surge would still be below HAT.

TROPICAL SQUALL LINES

Tropical squall lines are a feature of the northern Australian coastline in summer. They are less common on the north-eastern coast of Queensland. Often occurring at night or in the late afternoon, they typically move east to west on the northern coast or south–north or east to west along the north-western coast. They are accompanied by thunderstorms and often by severe squalls.

Warning signs

Tropical squall lines may be accompanied by 'roll' cloud (as are Southerly Busters) or convective activity. Watch for advancing lines of cloud or thunderstorms (see Figure 6.16).

BOAT HANDLING IN BAD WEATHER

'Wave Motion and Craft' in Chapter 4 is also relevant to this section. Its implications relate to all waves, not necessarily those associated with severe weather.

Few subjects could be more difficult to cover than how to handle a boat in bad weather. This is because no amount of theory can ever really prepare you for the fear that comes with the realisation that the weather and the seas are so severe that you may lose your craft — or even your life. The following advice is not a substitute for first-hand experience; you simply must learn for yourself how your boat will handle in big seas, beam seas, and strong wind — and, if called for, whether your boat prefers to heave to, or perhaps to run before the weather. The best and safest way to do this is to take your boat out in moderate conditions and experiment a little. This can be done in many places only a few nautical miles out to sea, or even within a harbour.

Most of the following section is applicable to boats generally. Specific information for powerboats and wind-driven craft follows. Normally, wind-driven and powerboats require different handling techniques in severe conditions. You should also (re)read Chapter 3 as well as this chapter.

Remember that avoidance is by far the most effective strategy against severe weather.

Avoid bad weather as much as possible. Remain in port an extra day or two if you have to. Be aware of safe anchorages along your route and places where you can take refuge. Listen to radio weather reports.

The boat

Before setting out ensure that you have a seaworthy craft and that all the charts and publications (for example, pilot books, tide tables, Almanac) that you require are on board. It is important that all navigation and radio equipment is working and, if possible, have an EPIRB (Emergency Position Indicating Radio Beacon) or a lifeboat radio on board. Make sure all safety equipment (for example, life jackets, distress flares) is on board and in good working order — even if the sun is shining and the barometer is stable. Never leave your moorings without it.

Regardless of the type of boat you are on, the person on watch must know the regulations for preventing collision at sea. These are available at virtually any boat shop or place that sells boating material. If in doubt contact one of the large boating associations and seek their advice. The rules of way are simple and intuitive; it's not good enough knowing these rules 'in theory', you must know them in practice and be able to apply them instinctively.

On encountering unavoidable bad weather

Make sure you can recognise the important cloud types and other signs associated with severe weather. If you have a warning that the weather is closing in, ensure that the boat is secured. Then do the following:
- Close all possible watertight openings and stow or secure all loose gear.
- Tie down any movable heavy weights so that they will not be able to shift.
- Show the safety equipment to the crew and demonstrate for all on board how to use the bilge pumps. Do you really think that you and your crew can bail out your craft by hand? No, if the weather is so bad that the boat is taking on water, all crew will be needed for other tasks.
- Alert either the authorities (for example, Water Police), or friends, of your position.

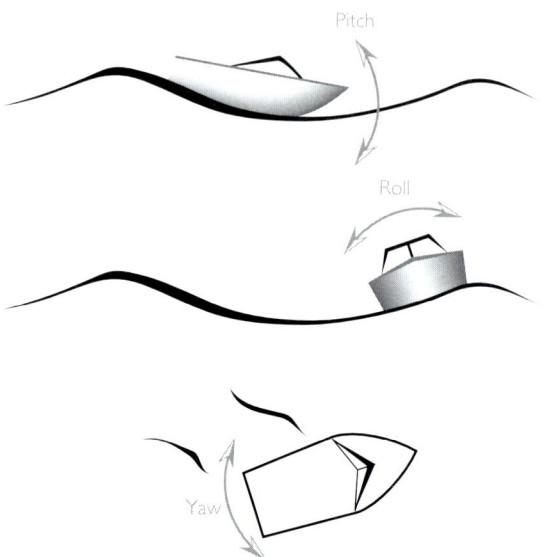

Figure 6.22
The three common movements of a vessel – pitch, roll and yaw.

The individuality of boats

No two boats behave in the same way. The particular behaviour of your boat will depend on many things, including the form and shape of the hull, the boat's construction and shape, its trim, speed, weight and load, the position and area of the rudder, the strength and direction of the wind and current, and the nature of the sea. The differences in boat behaviour are most noticeable in the steering characteristics. The response of boats that steer by changing their thrust direction (outboards and stern drives) is different from that of boats steered by rudders. The response of heavy displacement hulls to helm changes is quite unlike that of light planing hulls. When the going gets heavy, each hull design reacts differently and even individual boats of the same class may behave differently because of load and trim.

Effects of wind, waves and current

As a general rule, boats with low freeboard and superstructure but relatively deep draft (for example, sailboats) will be less affected by wind and more affected by currents than boats with light draft and high freeboard and deckhouses (motor boats), which float relatively high in the water.

The relatively higher bow and cabin forward of motor boats present the greatest exposure to wind. Thus their bows will be more affected by wind than the sterns. With wind abeam, the bow will tend to be driven to leeward more than the stern, so a motor boat will usually need a certain amount of weather rudder angle to compensate.

It would be useful here to define the three common movements of a boat in water: pitch, roll, and yaw (see Figure 6.22).

If you are facing the bow of the boat, pitch is the up-and-down movement perpendicular to the direction of travel. Roll is the side-to-side movement. And yaw is effectively the slip of the boat in the water — that is, any difference between the direction in which the boat is travelling and the direction in which it is facing. All these movements may be defined relative to an axis of rotation, as shown in the figure.

Bad visibility

Fog — defined as visibility reduced to less than 1000 metres by water droplets — is not very frequent around Australia (it features in less than 2 per cent of all ships' observations for all months). Fog over land is most frequent in winter; it usually occurs around dawn and disperses after a few hours. Near big cities, pollution increases the probability of fog. At times, patches of fog formed on land overnight drift out over coastal waters of estuaries and ports.

Fog at coastal stations is partly sea fog and partly land fog. The maximum occurs in winter on coasts sheltered from the sea (Melbourne) and in summer where coasts are very exposed (Gabo Island, Wilsons Promontory). Serious deterioration of visibility can occur during heavy rain but is usually over within an hour or less. Very rarely, more prolonged deterioration may occur due to thick haze from bush fires or large quantities of dust from dust storms.

If visibility is bad, you will need to detect other boats by sight, sound or radar early enough to avoid them. It is best to be able to stop in a short time, rather than resort to violent evasive manoeuvres to avoid a collision. Post a lookout — the more lookouts the better — as far forward as possible to watch for hazards such as rocks, breakers or buoys. Lookouts should not only be looking, but also listening, for other boats. Slow or stop engines as soon as your lookout has heard something. When underway in fog, at intervals slow your engines to idle or shut them off entirely and listen for fog signals of other vessels.

Radar comes into its greatest value when visibility is poor. If you have radar, use it, but not at the expense of posting a true lookout. Even if you do not have a radar, you should carry a passive radar reflector; this is the time to open and hoist it as high as possible.

Using oil on rough water

Oil that forms a slick on the surface can prevent seas from breaking; this is most effective in deep water. Although vegetable oil or fish oil tend to be used, the heavier and thicker the oil, the more effective it is. The best way to apply oil is to hang 1 or 2 canvas bags holding 1 to 2 litres of oil over the side or with the sea anchor and prick the bags with a sail needle to allow the oil to leak more easily.

The position of the bags depends on circumstances, as you want the oil to be to windward. If you are running before the wind, hang them on either bow and allow them to tow in the water. With the wind on the quarter, the effect seems to be less as the oil goes astern while the waves come up on the quarter. If you are lying-to, hang one from the weather bow and another further aft, using sufficient line for them to draw to windward while your boat drifts.

If you are in surf or in waves breaking on a bar, the effect of oil is uncertain, and nothing can prevent the larger waves from breaking under such circumstances. To cross a bar with a flooding tidal current, pour oil overboard and allow it to float in ahead of your boat; then follow, with another bag towing astern. The oil is less dependable in this situation, however.

> Clearly, discharging oil onto the sea is a technical violation of anti-pollution regulations even though only a small quantity is used. Use oil only in an emergency involving immediate danger.

Powerboat handling under adverse conditions

Many powerboats have run into trouble, or have been lost, at sea — some whilst relatively close to the shore. The following information is in no way intended to give powerboaters a complete guide to handling under adverse conditions. For this, consult References (page 207), which includes

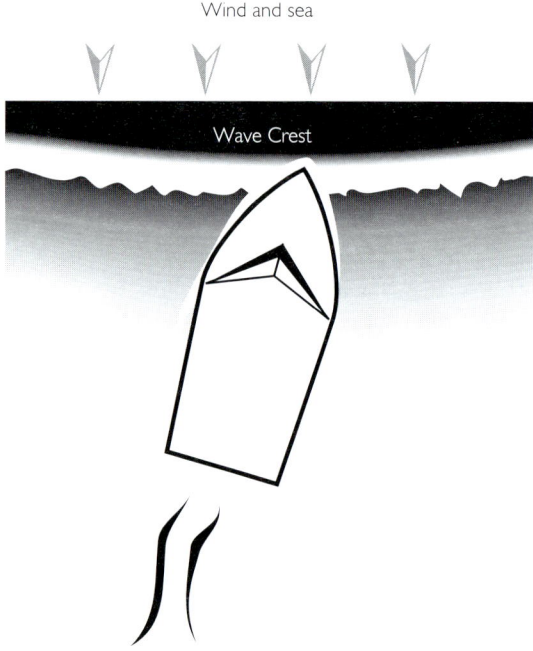

Figure 6.23
Steaming into the weather – keep the sea nearly dead ahead and reduce engine speed to just maintain headway and steerage.

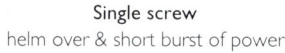
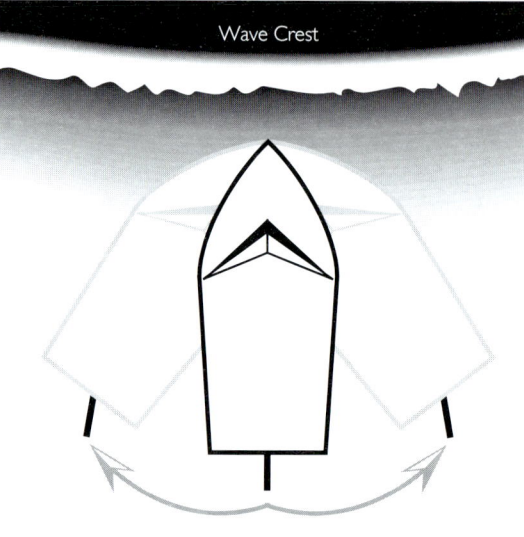
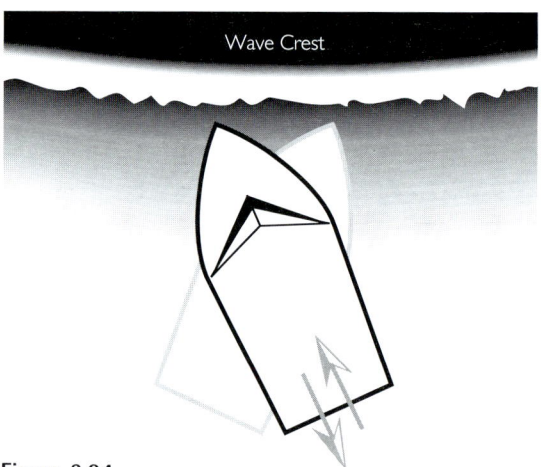

Figure 6.24
Manoeuvring a single and twin screw boat to minimise headway.

several works on the subject; you can also contact the Australian Powerboating Association (Sydney) for advice.

Unlike a sailboat, a motor boat cannot normally improve its weatherliness and can only minimise the effects of weather by altering course. Safety depends on being able to maintain steerage way. The safest course may not be the shortest one, and keeping to deep water for far longer than expected may be the only way to survive.

Rough weather is a relative term and unless the boat is big enough and powerful enough to shoulder aside the rising seas or drive through them, then your intended course must be abandoned, and instead you must steer a course that will prevent damage and make capsize unlikely.

There are two main options available in bad weather: either you stay at sea, or you run for cover.

Staying at sea

If staying at sea, you usually have three courses of action open to you:
- steam into the weather
- run before the sea
- heave to.

Steaming into the weather/head seas
Steaming into the weather should not be any trouble for the average, well-designed boat. If the boat starts to pound or the seas become too steep, ease the throttle and slow down. This gives the bow the chance to rise to each wave instead of being driven hard into it.

If conditions become really bad, slow the engine until you are making bare headway, holding your bow at an angle of up to 45° to the swell (Figure 6.23). By heading into the waves at a

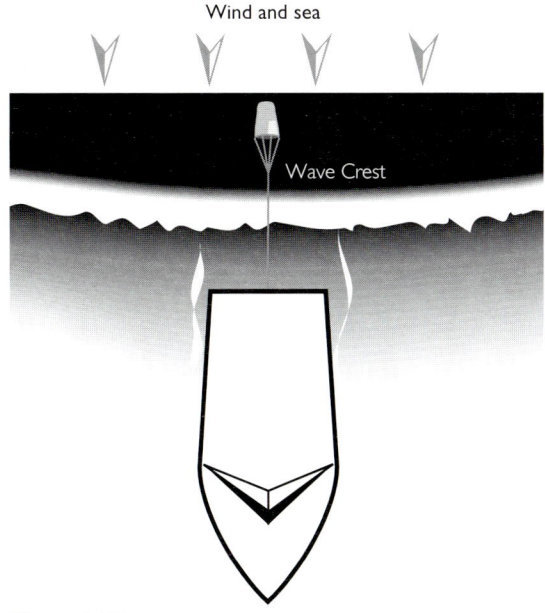

Figure 6.25

Reducing engine speed, or even reversing engines or towing warps, may help maintain steerage when running before the sea.

slight angle, wave impact will be reduced and the boat will be induced to lift to the waves and gently roll over each successive crest (the greater the angle to the waves, the more the boat will roll and this may increase sea-sickness). The more you reduce headway in meeting heavy seas, the less strain there will be on the hull. If the seas lift the screw clear of the water and the engine races, severe engine damage can result. This sounds dangerous, and it may be. There is a rapid increasing crescendo of sound as the engine winds up, then excessive vibration as the screw bites the water again. To avoid this, slow down and change your course until these effects are minimised.

Maintain steerage so the boat can be readily manoeuvred. In a head sea (swell approaching the bow), a vessel with too much weight forward will plunge (severe pitch) rather than rise. Under the same conditions too much weight aft will cause

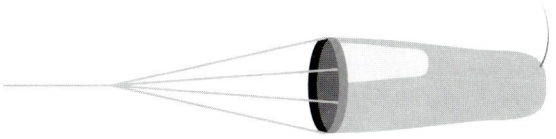

Figure 6.26

Sea anchors — canvas funnel type with a trip rope at the smaller end and a make-shift emergency sea anchor.

the boat to 'fall off' the swell and lose steerage. If the boat is poorly trimmed, change the weight aboard (tanks, other heavy gear) and, if necessary, ensure that passengers and crew maintain their given positions, to avoid swamping the boat. If your boat is in good trim and you nurse the wheel and the throttle (by varying your course and slowing or even stopping momentarily), you should weather moderate gales with little discomfort. Figure 6.24 shows the basic technique for manoeuvring single and twin screw boats while maintaining minimal headway.

Obviously, this assumes that you have adequate fuel to maintain your position for some hours. At the very least, if the seas are so large and dangerous that turning around and running for cover is not a viable option, steaming into the weather will buy you time to contact help.

Running before the sea

If the swell is coming from directly behind, running directly before it is an option if your boat is not thrown off course (see also 'Wave Motion and Craft', page 107). Turning to run before the sea, in an endeavour to absorb the wave impact on the stern rather than the bow, demands a great deal of skill and practice; it is an acquired feel. In heavy seas there is always the danger that the boat will accelerate down into the trough and yaw so badly that it may broach — that is, be thrown broadside, out of control, into the trough — and present its beam to the next oncoming wave. There is then a real danger of the boat rolling over and capsizing. Take every possible action to avoid broaching.

Modern powerboats are designed to provide a large, comfortable cockpit or afterdeck. The increased width at the transom increases the tendency to yaw and possibly broach. Slowing down to let the swells pass under your boat usually reduces this tendency to yaw, or at least reduces its extent. Any broaching tendency must be corrected before the swerve has time to develop. This is best done by a swift application of opposite helm to keep the stern of the boat as square on as possible to the wave that is approaching it. Each approaching wave will lift the stern, and only throttling back will stop the stern trying to overtake the bow into the hollow of the following trough. Cutting down engine speed will reduce the strain on the motor caused by alternate stern-down labouring and stern-high racing. While seldom necessary, you can consider towing a heavy line or drogue astern to help check your boat's speed and keep it running straight, as shown in Figure 6.25.

Excessive speed down a steep wave may cause a boat to 'pitchpole' — that is, drive its head under in the trough, tripping the bow, while the successive crest catches the stern and throws her end-over-end. If there is a risk of this, keep the stern down and the bow light and buoyant, by shifting weight aft as necessary. Adjust your boat's trim bit by bit rather than all at once, as too much weight aft may cause the boat to be swamped by the following sea. A zigzag track that puts the swells off your quarter, minimising their effects, can also be used.

Heaving to

If conditions get so bad that the boat cannot make headway and begins to take too much punishment, it is time to heave to, a manoeuvre whose execution varies for different vessels. Motor boats will usually be most comfortable if brought around and kept head to seas, using just enough power to make steerage while conserving fuel.

In extreme weather you may use a sea anchor — generally a heavy fabric cone or hemisphere with a hoop to keep it open at the mouth, which is streamed over the bow. A sea anchor is not meant to go to the bottom and hold, but rather will usually float just below the water surface and produce a drag that keeps the boat's bow into the wind (Figure 6.26).

Sea anchors work only if there is sufficient sea room, because the boat will steadily drift to leeward. When a boat is driven down onto a lee shore, the regular ground tackle must be used to ride out a gale. Use your engine to ease the strain during the worst of the blow, and have the anchor on the longest possible chain to give it the best chance to hold.

Running for cover

If running for cover, you may have to contend with:
• beam seas
• tacking
• running an inlet.

Beam seas

If your course for cover takes you directly or diagonally across the direction of swell, the swell waves will strike your boat side-on. These so-called 'beam seas' are the most dangerous, causing the boat to roll from beam to beam, possibly shipping water as it does so. Any water finding its way into the bilges will add its momentum to exaggerate the rolling. Water in the engine space can short-out the electric leads, immobilising a petrol engine or, in a diesel, saturating the air filters and choking the engine — most operators of diesel engines know that any water in the fuel usually spells disaster.

Tacking

If your course requires you to run broadside to the swells it is best to run a series of 'tacks', much like a sail boat. Set a course to take wind and waves at 45°, first broad on your bow then broad on your quarter. With the wind broad on the bow the boat's behaviour should be satisfactory; on the quarter, the motion may be less comfortable but at least it will be better than running in the trough. Make each tack as long as possible. To turn sharply, allow the boat to lose headway for a few seconds, throw the wheel hard over, then suddenly apply power. This is particularly effective with single screw boats, which will usually turn quickly. With twin screw powerboats the engine on the side in the direction of turn may be throttled back or even briefly reversed.

Running an inlet

Under adverse conditions, inexperienced powerboaters often run for shelter rather than remain safe at sea. In fact, an inlet or narrow harbour entrance is one of the worst places to be in violent weather. The shallower water builds up treacherous surf that often cannot be seen from seaward. If you do not have local knowledge, laying off or anchoring may be better.

If possible, follow a local boat, which will usually take the smoothest water (usually the deepest). If you must get through without local help, don't run directly in, but wait outside the bar and watch the waves as they pile up at the most critical spot in the channel (usually the shallowest). They generally come along in groups or sets. The last wave in the group will often be bigger than the rest; by watching closely you can pick it out.

When you are ready to enter, stand off until the last wave of a set has broken, then run through behind it. Watch the water both ahead and behind your boat; control your speed and match it to that of the waves. An ebb tide gives worse conditions on the bars than the flood tide because the outgoing water works against and

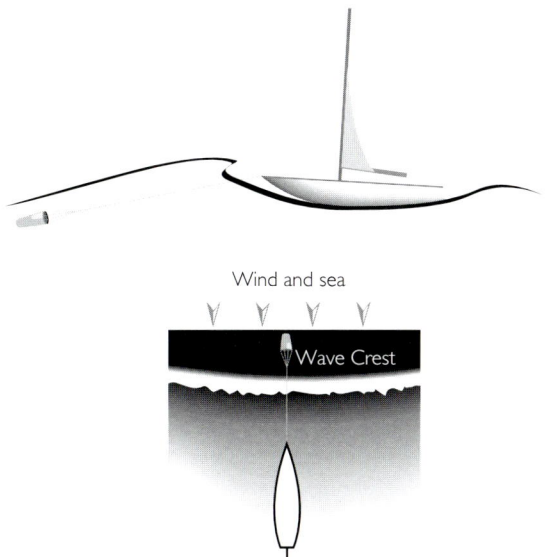

Figure 6.27
Lying to a sea anchor – note the aft storm sail to keep the bow into the weather and sea, and the helm lashed amidships.

under the incoming swells. If the sea looks too bad with an ebb tide, wait until the flood tide has had a chance to begin.

Sailboat handling under adverse conditions

Boats with ballast keels (sailboats) are better designed to ride out big seas. The two main problems for motor boats — capsize and swamping — do not occur as easily, since the ballast keels ensure that, if there is a capsize, it should be only temporary, and self-draining cockpits prevent swamping. However, many sailboats have been lost at sea, and the following information is not intended to provide sailors with a complete guide to handling under adverse conditions. It is merely intended to identify the main strategies — the sailor must seek instruction on how to set sail, under what conditions to heave to or lie to a sea anchor, and so on. References, page 207, lists several large works on the subject. You can also contact the Australian Yachting Federation (Sydney) for advice; the AYF runs courses and has a wealth of experience.

If the weather begins to deteriorate and the wind increases, your first strategy should be to reduce sail area. Normally this is done by reefing the mainsail just enough to maintain full control. If you carry too much sail, you risk causing struc-

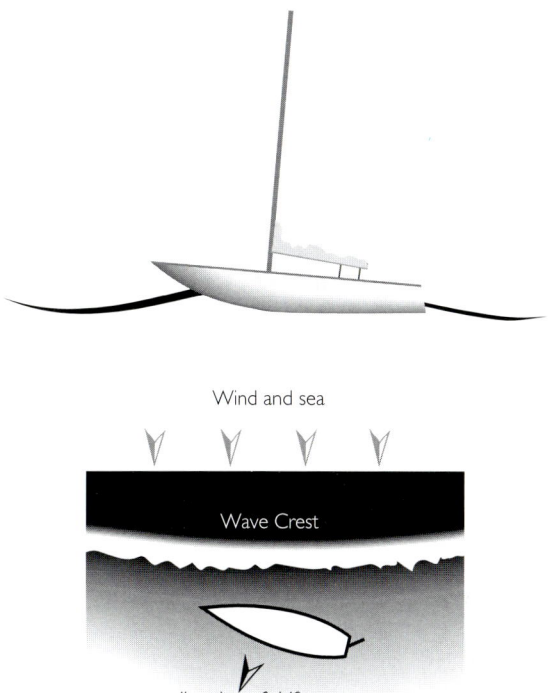

Figure 6.28
Lying ahull – all sails are furled and the helm lashed to leeward.

tural damage to the boat. If the wind continues to increase to severe levels, and you can't control the boat even with a fully reefed sail, four courses of action are normally available:
- lying to a sea anchor
- lying ahull
- heaving to
- running with the storm.

Lying to a sea anchor

The sea anchor will hold the boat's bow into the wind, exposing the strongest and most streamlined part of the hull to the seas. A small riding sail on the mizzen mast or a storm jib hanked to the mainmast backstay will also help keep the boat's head into the wind. Figure 6.27 shows a typical configuration. Lying to a sea anchor is probably most useful when the crew needs rest and the weather is bad but not extreme, but obviously the boat is going to drift to leeward and so you must have adequate sea room.

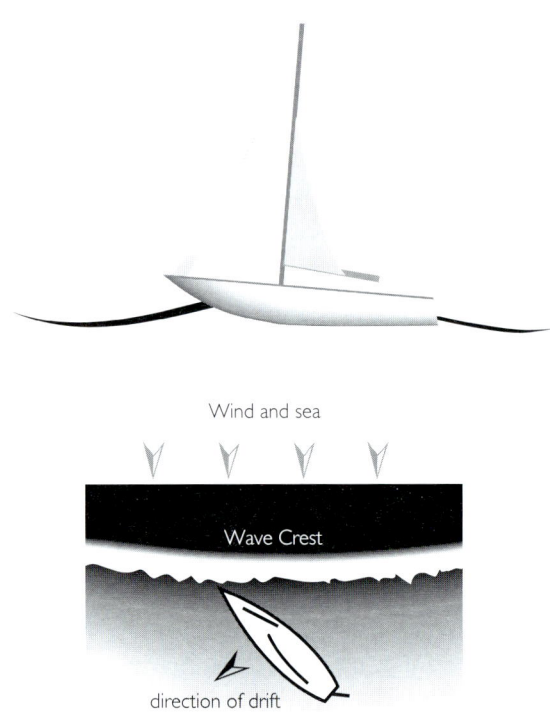

Wind and sea

Wave Crest

direction of drift

Figure 6.29
Heaving to — note the backed storm jib and opposing helm.

Lying ahull

When the boat is lying ahull, sail is reduced completely or at least to a small steering headsail, the tiller is lashed a little to leeward to keep the vessel's head from falling off, the hatches are battened down with the crew below decks, and the boat is left to ride it out (Figure 6.28).

This experience tends to be somewhat uncomfortable as the boat will lie at right angles to the wind and be buffeted at will by the seas, mostly broad to the beam. Lying ahull has the disadvantage of placing the boat in jeopardy should the beam seas begin breaking over it.

Although much has been written about the virtues of lying ahull, it is not recommend unless absolutely necessary. With all crew below decks, the boat is left without a watch, and thus is not in any position to avoid a collision. This is especially true of a collision with another boat — with all

the wind and spray, a large vessel is unlikely to spot a sailboat running bare poles (no sails up). With no sails up, the boat is a poor reflecting surface for radar and will be much harder to detect. In many ways, heaving to is a more seamanlike strategy as it does not preclude a watch, nor some of the crew from resting below.

Heaving to

This is a much more comfortable method for riding out a storm. The wind is brought on the weather bow and the vessel holds itself in that position, where it will ride most safely and easily (Figure 6.29).

To heave to, bring the boat about without releasing the leeward jib sheet and, once about, lash the helm to leeward. The backed jib and opposing rudder will cause the boat to jog slightly to windward, then fall off, then come up again, sailing forward slowly, but making a lot of leeway — which calls for plenty of sea room to leeward. Typically the boat will maintain an attitude of about 45–50° off the wind and move through the water at about one knot. Heaving to has the advantage of keeping the boat's speed under control and reducing the impact of breaking seas. One other advantage of heaving to, especially over lying ahull, is that you can maintain control of the boat and thus be in the best position to adjust to changing circumstances.

Running with the storm

Running before the seas with trailing warps to hold the boat steady is one of the best ways to ride out a blow. Unlike the sea anchor method, or lying ahull, it is unwise to leave the boat to its own devices when running with the storm. You must ensure that it does not broach, and must adjust the steering to counter the differing sizes and shapes of the seas. A trailed warp provides speed control and directional stability. A small storm jib usually makes steering easier.

Appendices

APPENDIX I: Conversion factors

To convert any factor, look down the far left column, then look across to the appropriate units. (See Appendix VIII for a list of symbols and abbreviations, page 198.) For example, what is 54.6 kilometres per hour in miles per hour? Go to the velocity table, look down the left column until you find km/h, then look across until you find mph. This shows that 1 km/h = 0.6214 mph. Multiply 54.6 by 0.6214 and the answer is 33.9 mph. In each case the appropriate cell shows the number to multiply by in order to convert from the units at the left to the other units.

Length

	centimetre	inch	foot	metre	kilometre	mile (statute)	mile (nautical)
centimetre	**1.0000**	0.3937	0.0328	0.0100	0.0000	0.00001	0.000005
inch	2.54	**1.0000**	0.0833	0.0254	0.0000	0.00002	0.000014
foot	30.48	12.0	**1.0000**	0.3048	0.0003	0.00019	0.000164
metre	100	39.4	3.28	**1.0000**	0.0010	0.00062	0.000540
kilometre	100,000	39,370	3,280	1,000	**1.0000**	0.6214	0.5397
mile (statute)	160,930	63,360	5,280	1,609	1.6093	**1.0000**	0.8684
mile (nautical)	185,300	73,000	6,080	1,853	1.8530	1.1516	**1.0000**

Weight

Weight is defined as mass (which is the same everywhere in the universe) times acceleration due to gravity; weight, if expressed in the old units of kiloponds, has the same numerical value as mass expressed in kilograms if you are on Earth (a reasonable assumption). Weight is thus a force.

	ounce	gram	kilogram	pound	ton (long)	tonne
ounce	**1.0000**	28.35	0.0283	0.0625	0.000028	0.000028
gram	0.0353	**1.0000**	0.0010	0.0229	0.000001	0.000001
kilogram	35.27	1,000	**1.0000**	2.2046	0.000984	0.001000
pound	16	43.59	0.4536	**1.0000**	0.000446	0.000454
ton (long)	35,840	1,016,000	1,016	2,240	**1.0000**	1.016057
tonne	35,270	1,000,000	1,000	2,205	0.9842	**1.0000**

Speed (velocity)

	knots	m/s	km/h	mph
knots	**1.0000**	0.515	1.853	1.152
m/s	1.943	**1.0000**	3.600	2.237
km/h	0.540	0.278	**1.0000**	0.621
mph	0.868	0.447	1.609	**1.0000**

Area

	sq. cm	sq. inch	sq. foot	sq. m	sq. km	sq. mile (statute)	acre	hectare
sq. cm	**1.0000**	0.1550	1.075E-03	0.000100	1.00E-10	1.00E-04	2.471E-08	1.000E-08
sq. inch	6.4516	**1.0000**	6.944E-03	0.000645	6.45E-10	6.45E-04	1.594E-07	1.000E-08
sq. foot	930	144	**1.0000**	0.092903	9.29E-08	9.29E-02	2.296E-05	6.452E-08
sq. m	10,000	1,550	10.76	**1.0000**	1.00E-06	3.86E-07	2.469E-04	9.294E-06
sq. km	10,000,000,000	1,550,003,100	10,763,600	1,000,000	**1.0000**	0.3861	2.471E+02	1.000E-04
sq. mile (statute)	25,898,464,900	4,014,489,600	27,878,400	2,589,800	2.5898	**1.0000**	6.400E+02	259.01
acre	40,469,000	6,272,640	43,560	4,050	0.0040	0.0016	**1.0000**	0.4050
hectare	100,000,000	15,499,700	107,600	10,000	0.0100	0.0039	2.47	**1.0000**

Temperature

The simplest way of converting between Celsius and Fahrenheit is by using the equations below. Taking the first one, 35°C would equal 95°F; and using the second one, 50°F would equal 10°C. Note that, in the following, C means degrees Celsius and F mean degrees Fahrenheit.

$$F = \frac{9}{5}C + 32$$

$$C = \frac{5}{9}(F - 32)$$

The following list shows some examples.

°C	-20	-15	-10	-5	0	5	10
°F	-4	5	14	23	32	41	50

°C	15	20	25	30	35	40	45
°F	59	68	77	86	95	104	113

Force

Force is defined as a mass times an acceleration (F = ma); below, force is defined in the metric system and its equivalent in the English system (foot pounds) is also given. As explained above, force and weight are equivalent on Earth (that is, weight is a force).

$$\begin{aligned} \text{Force} &= 1\,\text{kg.m.s}^{-2} \\ &= 1\,\text{Newton} \\ &= 0.224.\text{lb.force} \end{aligned}$$

Pressure (force per unit area)

Pressure can be thought of as a weight per unit area (for example, the pressure at the ground can be thought of as the weight of the column of air above the ground). Below, pressure is defined in the metric system and its equivalent in the English system (pounds per square inch) as well as in millibars.

$$\begin{aligned} \text{Pressure} &= 1\,\text{kg.m.s}^{-2} \\ &= 1\,\text{Pascal} \\ &= 0.000145\,\text{lb/sq.inch} \\ &= 0.01\,\text{mb} \end{aligned}$$

where mb is millibar.

Most weather maps today show pressure in hecto-Pascals, where 1 hPa = 1 mb (h is hecto, or 100). Normal atmospheric pressure is defined as 1013.25 hPa (1013.25 mb), which is equivalent to 101325 Pa, 760 mm Hg (mercury) or 14.7 lb/sq. inch.

Work (energy; force applied over a distance)

Work has the same units as energy and is defined below in several different systems, starting with the metric one.

Work = $1 \text{ kg.m}^2\text{s}^{-2}$

= 1 joule

= 0.738.ft.lb

Power (work per unit time)

Power can be thought of as work per unit time. If the same work is done in half the time, double the power is said to have been produced.

Power = $1 \text{ kg.m}^2\text{.s}^{-3}$

= 1 watt

= 0.00143 hp (horsepower)

ie 746 Watts = 1 hp

Miscellaneous

The density of water (normal conditions) is 1g/cm^3, which is equivalent to 1000 kg/m^3. Air is approximately 1.3 kg/m^3, nearly 800 times lighter. A US gallon of water is 3.79 litres; it weighs 3.8 kg and a cubic metre of water would contain 264 US gallons. A normal sized swimming pool might hold 150 tons of water. One litre is 33.8 fluid ounces.

Latitude — one degree of latitude is approximately equivalent to 111.1 kilometres, 69.1 miles and 60 nautical miles. See the Glossary for the derivation of the nautical mile.

APPENDIX II: How to Read and Understand Graphs

Many of the graphs in this book are what are termed 'xy' or x versus y graphs. These are used to show diagrammatically the relationship between two things. In Figure II.1, which is a diagram of potential wave heights in deep-water unlimited-fetch environments (see Chapter 4), the relationship is between wind speed on the x-axis and wave height on the y-axis. Normally, if the relationship is well understood, the thing being predicted or estimated is plotted on the y-axis and the thing from which it is being predicted is along the x-axis. Figure II.1 shows how to use the graph to estimate wave height for any given wind speed in deep, open water situations.

In this book there are also several three-dimensional figures, which, in some ways, are easier to understand than xy graphs. The 3D figures are organised in such a way that the thing being predicted is the third dimension — that is, 'height' above the page — and the things from which this is predicted are shown as width and depth. 3D figures are used when there is more than one dominant factor causing a particular effect.

Potential wave heights in deep-water unlimited fetch environments with 50 hour duration

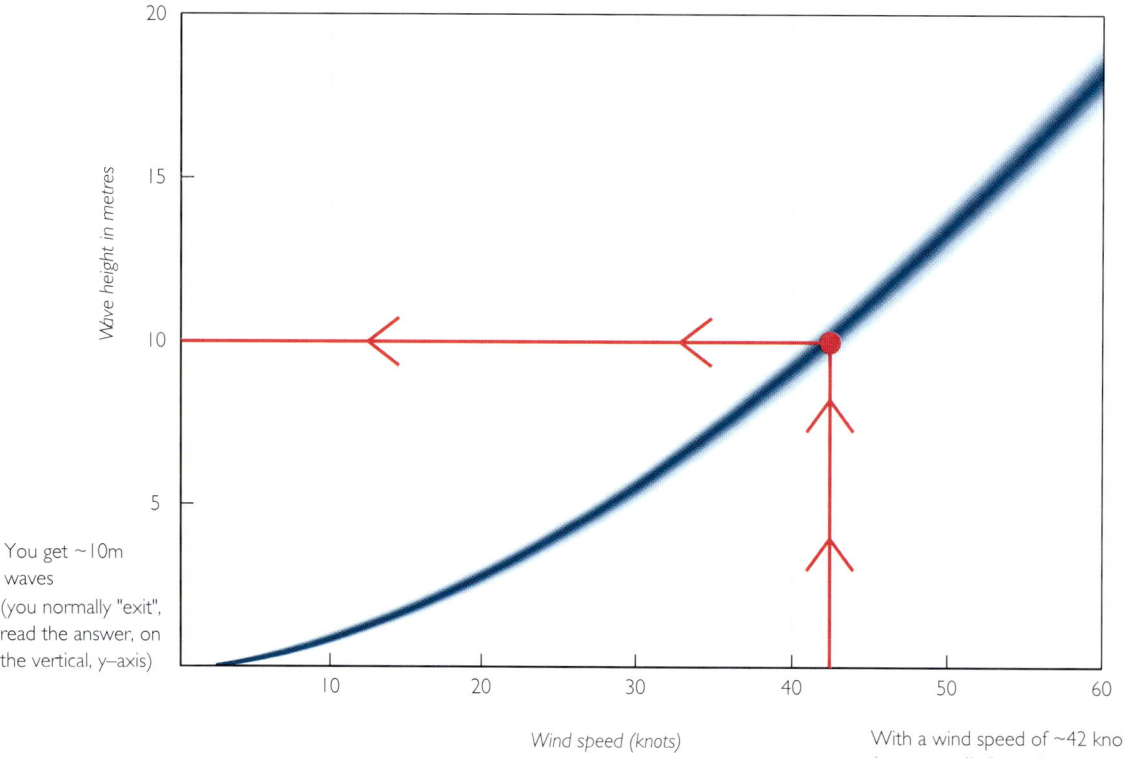

You get ~10m waves
(you normally "exit", read the answer, on the vertical, y–axis)

With a wind speed of ~42 knots ....
(you normally "enter" a graph on the horizontal or x–axis)

Figure II.1
How to read a normal xy graph.

APPENDIX III: Using Vector Addition for Estimating the Net Result of Two or more Winds, Currents, and so on

There are several references in this book to vectors. There is a simple method for resolving the net effect of the interaction of a number of different winds or currents (or any process which can be represented by speed *and* direction) operating together (that is, being superimposed). This technique, called vector addition, requires that speed be represented by the length of the arrow, and direction by the orientation of the arrow. By placing the different arrows 'tip to tail' and maintaining their orientations, you can draw a line from the tail of the first arrow to the tip of the last one and this will be the resultant or residual vector. The order of the arrows isn't important. In this book, vector addition is used to determine resultant wind and currents in situations where more than one source is operating.

Figure III.1 shows four examples of vector addition, applied to realistic situations, including the crossing of a flowing river — where the vector addition is the combination of the current and forward movement of the boat, and where a local lake breeze combines with a larger scale sea breeze, which in turn combines with a gentle synoptic (geostrophic) breeze to produce a resultant wind direction which might otherwise seem unexpected.

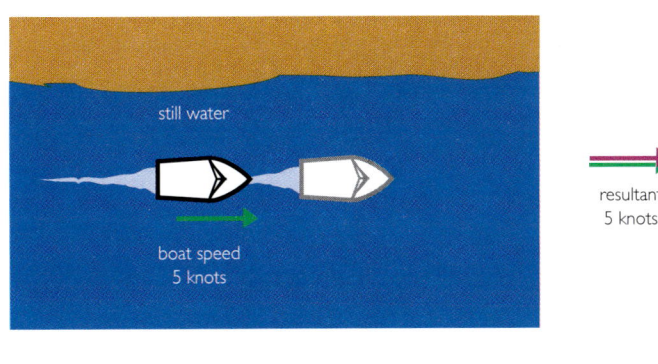

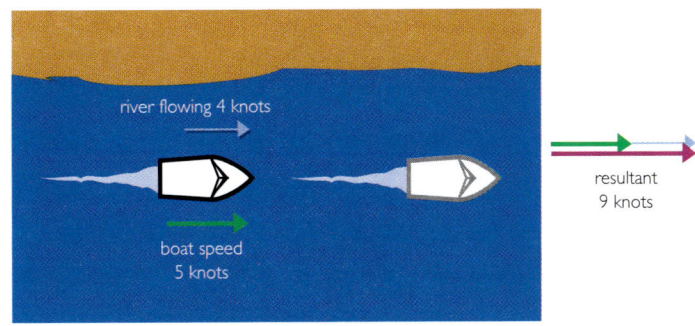

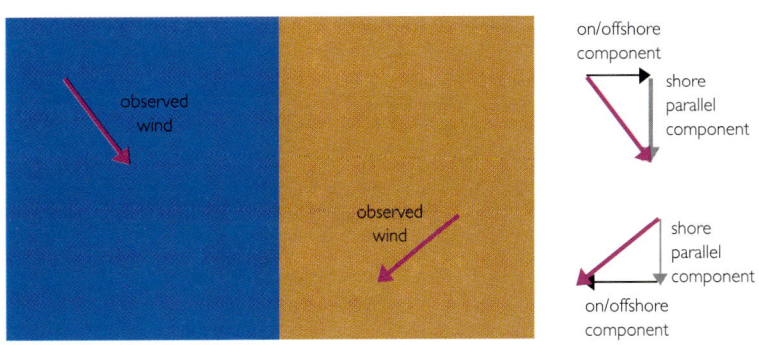

Figure III.1

A schematic representation of the use of vector addition to represent the combined effects of two or more qualities which can be represented by speed and direction.

APPENDIX IV: Quantitative Expressions for the Variation of Wind above Surfaces with Different Aerodynamic Surface Roughness

The magnitude of the wind shear — that is, the rate at which the wind speed increases as we move up from the ground, can be characterised by a number called the wind exponent, usually denoted by α. Knowing α enables us to relate the wind speed at a given height z_{ref} (in metres), where we might have a measuring instrument, to the speed at any other height, z, by the formula:

$$U_z = (U_{z_{ref}}) \times (z / z_{ref})^\alpha$$

A word of warning: this formula does not work too close to the actual surface roughness level. As a rule of thumb, don't apply it to less than three 'roughness heights' above the surface. For example, if the roughness is caused by swell 1 metre high, don't expect it to work below 3 metres. On the other hand, over 5 centimetre wavelets, it will be reliable down to 15 centimetres above the surface.

Some typical values for α are given below in Table IV.1.

Table IV.1

List of surface friction coefficients for use in estimating properties of wind profiles near the ground.

Description of terrain	α
Smooth hard ground, lake, ocean	0.10
Short grass, natural surface	0.14
Level country with long grass, odd tree	0.16
Tall crops, hedges, few trees or rough/heavy seas	0.20
Many trees, some buildings	0.22–0.24
Wooded terrain, small towns, suburbs	0.28–0.30
Urban areas with tall buildings	0.40

As an example, suppose the Beaufort Scale reading on a sea surface is 4, which is equivalent to a wind speed of 5.5 to 7.9 m/s (11 to 16 knots) at 10 metres above the surface. In this example, we will assume that the wind speed at 10m is 6.7 m/s (the middle of the Beaufort Scale 4). The question is what would the wind speed be at 2 metres above the ground? Using the equation we can say that:

$$U_z = (U_{z_{ref}}) \times (z / z_{ref})^\alpha$$
$$= 6.7 \times (2/10)^{0.1} \text{ (ie } \alpha = 0.1)$$
$$= 5.7 \, m/s$$

So, in this case the difference between the 10 and 2 metre wind speeds is only about 1 m/s, but over a rough land surface — say, ($\alpha = 0.28$) — the difference is much larger:

$$U_z = (U_{z_{ref}}) \times (z / z_{ref})^\alpha$$
$$= 6.7 \times (2/10)^{0.28}$$
$$= 4.3 \, m/s$$

As another example, if the wind speed at 2 metres above the sea on the deck of a boat is 20 m/s (39 knots), what would the wind speed be at 20 metres above the deck at the top of the mast, assuming a roughness of 0.12?

$$U_z = (U_{z_{ref}}) \times (z / z_{ref})^\alpha$$
$$= 20.0 \times (20/2)^{0.12}$$
$$= 26.4 \, m/s$$

It can be seen that the top of the mast experiences an increase of 6.4 m/s. Depending on how the sail behaves, this may be nearly 75 per cent more force.

The above equation was used in Chapter 3 to calculate the profiles shown in the graphics.

APPENDIX V: Estimating Wind Speed and Direction from Weather Maps

This appendix shows how to estimate wind speed and direction from normal newspaper weather charts, weather faxes and other sources. Due to the limitations of scale at which some of these are presented, the accuracy of the method may vary.

The basic idea is that wind speed and direction are the result of a balance between the following forces:

- the pressure gradient force — that is, the tendency of air masses to move from regions of high surface pressure to regions of low pressure
- the frictional force, which retards air masses as they move over the Earth's surface, and
- the Coriolis force, which is caused by the Earth's rotation.

Above the boundary layer, the frictional force can be ignored (this is one way of defining the boundary layer — that is, the layer of air affected by surface friction). That leaves two forces that are more or less in balance: the pressure gradient and the Coriolis force. The result of these is the somewhat counter-intuitive one that winds blow not from regions of high pressure to low pressure (that is, across the pressure lines on the map) but roughly along the isobars — the contours of constant pressure on weather maps.

In the Southern Hemisphere winds blow clockwise around regions of low pressure (cyclones) and anti-clockwise around regions of high pressure (anticyclones). In the following section you will see how to estimate surface winds from weather maps in two steps: first, by calculating the geostrophic winds above the boundary layer from the weather map, and then by adding in the effects of surface friction to derive near-surface winds using information about surface roughness.

Wind speed

The following method can be used to estimate U_g, the geostrophic wind. The geostrophic wind as estimated by the following method does not include the effects of curvature of the isobars, which may be important in cyclones and tightly curved anticyclones. U_g is estimated from the following formula, which expresses the balance between pressure gradient and Coriolis force:

$$U_g = \frac{A}{f} \times \frac{\partial P}{\partial x}$$

where A is a constant related to the density of air and has a value of 77.5, f is the Coriolis parameter,

and $\frac{\partial P}{\partial x}$ is the horizontal pressure gradient,

usually expressed as hecto-Pascals per metre (1 hecto-Pascal = a milli-bar).

Step 1: Firstly, determine the correct value of the Coriolis parameter, f. To do this, you will need to know the latitude of your area in degrees. Then look in Table V.1 below and read off the 'Coriolis parameter' (s^{-1}). Its value is constant for any given latitude, but for a country as big as Australia, it is a very important variable. Although it may not at first seem obvious, the data in Table V.1 indicates that to produce any given wind speed, a smaller pressure gradient is required near the equator than near the poles. In other words, any given pressure gradient will produce stronger winds near the equator than near the poles. Table V.2 shows the values of f for the capital cities around Australia.

For anyone interested in calculating f:

$f = 2\Omega Sin\ (latitude)$

where $\Omega = 7.29 \times 10^{-5}$ radians per second and is the angular velocity of the Earth's rotation, and for completeness F_c, the Coriolis force = fv, where v is velocity in ms^{-1}.

Step 2: Calculate the scale of your weather (surface pressure) map. A simple way to do this is to measure with a ruler the distance in millimetres between known points on the map (for example, Cape York to Melbourne), then divide the known (real) distance by this figure. Make sure all measurements are in the same units such as millimetres or inches. If, for example, you measure on the map 10 millimetres between two points which you know are, in reality, 400 kilometres apart, then the scale (remembering that kilometres must be multiplied by 1,000,000 to get millimetres) is:

= Known (real) Distance / Distance on map

= 400 (km) / 10 (mm)

= 400 (km) $\times$ 1000 (m) $\times$ 1000 (mm) / 10 (mm)

= 1: 40,000,000

Typically, weather maps are published in newspapers at a scale of about 1:40,000,000 (or 1 mm = 40 km), but they are sometimes at other scales. Several examples are given below. You will need to know the scale in order to calculate the horizontal pressure gradient (Step 3).

Step 3: Calculate the horizontal pressure gradient $\left(\dfrac{\partial P}{\partial x}\right)$ from the surface pressure map for a spot near you. Look at the two isobars nearest you. Check whether the isobars are 2 or 4 hecto-Pascals (that is, 2 or 4 mb) apart. The pressure gradient is simply the difference in pressure (2 or 4 hPa normally) divided by the horizontal distance in metres between the two same isobars. Obviously, the closer the isobars, the stronger the pressure gradient, and the stronger the wind.

Step 4: Now enter your variables, horizontal pressure gradient and Coriolis parameter (f) into the wind speed formula (noting that this will be the same for your locality, regardless of the pressure gradient):

For example, if your latitude is 30 degrees, and the isobars are 4 hPa (that is, 4 mb) and 300 kilometres apart (300,000 metres), then the geostrophic wind above the boundary layer is approximately:

$$U_g = \frac{77.5}{0.0000729} \times \frac{4}{300,000}$$

$$= 14.2\ ms^{-1}$$

$$= 27.5\ knots$$

Now, don't forget to correct this value for the type of surface (using Table V.3). If you were sailing on a fairly smooth sea you might expect the wind to be 65 per cent of its theoretical value (14.2 m/s), which is 9.2 m/s (17.9 knots). Note that the value 77.5 is a 'constant' and does not change for different areas.

Table V.3 lists corrections that must be made to the calculated geostrophic wind and direction so as to estimate the surface wind correctly.

Once you get accustomed to using this method you can quickly estimate the synoptic

Table V.1

Values of the Coriolis parameter (f) for different latitudes for use in estimating wind speed from weather maps.

Latitude (degrees)	Coriolis parameter (f) (s^{-1})
10.0	0.0000253
12.5	0.0000316
15.0	0.0000377
17.5	0.0000438
20.0	0.0000499
22.5	0.0000558
25.0	0.0000616
27.5	0.0000673
30.0	0.0000729
32.5	0.0000783
35.0	0.0000836
37.5	0.0000888
40.0	0.0000937
42.5	0.0000985

Table V.2

Values of the Coriolis parameter (f) for different capital cities in Australia.

Perth	0.0000772
Darwin	0.0000315
Adelaide	0.0000831
Brisbane	0.0000672
Sydney	0.0000812
Canberra	0.0000843
Melbourne	0.0000894
Hobart	0.0000992

Table V.3

Correction factors to be applied to calculated wind speed and direction values.

Type of surface	Wind speed at surface as % of calculated geostrophic wind speed	Correction for geostrophic wind direction in degrees
Smooth sea	**65**	10
Rough sea	60	**15**
Smooth land	55	20–25
Rough land	45–50	25–35
Land at night	< 50 (as low as 25)	40–50

Note: In the nomograms for estimating wind speed and direction from weather maps in Chapter 3, the figures in bold have been assumed.

wind by simply measuring the distance between isobars on the weather map and producing a 'lookup table' for your area, in other words, by working out that x mm between isobars equals y m/s wind speed.

> The nomograms in Chapter 3 assume the wind at the surface water and land surface respectively to be 65 per cent and 50 per cent of the geostrophic speed.

Wind direction

Wind direction is a little easier to estimate than wind speed. At some height above the earth's surface, the wind normally blows parallel to the isobars. As one gets closer to the surface, one observes that the wind direction changes markedly from this. Because winds blow anticlockwise around high pressure systems and clockwise around low pressure systems (in the Southern Hemisphere), you can estimate what the wind direction would be above the earth's surface by drawing an arrow parallel to the isobars and in the correct direction (that is, depending on whether you are looking at a high or a low). However, by looking at the correction table (Table V.3) one can see that, over rough surfaces, there can be a very large deflection from the theoretical wind direction.

The question is, which way does the wind deflect? The simple rule is that the wind always deflects in the direction from high to low pressure. That is, if you estimate the theoretical direction (parallel to isobars), then, as you get closer to the ground, the head of the arrow indicating direction will point more and more in the direction of the pressure gradient (from high to low). The amount of deflection can be estimated from the correction table. Generally speaking, using this method will allow you to estimate wind direction with good accuracy; but obviously, because you're using such a small scale pressure map, wind speed will be an estimate only, and the 'real' wind speed could be considerably higher or lower, depending on the accuracy of the map. However, most of the time this method will provide a useful estimate.

> Chapter 3 uses the above model to present an 'equation-free' way of estimating wind speed and direction for the Australian region. In the solution it was assumed that wind speeds over water and land surface were 65 per cent and 50 per cent, respectively, of the estimated geostrophic values.

APPENDIX VI: The Froude Number, F_r

The airflow around any physical feature is actually determined by the balance between inertia — the tendency of the air (wind) to keep on moving in the direction in which it is already going — and buoyancy — the fact that lighter fluids (air, water) rise over denser fluids. Scientists express this balance in a single number, the Froude Number, F_r for short. Students of boat design will recognise the name. This is the same Froude and the same Froude Number used to predict the wave drag effect on boats.

One of several equivalent ways of writing the Froude Number is:

$$F_r = \frac{U}{Nh}$$

U is the wind speed in m/s, h is the hill (or feature) height in metres and N is another new quantity called the 'Buoyancy Frequency' or 'Brundt-Vaisala Frequency'. Before going into what N means, the equations says that F_r will increase as U increases, and will decrease as N increases or as h, the height of the feature, increases.

N is directly related through another formula to the atmospheric temperature gradient — that is, the rate of change of temperature with height:

$$N = \sqrt{\frac{g}{T} \times \frac{\partial T}{\partial z}}$$

where g is the gravitational acceleration (9.81 m/s^2), T is the Potential Temperature in degrees Kelvin (ordinary measured temperature plus 1°C for every 100 metres' height gain or fraction thereof) and $\frac{\partial T}{\partial z}$ is the rate of change of temperature with height in °C per metre of height gain. We use Potential Temperature to correct for the effect on buoyancy of the natural cooling of the air as it expands with height. The units of N are s^{-1}, which is the same unit as frequency — that is, cycles or oscillations per second).

The equation shows that the value of N will increase as $\frac{\partial T}{\partial z}$ increases — in other words, as the atmosphere becomes more stable. Note that we have to take the square root of the expression in brackets on the right hand side of the equation, and if this is negative we can't do so. What a negative right hand side means is that the atmosphere is unstably stratified, and the kind of buoyancy effects we are about to discuss don't occur; instead we get turbulent mixing.

Often things are not simply one thing or another; in the daytime it is common to have unstable stratification near the ground in the boundary layer and stable stratification with a positive buoyancy frequency at higher levels, as in Figure 3.17.

Going back to the expression for the Froude Number, we can see that we can obtain the same F_r value by varying the wind speed, the stratification or the hill height. Large values for F_r means that inertia dominates and buoyancy has little effect on flow patterns (recall that in 'Wind and Barriers' our aim was to explain conditions where extreme gusts can occur and this requires a large scale feature and buoyancy dominating inertia). Large F_r occurs where there are strong winds, weak stratification and small hill heights. (Strong winds and weak stratification usually go together near the ground, as the turbulence caused by the friction of the strong winds on the ground effectively mixes up the lower part of the boundary layer and smooths out the temperature gradient.)

Small values for the Froude Number mean that buoyancy forces dominate the flow patterns. Small Froude Numbers occur where there are weak winds, strong stratifications (that is, large buoyancy frequency) or large hills. The reason we have chosen to concentrate on the last factor, the hill height, and have talked about large scale and small scale effects rather than large and small Froude Number effects is that, at small scale, for buoyancy effects to be important the winds must be so weak that they pose no danger. Conversely, flows over very large hills cannot fail to have low Froude Numbers and be strongly affected by buoyancy, even when winds are strong, because they disturb the stably stratified atmosphere above the boundary layer. It can now be seen that the combination of a strong wind, large hill and very stable atmosphere can produce wind conditions of the kind shown in Figure 3.18.

A final complication that we must deal with is where to measure the quantities — wind speed, temperature and temperature change with height — that go into the Froude Number formula, as these all vary with height. In fact, this height variation is important and will affect the flow patterns. For example, lee waves (which were discussed on page 74), occur only with certain special height variations of these quantities. To answer the basic question of whether buoyancy effects are going to be important, we need only look at the Froude Number formed from average values of these quantities over a height of two to three hill heights above the ground. Since the atmosphere above the boundary layer is always stably stratified we will get a positive value for F_r for any hill occupying a significant fraction of the boundary layer, but not for a hill that is only a small fraction of a deep, well mixed boundary layer (Figure 3.17).

APPENDIX VII: Weather Lore — Should We Believe It?

Why is the figure of a cock used in many wind vanes?

Historically, by papal enactment made in the middle of the ninth century, all church steeples had to have a cock set up as an allusion to St Peter's 'denial of Christ thrice before the cock crew twice'.

Can weather be predicted from birds, insects, or animals?

There are literally dozens of weather proverbs involving animals, insects and birds. For example, it was once believed that the severity of a coming winter could be predicted from the width of the stripes on a woolly-bear caterpillar (an American species). Other proverbs involve muskrats, beavers, squirrels, albatrosses, ground hogs, frogs, crickets (chirrups per minute)... even the lowly leech has rated a mention. So far none of these has been proven to be a reliable indicator, and even the once trusted migratory birds have been observed *en masse* to fly directly (and fatally) into severe storms. However, before we all get too blasé, experiments reported in *National Geographic* magazine have indicated convincingly that certain bird species can 'hear' extremely deep subsonic sounds, can 'see' ultraviolet light, and furthermore can sense the Earth's magnetic field and orientate themselves by the stars. Perhaps it would be more correct not to assume that other life forms are as insensitive to the weather as we are, but that we have either asked the wrong questions or that we have failed to see the answers when they are there. For example, bird banding studies have confirmed that individual Arctic terns have returned to the very same tree in which they last nested, having flown half way around the world!

Is there any validity in the following proverbs?

Red sky at night, sailor's delight; red sky in the morning sailors take warning *and* Rainbow in the morning gives you fair warning.
Often the appearance of a red sky indicates the presence of essential rain elements, dust and moisture, and high cloud; the presence of a rainbow indicates rain and shower activity. Normally a rainbow is seen when the observer's back is toward the sun and, similarly, the colour of the sky is far more apparent when the sun is behind the observer. Thus, given that weather systems move from west to east in the middle latitudes, a red sky (or rainbow) at night indicates that the poor weather has passed the observer, whereas if the same is observed in the morning (red western sky or rainbow in the west), there is a chance that poor weather is moving towards the observer.

Mares' tails and mackerel scales make tall ships take in their sails.
The mackerel sky, as it is called, consists of tiny ripple-like formations of cirrocumulus clouds. Mares' tails (cirrus) often precede an approaching warm front with veering winds and impending rain. If the clouds thicken and fuse, then the proverb tends to be particularly reliable.

Clear moon, frost soon.
If the atmosphere is clear and cloud free, the surface of the earth will cool by radiating to outer space, particularly if there is little wind. These conditions are most common under anticyclonic conditions and frequently result in frost.

Halo around the Sun or Moon, rain or snow soon.
Lunar or solar halos are evidence of cirriform clouds, which are frequently formed ahead of a warm front (fairly uncommon in Australia) and thus there is a good chance of rain or snow over the following 2 to 3 days.

Sunshine and showers, rain again tomorrow.
Rain interspersed with sunny periods is indicative of a moist airstream and strong surface heating and the formation of cumulus clouds. This type of weather may last for several days at a time.

Sharp rise after low foretells of a stronger blow.
A quick pressure change may be regarded as, say, 10 millibars in 3 hours. Although high pressures are usually a sign of fair weather, an abrupt pressure increase from a low value, or from a 'normal'

seasonal average, will generally indicate that bad weather is on the way. If you happen to be unfortunate enough to be sitting in the middle of a tropical cyclone, a rapid pressure rise will indicate that the cyclone has passed over you (cyclones are often relatively calm in the centre) and thus you should expect a dramatic change in wind direction accompanied by an increasing wind. (See Chapter 2.)

When mist takes to the open sea, fine weather shipmates, it will be.
Early morning coastal mist and fog are common occurrences under stable anticyclonic conditions; if these clear fairly quickly with the rising of the sun and an increase in air temperature, particularly if aided by a land (offshore) breeze, it usually foretells of continuing fine and settled weather.

If the sun goes pale to bed, there's rain by dawn, so 'tis said. The moon the governess of the floods, pale in anger, washes all the air. When round the moon she wears a brough (halo), the weather will be cold and rough.

If the moon is surrounded by an iridescent halo (usually cirrostratus) and the stars have a watery appearance it is quite likely that a front is approaching. One should expect rain and wind to follow the next morning.

Heavy dews in hot weather foretell fair weather. But, no dew after sun, fine weather's on the run.
Dews and frost generally form during clear and calm nights, which are commonly associated with anticyclonic weather. If the weather is also warm you can be fairly certain that good weather will persist for some days. If dew does not form it probably indicates overnight cloud and changing conditions.

Seagull, seagull stay out from the land. We'll ne'er have good weather while you're on the sand.
Sea birds that normally forage far out to sea have often been observed flying 'haphazardly' and noisily toward land well before bad weather sets in. Similarly, if sea birds seem to swarm inshore to fish, bad weather may be on the way.

APPENDIX VIII: Symbols and abbreviations

Symbol or abbreviation	Glossary entry
α	wind exponent
D	distance
d	depth
F	frequency
f	Coriolis parameter
F_c	Coriolis force $(=fv)$
g	acceleration due to gravity, 9.8ms^{-2}
H	wave height (also height of water surface in Chapter 5)
h	hecto
hPa	hecto-Pascal
H_s	significant wave height
km	kilometre
km/h or kmh^{-1}	kilometres per hour
kt	knot

Symbol or abbreviation	Glossary entry
l	length (m)
L	wavelength (m)
m	metre
m/s	metres per second (same as ms^{-1})
mb	millibar (sometimes mB)
nm	nautical miles
Pa	Pascal
s	Seconds (time)
T	Wave period
U, u*	Wind velocity
U_g	Geostrophic wind (speed)
V	Wave velocity (m/s, km/h)
v	Geostrophic current, flow (water, currents)

*(lower case used in some mathematical equations)

Glossary

adiabatic lapse rate
(atmosphere) the theoretical decrease in air temperature with increasing height — if no energy is added or removed, air will expand and cool as it rises at a rate of 0.0098°C/m.

amphidromic point
used in Chapter 5 to describe a theoretical location, usually in open oceans, about which the tidal wave rotates in response to the tide-producing forces.

atmospheric stability — *see* **stability**

backing (the opposite of veering: *see* Table 3.2)
(wind) change in wind direction, independent of hemisphere and independent of whether wind direction is expressed from the point of view of an observer facing the wind or facing away from it. Backing is anticlockwise in the sense that an observer facing into the wind would have to turn in an anticlockwise direction in order to remain facing directly into the wind. (Note that the observer would still have to turn anticlockwise if the wind was on his or her back.) For example, if the wind direction changes from easterly to north-easterly, this is backing (recall that easterly means 'coming from the east'). In Figure 3.2 it is shown that wind passing from oceans to land in the Southern Hemisphere will normally veer and wind passing from land to oceans will normally back. In the Northern Hemisphere the opposite happens: wind backs from water to land and veers from land to water (due to a force associated with the Earth's rotation which acts to the left south of the Equator and to the right north of it).

Symbolically, backing could be shown as:

   from ← to ↙ .

Symbolically, veering could be shown as:

   from ↙ to ← .

barometric pressure (usually just 'pressure')
[hPa or mb]
is the weight of the atmosphere per unit area at a particular location, usually measured in hecto-Pascals (hPa), noting that 1 hPa = 1 mb. The lines (termed 'isobars') on weather maps join places with equal barometric pressure.

bathymetry
the shape of the sea or ocean floor; analogous to topography (of land).

beat (beating) (*see also* **surf beat** and Figure 5.24)
what happens when two (or more) signals that are slightly out of phase with each other are combined. The figure relates to the different gravitational pulls of the Sun and Moon, but the principle relates equally to swell waves, sound and other wave forms. In the case of the Sun and Moon this phase difference causes the familiar spring–neap cycle.

Beaufort Scale (*see* table 3.1)

blue water (*see also* **upwelling**)
refers to the clear, almost translucent oceanic waters found where, in the absence of major upwellings, there are relatively low concentrations of nutrients and plankton.

boater (*see also* **fisher**)
a person who goes boating.

boundary layer (or planetary boundary layer)
the layer of the atmosphere that is influenced directly by the ground surface (*see* Chapter 3).

breeze
a 'light wind'. In this book 'wind' is used to describe air movements caused by convection and differential cooling of surfaces such as land, water or hilltops and valleys (loosely sea, land, mountain and valley breezes). Table 3.1 (the Beaufort Scale) shows that the term *breeze* is used for Beaufort 2–6 inclusive.

Buys Ballot's law
(wind/orientation) Buys was a Dutch meteorologist who, in 1857, worked out the following relationship between wind direction and the relative

position of highs and lows: if an observer stands with his or her back to the wind, the pressure is lower on the left in the Northern Hemisphere and on the right in the Southern Hemisphere. Extending this for Australia (and elsewhere in the Southern Hemisphere): if you are facing into the wind (normally wind direction is defined as the direction from which the wind is coming) the lower pressure area (or low) is to your left and the high to your right. Well above you (perhaps 1000 metres) the wind will be blowing parallel to the isobars. Closer to the surface, friction effects will cause the direction to veer (observer will have to turn clockwise in order to have wind still directly in the face) by perhaps 15° over water and 35° over land. See Table 3.2, which presents a range of simple orientation rules.

climate (*see also* **weather**)
long term weather; there is no real cut-off between weather and climate.

Coriolis force [F_c]
used in the equation for predicting wind speeds from weather maps (see Appendix V). The Coriolis force is equal to the Coriolis parameter times the velocity of the wind (or current). In the Southern Hemisphere the Coriolis force causes a deflection to the left and 'explains' why highs rotate anticlockwise and lows clockwise in our part of the world. In mathematical terms the Coriolis force = Coriolis parameter (f) × velocity, meaning that the faster an object moves, the more it will be deflected by the Coriolis force.

Coriolis parameter [f]
used in the equation for predicting wind speeds from weather maps (see Appendix V). Units are s^{-1}. The Coriolis parameter arises from the rotation of the Earth and is proportional to latitude.

current (*see also* **tidal current**)
– (direction of) normally specified in terms of the direction *towards* which water is moving; unlike wind, which is normally specified in terms of the direction *from* which it is blowing.
– (water) unless specified otherwise, the term (usually plural) is used in this book to describe ocean currents — the horizontal movement of water due to wind acting over oceanic waters for long periods of time (for example, the Trade Winds, which may blow for half the year). Not to be confused with *tidal currents*, which refers to horizontal

motion of water associated with the gravitational attraction between the Earth, Moon and Sun.

declination (tides)
the angle of the Sun or Moon relative to the equator (zero declination means directly over the equator).

depth [d]
(usually of water) height from top to bottom, normally measured in metres.

distance [D]
unit of length, normally in metres, kilometres or nautical miles.

diurnal (tide) (*see also* **semidiurnal**)
a tide that occurs once per twenty-four hours.

diurnal (time scale)
commonly used in climatology to describe processes or patterns that vary on a daily (24 hour) basis, such as maximum and minimum temperature; also used in oceanography to describe tides — in this case, one tide per twenty-four hours.

duration (wind)
the length of time for which wind has been blowing.

Ekman spiral — *see* Figure 5.2 and associated text

Ekman transport
(large scale oceanic waters) When wind blows constantly over large expanses of water, the surface water is set in motion. This movement in turn sets deeper water in motion and so on down to about 100 metres or so. As explained in Chapter 5, the net movement of the water column ends up almost at right angles to the prevailing wind. In the Southern Hemisphere, the direction of movement is to the left.

equilibrium tide
the name given to an idealised hypothetical circumstance which would occur on an Earth with real dimensions but entirely covered by a deep ocean, and in the presence of the real Moon and Sun. Under such conditions, the response of the ocean to the changing gravitational fields would be complete and virtually immediate.

fetch (kilometres, nautical miles)
the distance the wind has been blowing *over water*. The term is used throughout Chapter 4.

fisher (*see also* **boater**)
a person who goes fishing.

freak wave — *see* **rogue wave**

frequency [F]
used in Chapter 4; number of cycles (e.g. waves) per second. The inverse is period (T), the number of seconds per cycle (wave).

fully developed sea
for a given wind speed in open and deep waters, there is a condition in which energy lost through spray is approximately equal to energy gained from the wind. At this point the waves have reached their maximum size. The stronger the wind, the greater this height, but the longer it takes to reach the fully developed condition. See Table 4.4.

geostrophic current, flow (water, currents) [v]
as explained in Chapter 5, the geostrophic current is a theoretical current that moves around hills of water in the major oceanic basins. Fanciful as this sounds, these 1 to 2 metre 'hills' are created by the action of prevailing wind acting over the vast oceans, via Ekman transport. Their existence has been verified only recently by accurate satellite height data.

geostrophic wind speed [U_g] (*see also* wind velocity)
U_g is a theoretical wind observed well above the Earth's surface, where friction effects are negligible and the pressure gradient (as measured by isobars on weather maps) is balanced by the Coriolis force. This is explained in detail in Appendix V. At 1000 metres or more above the Earth's surface, observed winds are frequently in close agreement with U_g; the term *theoretical* is used here to mean based on a theoretical equation. Occasionally in this book, regional or synoptic wind is used to mean the geostrophic wind.

global scale — *see* **map scale**

green water
less common than 'blue water'; areas of water where upwelling — usually associated with large scale currents — brings nutrients close to the surface and supports large populations of plankton.

hecto [h]
literally, 100 (hecto-Pascal = 100 Pascals; hecto-litre, 100 litres; and so on).

hecto-Pascal [hPa]
unit of pressure (100 Pascals); hence 1 hecto-Pascal = 1 millibar.

isobar
lines on weather maps (usually) joining places with equal pressure. In this book isobars usually refer to ground level pressure.

kilometre [km]
unit of length (1000 m).

kilometres per hour [km/h or kmh^{-1}]
unit of velocity (speed).

knot [kt]
unit of speed (1 nautical mile per hour). The 'normal' abbreviation for unit of wind speed over water.

lapse rate
the rate of change of temperature with height above ground. Under normal daytime conditions, temperature decreases with height by about 0.6°C per 100 metres.

length [l] (of a pool, basin or other body of water)
the factor used in Merian's formula to predict T, the periodicity of resonance (*see* Chapter 5).

littoral zone
the area of land/sea interface (roughly, the coast) affected by waves or alternately exposed and submerged by tides or both.

map scale
mapping convention which can be confusing since 'small' means a broad, rather than detailed, scale. 'Small scale map' means broad coverage (say, 1:1,000,000), where 1 millimetre on the map equals 1 kilometre on the ground. Large scale map means detailed coverage (say, 1:10,000), where 1 millimetre on the map equals 10 metres on the ground.

maximum condition (waves) — *see* **fully developed sea** and Table 4.4.

metre [m]
unit of length.

millibar [mb]
unit of pressure (1 mb = 1 hPa).

nautical mile [nm]
one minute of earth arc, i.e. 1 degree of latitude is 60 nautical miles at the Equator.

neap and spring tides
in the course of a calender month the Sun, Moon and Earth are aligned in such a way that the gravitational forces reinforce each other and produce larger than average spring tides. Spring tides occur near to the time of the new and full moons and may be 20 per cent higher than average. Correspondingly, neap tides occur at the first and third quarter of the Moon and may be 20 per cent lower than average. In between the two spring and two neap tides the tides may vary smoothly (i.e. becoming smaller or larger day by day) or may show complex patterns. (Figure 5.22.)

nomogram (a form of graph)
term used in Chapter 3, 'How to estimate wind speed and direction from weather maps'. A nomogram is a 'cross' between a graph and a series of mathematical curves; in this case the nomogram shows surface pressure gradients (isobars) superimposed on a map of Australia, over which estimated wind — shown as colours — is also draped.

Pascal [Pa]
unit of pressure. 1 hecto-Pascal (100 Pa) = 1 millibar (mb).

pattern (patterns and processes)
the focus of this book is on understanding and mapping the patterns of wind speed, wave height, solar radiation etc rather than the physics of the processes causing them.

potential wave height
same as maximum condition or fully developed sea; describes the maximum wave height for a given wind speed; note that the higher the wind speed the longer it takes to reach the maximum condition (and the larger the waves). See Table 4.4.

predominant wind direction
the most common (frequent) directions from which the winds blow.

process (patterns and processes)
the focus of this book is on understanding and mapping the patterns of wind speed, wave height, solar radiation etc rather than the physics of the processes causing them.

pycnocline (oceans)
a strong gradient in the density of the water column which occurs somewhere at about 200 to 300 metres' depth in the oceans.

qualitative
used in this book to refer to non-mathematical description of a process or pattern.

quantitative
used in this book to refer to mathematical (usually involving symbolic equations) descriptions of a process or pattern.

rainday
a day with measurable rainfall, normally defined as ≥ 0.2mm.

rogue wave
an unfortunate and misleading term. As explained in Chapter 4, approximately 1 in 1000 waves will be almost twice the significant wave height and there is nothing roguish about it. Like trying to toss five heads in a row, it is a relatively rare event but will happen sooner or later. In this case *sooner* usually means within three hours under normal swell conditions.

scale (in maps) — *see* **map scale**

seas (*see also* swell)
seas or sea waves (sometimes referred to as wind seas) are waves that have been generated more or less locally under the influence of a wind that is still acting upon them; they are characteristically choppy and disorganised. Eventually, seas become swell — recognisable by a smooth, round and regular appearance.

semidiurnal (tide) (*see also* diurnal)
happening twice per twenty-four hours.

shoaling (waves)
breaking (plunging). Waves approaching shallow waters normally begin to 'feel' the bottom when the depth is less than half their wavelength. For example, a wave with a 100 metre wavelength will feel the bottom in approximately 50 metres of water and will shoal when the depth is about 80 per cent of the wave height. Thus, a 3 metre swell wave might break in about 2.4 metres of water.

significant wave height [H$_s$]
a statistical concept referring to the height of the highest one-third of waves in a given time period.

solar radiation (aka short wave radiation)
loosely, 'sunlight'. Most of the Sun's energy is in the range (all in units of microns or 10^{-6}m) $0.2 < \lambda < 3$, with virtually no overlap with the heat (terrestrial) radiation of the Earth (land, water, snow, etc), which is concentrated in longer wavelength as $4 < \lambda < 50$.

spring and neap tides
twice in the course of a calender month the Sun, Moon and Earth are aligned in such a way that the gravitational forces reinforce each other and produce larger than average spring tides; these occur at about the time of the new and the full Moon and may be 20 per cent higher than average. Correspondingly, neap tides occur at the first and third quarters of the Moon and may be 20 per cent lower than average. In between the two spring and two neap tides, the tides may vary smoothly (ie by becoming smaller or larger day by day) or may show complex patterns — *see* Figure 5.22.

stability (atmospheric) — *see also* **lapse rate** and Chapter 3, pages 71–6.
Under normal well mixed conditions (loosely, windy), the temperature of the atmosphere decreases with height by an average of about 0.6°C per 100 metres. The theoretical decrease — if heat is neither added to nor lost from the atmospheric column — is called the adiabatic lapse rate, and its value is nearly 1°C per 100 metres and is shown with the symbol Γ. Recalling that 'lapse' means decrease in temperature with height, we can define the states of stability thus:

atmospheric lapse rate
  $< \Gamma$ is a stable atmosphere;

atmospheric lapse rate
  $= \Gamma$ is a neutral atmosphere, and

atmospheric lapse rate
  $> \Gamma$ is an unstable atmosphere.

Atmospheric mixing is suppressed under stable conditions, and wind conditions on the ground may be independent of those aloft. Pollution may become trapped near the ground. Unstable conditions enhance mixing, vertical development of clouds and other thermally driven phenomena. Winds aloft are well transported to the ground. Wind veering is more pronounced under stable conditions.

streamline
a line showing where wind or water is travelling; not to be confused with direction, which, in the case of wind, is defined as the direction *from* which the wind is blowing.

surf beat (*see also* beat)
the interaction of two or more wave trains, causing some waves to be much higher, and some to be much lower, than either train separately. From the perspective of sitting on the beach, wave beat is most obvious when two sets of swell waves approach from different directions. (Figure 5.24.)

swell (*see also* seas)
smooth, regular, organised sets of waves, most common in open waters; swell waves form from seas after considerable time (and distance away from where they were generated by the action of wind on water).

synoptic scale — *see* **scale**

temperature inversion (*see also* lapse rate)
reversal of the normal decrease in atmospheric temperature with increasing height; common on still clear winter nights. By definition, inversions result in a stable atmosphere, one in which little vertical mixing occurs.

thermal
(wind) in a way that is closely analogous to how horizontal pressure induces the geostrophic wind, horizontal temperature gradients also induce wind that blows parallel to the isotherms rather than to the isobars. In the Southern Hemisphere, this 'thermal wind' blows along the isotherms with the low temperature to the right (ie if you face the thermal wind, the higher temperature will be on the

right). The stronger the temperature gradient, the stronger the thermal wind (*see* Chapter 3).

tidal current (*see also* **current**)
the vertical movement of water, associated with gravitational attraction of the Sun and Moon and their interactions, can induce horizontal movements as well, especially in shallow waters. In some places the tidal currents can be very strong.

tide range
the vertical extent of tide — that is, the difference (usually measured in metres) between the water level at high and low tides, caused by astronomical factors.

tide wave
(also 'tidal wave') the tide itself. As described in Chapter 4, the tide is a long-wavelength, relatively slow-moving wave, which behaves like other waves. Should not be confused with a tsunami (also erroneously called a 'tidal wave'), which is caused by undersea tectonic disturbances and 'landslides'.

upwelling (*see also* **blue water**)
(oceans, seas) cold — usually nutrient-rich — water being brought to the surface (see page 114).

vector (**mathematical meaning**)
the representation of a quantity in terms of its speed and direction (see Appendix III).

veering (**the opposite of backing:** *see* **Table 3.2**)
(wind) change in wind direction, independent of hemisphere; veering is clockwise in the sense that an observer facing into the wind would have to turn in a clockwise direction in order to remain facing directly into the wind. For example, if the wind direction changes from north-easterly to easterly, this is veering (recall that north-easterly means 'coming from the north-east'). In Chapter 3 it is shown that wind passing from oceans to land will normally veer and wind passing from land to oceans will normally back. In the Northern Hemisphere the opposite happens: wind backs from water to land and veers from land to water (due to a force associated with the Earth's rotation which acts to the left south of the equator and to the right north of it).

Symbolically, backing could be shown as:

from ← to ↙

Symbolically, veering could be shown as:

from ↙ to ←

wave height [H] (*also* **height of water surface** in Chapter 5)
distance (usually in metres) from the trough to the crest of a wave; also used to describe the height of local water above the general sea level height.

wave period [T]
(usually in seconds) the time between successive waves, normally measured between successive troughs or crests; also used as symbol for resonance period in relation to lakes and oceans, or to define period of resonance (in seconds).

wave velocity (m/s, km/h) [V]
the speed at which a wave is moving.

wavelength (m) [L]
(used in wave equations) the distance between successive troughs or crests.

weather (*see also* **climate**)
weather refers to the day-to-day levels and fluctuations in temperature, humidity, rainfall, wind, solar radiation, and so on.

wind
in this book wind is used to describe air movements caused by difference in pressure over regions and large areas (loosely regional and large scale pressure gradients).

wind duration
the amount of time wind has been blowing.

wind exponent [α]
Appendix IV presents a simple mathematical model for estimating wind speeds above land and water surfaces. This model is based on observations of the rate at which the wind speed increases as we move up from the surface. Close to the ground this rate is controlled mainly by the nature of the surface, which can be characterised by a number called the wind exponent (usually denoted by the Greek letter 'alpha', or α). Knowing α enables us to relate the wind speed at a given height, where we might have a measuring instrument, to the speed at any other height.

wind rose
a simple graphical representation of wind direction (see, for example, Figures 2.40 to 2.43). In this book the wind rose arrows face into the wind.

wind shear
change in wind speed with height; derived from the shearing or sliding effect of fast-moving air (above) slipping over slower moving air (near the surface). The friction between the moving air and the ground leads to much lower wind speeds near the surface.

wind velocity (m/s, km/h, knots) [U, u] (See also geostrophic wind speed.)
the speed of wind.

x-axis (a graphing convention)
the horizontal axis on a normal two-dimensional graph.

y-axis (a graphing convention)
the vertical axis on a normal two-dimensional graph.

The Photographs

Robert Keeley has been a journalist and photographer for the last 21 years. He is currently editor of the yachting magazine *Cruising Helmsman* and has been photographing marine subjects for the last 10 years. Until recently he was a regular dinghy sailor and competed in many state, national and world titles. He has sailed offshore in a range of yachts.

page x
A trailable yacht sits lifeless in a winter drifter on Sydney's Botany Bay.

page 6
Footprints in the sand. Australia's coasts offer unlimited potential for those seeking solitude.

page 13
A Skiff Moth on Sydney's Middle Harbour. These intermediate training dinghies are fast and flighty. The tricky waters of Middle Harbour, with shifty breezes and tangled currents, offer an extremely tough challenge for young sailors.

page 14
A fine day at Tasmania's Port Davey, looking east towards Bathurst Harbour. This spectacular but isolated harbour on the southwest coast of the island State is subject to frequent Southern Ocean gales.

page 46
An Eighteen Foot Skiff slices upwind on Sydney Harbour. Each summer the skiffs dominate the dinghy action on Sydney Harbour, with their big rigs carrying sponsor's insignias. The summer nor'easter on Sydney Harbour can create a short chop, which is exacerbated by yacht traffic, ferries and ships.

page 89
Sailing downwind with a spinnaker flying; a yacht makes its way down Broken Bay, north of Sydney.

page 90
A couta boat races on the waters of southern Port Phillip Bay. In summer the bay is unpredictable. Whilst the prevailing breeze is a sou'westerly of moderate strength, sudden gales can sweep in from the south, west, or east. When the breeze swings to the north it can also build to gale force over a couple of days.

page 102–3
Summer couta boat racing at Portsea in the shallow waters of the southern reaches of Port Phillip Bay.

page 110
Dark clouds close in at a late afternoon anchorage.

page 122–3
Late in the day a coastal cruising yacht attracts a school of playful dolphins. They stayed with the yacht for almost an hour.

page 149
Friends enjoying a beautiful surf beach with some early morning fishing.

page 150
A wintery, glassy Port Phillip Bay. In early and mid-winter the bay usually has very little wind, but later in winter and into spring a strong nor' westerly pattern is not uncommon.

page 162–3
Darwin's Fanny Bay. This stretch of water regularly has sensational sunsets. A large cruising fleet gathers in Darwin each July in preparation for the annual Darwin to Ambon (Indonesia) yacht race.

page 183
The Whitsunday islands offer some of the best sailing waters in Australia and attract large numbers of cruising yachts. Their best conditions are usually experienced during the winter months, especially July, August and September, when lighter winds and drier weather generally prevail.

page 184
Coasters Retreat on the western shore of Sydney's Pittwater is a magnet for weekend sailors. It is well protected from southerlies and westerlies, but exposed to Sydney's predominant summer breeze, the nor'easter.

References

Accad, Y., and Pekeriss, C.L. (1978) 'Solutions of the tidal equations for the M_2 and S_2 tides of the world ocean from a knowledge of the tidal potential alone', *Phil. Trans. R. Soc. London*, Series A, 290, 235–66.

Allan, R., Lindesay, J., and Parker, D. (1996) *El Niño Southern Oscillation and Climate Variability*, CSIRO, Victoria, Australia.

Baines, P. G. (1989) 'The physical oceanography of Australian waters – a review', *Australian Meteorological Magazine* 37, 155–65.

Bureau of Meteorology (1977) *General Meteorology Part I* and *Aviation Meteorology Part II*, AGPS, Canberra.

Bureau of Meteorology (1984) *Observing the Weather*, AGPS, Canberra.

Bureau of Meteorology (1989) *Wind Waves Weather: Victorian Waters*, AGPS, Canberra.

Bureau of Meteorology (no date) 'Understanding cyclones: Western Australia', AGPS, Canberra. Cat. no. 92 2740 1.

Colls, K., and Whittaker, R. (1990) *The Australian Weather Book*, Child and Associates, Sydney.

Department of Transport (1981) *Safety for Small Craft*, AGPS, Canberra.

Doodson, A. T. (1921) 'The harmonic development of the tide generating potential', *Proceedings of the Royal Society*, A, Vol. 100.

Easton, A. K. (1970) *The Tides of the Continent of Australia*, The Horace Lamb Centre, Flinders University (South Australia).

Haughton, D. (1988) *Understanding Weather at Sea* Parts 1, 2. [video], Carpenter Videos, available in Australia through Mastercraft Videos, Sydney (tel. 02 9958 1724).

Haughton, D. (1988) *Understanding Wind Strategy* Parts 1, 2 [video], Carpenter Videos, available in Australia through Mastercraft Videos, Sydney (tel. 02 9958 1724).

Hydrographer of the Navy (1979) *The Mariner's Handbook* Hydrographer of the Navy (5th edition), Hydrographer of the Navy, Somerset.

Hydrographer of the Navy (1982) *Australia Pilot Vol. III*, Hydrographer of the Navy, Somerset.

Linacre, E., and Hobbs, J. (1977) *The Australian Climatic Environment*, Wiley and Sons, Brisbane.

Macmillan Atlas of the Oceans (1977), Macmillan, New York.

Maloney, E. S. (1985) *Chapman Piloting Seamanship and Small Boat Handling*, Hearst, New York.

NATMAP (1986) *Atlas of Australian Resources: Climate*, edited by Geoff Parkinson, produced by the former NATMAP (Division of National Mapping, Canberra), Commonwealth Government Printer (now AGPS), Canberra.

Neiburger, M., Edinger, J. G., and Bonner, W. D. (1982) *Understanding our Atmospheric Environment*, Freeman, San Francisco.

Open University Course Team (1989) *Waves, Tides and Shallow-Water Processes*, Open University Course Team (UK), Pergamon, Oxford.

Physick, W.L., and Byron-Scott, R.A.D. (1994) 'Observations of the sea breeze in the vicinity of a gulf', *Weather* 32, 373–81.

Rossiter, J. R., and Lennon, G.W. (1968) 'An intensive analysis of shallow water tides', *Journal of the Royal Astronomical Society*, 16 (3).

Simpson, J.E. (1994) *Sea Breeze and Local Winds*, Cambridge University Press, UK.

Tapper, N., and Hurry, L. (1993) *Australia's Weather Patterns*, Dellasta, Victoria.

Thurman, H. V. (1989) *Introductory Oceanography*, Macmillan, New York.

Index